Progress in Nonlinear Differential Equations and Their Applications
Volume 28

Editor
Haim Brezis
Université Pierre et Marie Curie
Paris
and
Rutgers University
New Brunswick, N.J.

Editorial Board
Antonio Ambrosetti, Scuola Normale Superiore, Pisa
A. Bahri, Rutgers University, New Brunswick
Luis Cafarelli, Institute for Advanced Study, Princeton
L. C. Evans, University of California, Berkeley
Mariano Giaquinta, University of Florence
David Kinderlehrer, Carnegie-Mellon University, Pittsburgh
S. Klainerman, Princeton University
Robert Kohn, New York University
P. L. Lions, University of Paris IX
Jean Mawhin, Université Catholique de Louvain
Louis Nirenberg, New York University
Lambertus Peletier, University of Leiden
Paul Rabinowitz, University of Wisconsin, Madison
John Toland, University of Bath

Augusto Visintin

Models of Phase Transitions

Birkhäuser
Boston • Basel • Berlin

Augusto Visintin
Dipartimento di Matematica
Università Degli Studi di Trento
38050 - Povo
Italy

Library of Congress Cataloging-in-Publication Data
Visintin, A. (Augusto)
  Models of phase transitions / Augusto Visintin.
    p.   cm. -- (Progress in nonlinear differential equations and
  their applications : v. 28)
    Includes bibliographical references and indes.
    ISBN 0-8176-3768-0 (acid-free paper). -- ISBN 3-7643-3768-0 (acid-
  free paper).
    1. Phase transformations (Statistical physics) -- Mathematical
  models. 2. Transport theory--Mathematical models. 3. Differential
  equations, Partial--Numerical solutions. I. Title. II. Series.
  QC175.16.P5V57   1996
  530.4'14'015118--dc21   96-45299
                                            CIP

Printed on acid-free paper
© 1996 Birkhäuser Boston   *Birkhäuser*

Copyright is not claimed for works of U.S. Government employees.
All rights reserved. No part of this publication may be reproduced, stored in a retrieval system,
or transmitted, in any form or by any means, electronic, mechanical, photocopying, recording,
or otherwise, without prior permission of the copyright owner.

Permission to photocopy for internal or personal use of specific clients is granted by
Birkhäuser Boston for libraries and other users registered with the Copyright Clearance
Center (CCC), provided that the base fee of $6.00 per copy, plus $0.20 per page is paid directly
to CCC, 222 Rosewood Drive, Danvers, MA 01923, U.S.A. Special requests should be
addressed directly to Birkhäuser Boston, 675 Massachusetts Avenue, Cambridge, MA 02139,
U.S.A.

ISBN 0-8176-3768-0
ISBN 3-7643-3768-0
Typeset by the author in TeX.
Printed and bound by Quinn-Woodbine, Woodbine, NJ
Printed in the U.S.A.

9 8 7 6 5 4 3 2 1

Contents

Preface ix
Introduction 1
Reader's Guide 4

Part 1. Some Nonlinear P.D.E.s

Prelude 5

I. Models and P.D.E.s 6
1. Modelling and Analysis 6
2. Nonlinear P.D.E.s and Minimization Problems 12
3. Examples of Nonlinear P.D.E.s 25
4. Comments 30

II. A Class of Quasilinear Parabolic P.D.E.s 31
1. Variational Techniques of L^2-Type 31
2. Further Results via L^2-Techniques 39
3. Techniques of L^1- and L^∞-Type 47
4. Local Regularity Results 54
5. Integral Transformations 58
6. Semigroup Techniques 61
7. Comments 65

III. Doubly Nonlinear Parabolic P.D.E.s 68
1. Doubly Nonlinear Parabolic Equations of First Type . . 68
2. Doubly Nonlinear Parabolic Equations of Second Type . . 74
3. Other Nonlinear Parabolic Equations 78
4. Use of Compactness by Strict Convexity 82
5. Comments 87

Part 2. Phase Transitions

IV. The Stefan Problem 90
1. Strong Formulation of the Stefan Problem 90
2. Surface Tension 96
3. Length Scales and Mushy Region 98
4. Weak Formulation of the Stefan Problem 100

5. On the Analysis of the Stefan Problem 104
6. Comparison between Strong and Weak Formulations 106
7. The Muskat and Hele-Shaw Problems 110
8. A Stefan-Type Problem Arising in Ferromagnetism 113
9. On the History of the Stefan Problem 117
10. Comments 121

V. Generalizations of the Stefan Problem 123

1. Kinetic Undercooling and Phase Relaxation 123
2. Phase Transition in Two-Component Systems 131
3. Approach via Nonequilibrium Thermodynamics 137
4. Analysis of the Model of Section V.3 141
5. General Nonequilibrium Thermodynamics 143
6. The Evolution of the Free Energy 147
7. Comments 152

VI. The Gibbs-Thomson Law 155

1. Free Energy 155
2. Entropy 160
3. Phase-Dependent Conductivity 165
4. The Gibbs-Thomson Law 167
5. The Phase Field Model 174
6. Comments 176

VII. Nucleation and Growth 178

1. Local and Global Minimizers 178
2. Nucleation 181
3. Stable and Metastable States 185
4. Pure Phases 187
5. From Nucleation to Growth 190
6. Mean Curvature Flow 193
7. Nonlinear Mean Curvature Flow 196
8. Hysteresis in Front Motion 198
9. Comments 200

VIII. The Stefan-Gibbs-Thomson Problem with Nucleation 203

1. Modes of Phase Transition 203
2. Formulation of the Problem 208
3. Some Auxiliary Results 213
4. Existence Result 216
5. The Mullins-Sekerka Problem 225
6. Comments 227

IX. Two-Scale Models of Phase Transitions 229
1. Two-Scale Stefan Problem and Nonadiabatic Nucleation 229
2. Another Model with Surface Tension 237
3. A Mean Field Model 239
4. Micromagnetics 242
5. Some Comparisons 244
6. Comments 247

Appendix

X. Compactness by Strict Convexity 248
1. Extremality and Compactness 248
2. Strictly Convex Functionals 252
3. Applications 255
4. Comments 259

XI. Toolbox 260
1. Some Function Spaces 260
2. Sobolev Spaces 264
3. Compactness 268
4. Convexity 273
5. Monotonicity 279
6. Accretiveness 285
7. Minimization 288
8. Geometric Measure Theory 290
9. Other Results 293

Book Selection 295

Bibliography 297

Index 319

Preface

> ... *"What do you call work?"*
> *"Why ain't that work?"*
> Tom resumed his whitewashing, and answered carelessly:
> *"Well, maybe it is, and maybe it ain't. All I know, is, it suits Tom Sawyer."*
> *"Oh come, now, you do not mean to let on that you like it?"*
> The brush continued to move.
> *"Like it? Well, I do not see why I oughtn't to like it. Does a boy get a chance to whitewash a fence every day?"*
> That put the thing in a new light. Ben stopped nibbling the apple. ...
> (From Mark Twain's *Adventures of Tom Sawyer*, Chapter II.)

Mathematics can put quantitative phenomena in a new light; in turn applications may provide a vivid support for mathematical concepts.

This volume illustrates some aspects of the mathematical treatment of phase transitions, namely, the classical Stefan problem and its generalizations. The intended reader is a researcher in application-oriented mathematics. An effort has been made to make a part of the book accessible to beginners, as well as physicists and engineers with a mathematical background. Some room has also been devoted to illustrate analytical tools.

This volume deals with research I initiated when I was affiliated with the Istituto di Analisi Numerica del C.N.R. in Pavia, and then continued at the Dipartimento di Matematica dell'Università di Trento. It was typeset by the author in plain T$_E$X.

I express my gratitude to E. Magenes, who introduced me to the Stefan problem; to A. Damlamian, who helped me at the beginning of that research; to S. Luckhaus, to whom I owe some of the ideas which are developed here in Chap. VIII; to P. Colli, N. Kenmochi, and P. Krejčí, who kindly read some parts of the draft and contributed several suggestions.

I am also indebted to H. Brézis, who included this volume in the series he directs; and to the staff of Birkäuser, who assisted me in the redaction of this work.[1]

Povo di Trento, August 1996

Author's address: Dipartimento di Matematica dell'Università di Trento - I 38050 Povo (Trento) - Italy. Tel.: (39) 461-881635, Fax: (39) 461-881624. e-mail: visintin@science.unitn.it (Internet)

---

[1] *Work?* Does a researcher get a chance to write a book every day?

Introduction

This volume deals with modelling and analysis of phase transitions in two-phase systems. It aims to offer an introduction to this subject, and also to present some research problems. The models that we construct represent a compromise between the description of physical phenomena and analytical requirements; accordingly our presentation is characterized by a strict interplay between physics and mathematics.

This essay expresses a point of view, rather than an overview of the large existing literature.

Phase Transitions. The *Stefan problem* is the classical model of phase transition in solid–liquid systems, which it describes in a rather simplified way, by accounting for heat diffusion in each phase and exchange of latent heat at the phase interfaces. This is an example of *free* (or *moving) boundary problem,* as the evolution of those interfaces is a priori unknown.

Several thousand papers have been devoted to this problem and to its multifold generalizations. Why such popularity? A first answer refers to applications. Phase transitions occur in many industrial and nonindustrial processes, for example, metal casting, steel annealing, crystal growth, thermal welding, freezing of soil, food conservation, and so on. But there is also another reason. The Stefan problem is a typical *model problem:* although its formulation is rather simple, it exhibits some attractive mathematical features, which also underlie more sophisticated models.

We shall move from this problem to formulate a more refined model, which accounts for *undercooling* and *superheating, surface tension, and phase nucleation.*

Two Approaches. At first, we introduce a *convex macroscopic* model, which accounts for local equilibrium and for possible occurrence of the so-called *mushy region* (namely, a very fine mixture of solid and liquid). We formulate a problem in the framework of Sobolev spaces, and show its well-posedness via monotonicity and compactness techniques. These developments are fairly standard.

Then we consider a *mesoscopic* (i.e., small length scale) model, whose analysis appears more challenging. At such a scale, only pure phases (without mushy regions) can be represented. At variance with the convex macroscopic model,
 • the requirement of pure phases corresponds to a nonconvex constraint;
 • the nonconvexity is a source of *metastability,* in particular of the *undercooling* which is required for solid nucleation; [1]

---

[1] This undercooling may even be of the order of hundreds of degrees for some materials, e.g. about $400K$ for platinum.

- the surface tension is represented by a *space-interaction* term in the free energy, which penalizes space derivatives;
- this additional term forces the phases to be separated by smooth interfaces, and gives rise to a *fine scale structure;*
- that term also provides the compactness needed to *sustain* the nonconvex constraint.

Similar features are exhibited by models arising in other phenomena as well, for example, in ferromagnetism.

Outline of the Book. This volume consists of two main parts, which respectively deal with some classes of nonlinear P.D.E.s and phase transitions. An appendix illustrates some mathematical tools.

In Chap. I we present an informal discussion about modelling of space-distributed systems and analysis of P.D.E.s. This is a review of well-known methods and results; it is intended to introduce the reader to some points of view that are at the basis of this essay, and is especially addressed to less experienced researchers.

Chaps. II and III are then devoted to initial and boundary value problems for some classes of (possibly degenerate) parabolic P.D.E.s, in view of successive applications to problems of phase transitions. Chap. II deals with differential inclusions of the form $\partial \alpha(u)/\partial t - \Delta u \ni f$ in the sense of distributions, where α is a (possibly multi-valued) maximal monotone function, and Δ is the Laplace operator. This includes the weak formulation of the *Stefan* and *Hele-Shaw problems,* introduced in Chap. IV. *Doubly nonlinear* (possibly degenerate) parabolic inclusions are then studied in Chap. III, in view of the models of Chap. V.

Existence results are proved for corresponding boundary and initial value problems via standard methods: time-discretization, derivation of a priori estimates by L^p-techniques, and passage to the limit by compactness and monotonicity procedures. Well-posedness and regularity properties are also proved. Several of these results can be extended to systems of nonlinear P.D.E.s.

We then move to phase transitions. In Chap. IV we introduce and discuss the strong and the weak formulation of the standard Stefan problem, either in a single or in several space dimensions. We also mention some related models: the Muskat and Hele-Shaw problems, and a vectorial Stefan-type model representing ferromagnetic processes. In Chap. V we discuss some generalizations, including so-called *phase relaxation* and phase transitions in multi-component systems. We also outline the viewpoint of *nonequilibrium thermodynamics,* which provides a powerful unifying approach for several phenomena, on the basis of the second principle. These problems can be formulated in terms of equations of the form we studied in Chaps. II and III, and those results can be applied. The variety of problems that can be studied is fairly large; so we refrain from dealing with specific examples in detail, and rather illustrate the main techniques that can be used in the analysis.

We then deal with surface tension effects in solid–liquid systems; this requires the use of a mesoscopic length scale and the formulation of nonconvex models. In

Chap. VI we introduce free energy and entropy potentials that include a perimeter term, and derive the classical *Gibbs-Thomson law*.

In the next two chapters we introduce the reader to more advanced problems under current research. Chap. VII deals with stable and metastable equilibria in two-phase systems, as well as phase nucleation phenomena. In Chap. VIII we generalize the Stefan model, to account for the Gibbs-Thomson law and phase nucleation. We present a first formulation in the framework of Sobolev spaces, and prove existence of a solution by an argument that is largely due to Luckhaus. In Chap. IX we then amend that model on the basis of physical motivations, and introduce what might be called a *two-scale Stefan problem,* to which we extend the existence argument. We then outline some related models.

The two final chapters constitute an appendix; they deal with some analytical tools used in this volume, and include some complements. In Chap. X we briefly illustrate how in certain cases strict convexity and strict monotonicity may entail strong convergence properties. Finally, in Chap. XI we review some definitions and basic results about function spaces, compactness, convex analysis, maximal monotone operators, nonlinear semigroups, functional minimization, and geometric measure theory. Here some complements have been inserted.

Several exercises have been unevenly distributed throughout the text.

Note. Chapters are labelled by Roman numbers. They are indicated in referring to theorems and formulae of other chapters, and to sections; in other cases they are omitted. Sections are labelled by Arabic numbers.

Reader's Guide

Here we indicate two possible paths for reading this book, and provide some general references.

Analysis of Some Nonlinear P.D.E.s.: Chaps. I, II, III, X, XI. Chap. I is a digression about modelling and analysis of nonlinear P.D.E.s. Some fundamental analytical concepts and tools are reviewed in Chap. XI, which is aimed at providing the reader with the tools needed to read Chaps. II and III. These two chapters deal with certain nonlinear parabolic P.D.E.s, including some of the models that are derived in the rest of the book. *Compactness by strict convexity* is illustrated apart in Chap. X.

Modelling of Phase Transitions: Chaps. IV through IX. Chaps. IV and V are an introduction to the Stefan problem and its generalizations, and do not exhibit any mathematical difficulty. Chaps. VI and VII deal with surface tension effects, and are also fairly accessible. Chap. VIII and IX generalize the Stefan model, accounting for phase nucleation and surface tension effects. The two latter chapters are aimed at introducing the reader to a topic under current research, and are based on some nonelementary techniques of analysis of nonlinear P.D.E.s in Sobolev spaces.

These two paths can be read independently of each other.

A Guided Tour. The following sections might be chosen for first reading, or form the basis of a Ph.D.-level course:

I.1, I.2, I.3;
II.1, II.2, II.3, II.5;
IV.1 through IV.6, IV.9;
V.1, V.2;
VI.1, VI.4;
VII.1, VII.2, VII.3;
VIII.1, VIII.2;
X.1, X.2, X.3.

Prelude

To start this book I intended to devote a few words to mathematical research. A nice opportunity was then provided by a talk which professor Ennio De Giorgi recently gave at the University of Trento. I partly translate it here with De Giorgi's kind permission.

To make mathematics, one should have in mind the famous sentence of William Shakespeare: "There are more things in heaven and earth, Horatio, than have been dreamt of in our philosophy." [1] *One should also be pleased to know that more things expect to be discovered than have already been discovered.*

One needs some freedom to dream, to imagine visible and invisible worlds; and a certain taste to communicate, otherwise the fantasy would remain chaotic. When one tries to communicate the ideas he has conceived, he must give them a precise form. Here is the source of the mathematical rigour: the distinction between what follows from the assumptions and what may be true but was not derived from them, this stems from a desire for communication and clarity.

My mathematical culture is largely due to friendship, to the fact that I met some good friends who talked to me with much simplicity about what they knew and also — this is essential — about what they ignored.

It is also important not to be confined by specializations, to maintain communication with scholars of different branches of culture: I talk with the physicist, the biologist, the linguist, the philosopher, the theologian, to keep contact with other persons who love learning. This ideal goes back to the great Greek philosophers and to the Bible: Socrates died for the sake of knowledge, and the Book of Proverbs tells us about the knowledge which was with our Lord when He created the world, and which goes towards the one who searches and longs for it, being conscious of his own ignorance.

Science is part of that knowledge, and a disinterested exchange provides an enrichment which can also improve scientific activity. If one loses the ability and the interest to communicate, even his comprehension of the topics he teaches will be weakened.

---

[1] Hamlet, Prince of Denmark. Act I.

Chapter I. Models and P.D.E.s

Outline

This chapter deals with some general features about modelling and analysis of nonlinear partial differential equations (P.D.E.s) and minimization problems. The presentation is nontechnical and expository.

Prerequisites. Some background in functional analysis, and some tolerance towards nontechnical digressions.

I.1 Modelling and Analysis

> *Die Mathematiker sind eine Art Franzosen:*
> *redet man zu ihnen,*
> *so übersetzen sie es in ihre Sprache,*
> *und dann ist es alsobald ganz etwas anders.*
> (Johann Wolfgang Goethe) (1)

... indeed mathematical models are not always very transparent, and *modelling* is not an obvious concept, either. In this chapter we discuss it, and outline some of the viewpoints that underlie this book. We also overview some general techniques of analysis of P.D.E.s and variational problems, for the benefit of the less experienced reader.

We begin with a general discussion about modelling and mathematical analysis of problems related to applications.

A Play with Different Characters. At first, we try to represent the way applied researchers and mathematical analysts interact (or might interact ...). This picture is oversimplified: it is just a *model,* and hopefully will only be regarded as the starting point for a more thorough discussion.

Let us distinguish the following steps in the study of *applied problems* (namely, problems issued from applications) which are susceptible to a mathematical treatment.

---

(1) Maximen und Reflexionen 1279 in der Artemis Gedächtnisausgabe der Werke Goethes, Zürich 1950, vol. 9, p. 660. (This reference was kindly indicated to this author by Prof. H. Heuser.)

Translation: *Mathematicians are like Frenchmen: whatever you say to them, they translate into their own language, and then it is something entirely different.*

(i) *Study of the Phenomenon.* This is left to the applied researcher — be (s)he a physicist, a biologist, an engineer, an economist, or the like. Henceforth we often refer to physicists and physical problems, since in this essay we deal with such problems; however, our discussion is intended for a more general scope.

As a pictorial image, one might think that a physicist brings a problem to the attention of a mathematician, although sometimes the mathematician has the initiative. This leads to our second step.

(ii) *Formulation of a Quantitative Model.* [2] This is interdisciplinary work between physics and mathematics. Here a compromise must be reached between the exigencies of the two disciplines. Variational principles and differential equations are part of the typical outcome of this pursuit, and are our main concern in this volume.

In formulating a model, an effort should be made to determine its limits of validity. It should also be realized that the same phenomenon might be described by completely different, even contrasting, models — typically with different limits of validity. For instance, in multi-phase systems the different phases may be thought of as separated either by *thin* layers or (and this is the viewpoint we adopt in this essay) by surfaces. The first model requires the use of a very small length scale, whereas the second one can be applied at the macroscopic scale.

(iii) *Analysis of the Problem.* This is essentially left to the mathematical analyst. Obviously this step requires the preliminary formulation of an analytical problem. This issue is less trivial than it might appear: in P.D.E.s and in many variational problems the choice of the functional environment is a question of major importance. For the cases we are interested in, a *continuous* problem is formulated, in which the unknowns are fields defined on a domain in space (or space-time).

We regard the model and its analytical formulation as different concepts: a *precise* analytical problem does not just consist of a set of *formulae*. In this book we systematically distinguish between two phases of analysis. At first we describe the quantitative features of a phenomenon with little concern for mathematical rigour. Then we provide a precise formulation and analyse it rigorously, without any reference to the underlying applications.

In the next section we briefly discuss the main issues of the mathematical analysis of a large class of problems. These include existence of a solution, its uniqueness, *qualitative* properties, and so on.

(iv) *Numerical Approximation.* Computations can be accomplished only after the continuous problem has been discretized, which corresponds to passing from infinite to finite dimension. [3] Therefore an approximation procedure must be

---

[2] By the term *model* we mean a representation, a metaphor, a picture that describes a phenomenon, or (more pompously) *a way to organize the reality.*

[3] *Caveat:* Ordinary differential equations (O.D.E.s) are *infinite* dimensional problems but are often labelled as *finite* dimensional, since the unknown functions are time-dependent Euclidean vectors.

chosen. The final product is an *algorithm* for the computation of the approximate solution.

(v) *Implementation of the Algorithm.* Usually numerical simulations are accomplished at first on some simplified settings, to test the soundness of the method, and then on data of interest for applications. They represent an experimental counterpart to the previous analysis, and may suggest modifications of the latter. [4]

(vi) *Interpretation of the Mathematical Results.* With reference to the preceding picture, we might say that at this point theoretic and numerical results are offered for the evaluation of the physicist. Of course if they are at variance with the physical evidence, the previous process must be revised.

In this essay we mainly deal with steps (ii) and (iii) for models of phase transitions.

Some Methodological Remarks. Obviously the above picture is oversimplified. Some *feedback* between the above steps is in order, and the protagonists are expected to have some interdisciplinary inclination. [5] For instance, the mathematician may contribute to the formulation of the model — this is a point of view which we try to foster in this essay. [6]

By the same token, approximation procedures may be of both numerical and theoretic interest; for instance, existence arguments can be based on convergence of discretizated schemes (in this volume we often use approximation by time discretization). Moreover, numerical simulations performed on appropriate mathematical models may replace experiments (to the dismay of wind tunnel builders, for instance ...).

In several cases one proceeds to numerical simulations without the comfort and guide of analysis. Of course this bold behaviour has its own risks, but we must realize that this occurs for most applied problems. [7]

Usually the original setting is far too complicated to be treated as it is. Therefore simplifications are in order, and the question arises of the relevance and relative importance of the various features of the problem. For instance, one should not neglect a term that is quantitatively more important than one which is left in (an issue which often is not easily established prior to effective computation).

---

[4] The presence of an experimental moment in the mathematical investigation is of high significance, and has radically influenced most of scientific disciplines in the last half century.

[5] The more a researcher is interested in the solution of *problems*, rather than in the academic development of knowledge, the more (s)he must be disposed to compare methods and results originated from different contexts — without ignoring the difficulties this may imply. Actually, physicists and other applied researchers are used to implementing by themselves the numerical solution of many of the problems they study.

[6] Indeed if mathematics is the language of sciences, then it should be capable of conveying concepts endowed with an applied content. In fact the roots of many mathematical theories can be found in applications. Actually, the distinction between pure and applied (or rather *application-oriented*) mathematics used to be definitely less marked in the past.

[7] Engineers tend to become impatient, and then start computing. They hardly stand waiting until mathematicians eventually *solve* some of their problems.

As may be expected, in many cases a compromise is hardly reached between what the physicist demands and what the mathematician is able to accomplish rigourously. It is then convenient to deal with so-called *model problems*. [8] These are oversimplified problems, which may even appear as *caricatures* of the phenomenon under study, but which have some mathematical interest while retaining some relevant features of the original setting. [9] For instance, in most of either physical or nonphysical phenomena, several effects coexist; to deal with difficulties separately, it is expedient to formulate models that account for just some of these aspects, as we often do in this essay.

Other Problems. At this point the reader might have got the impression that computation is the final goal of modelling. This is not always true: qualitative results may also be useful in themselves, providing *insight* into the phenomenon under study.

Besides *modelling*, other mathematical issues are of major practical importance; for instance, *parameter identification, optimization,* and *control* are the focus of intense mathematical research. A class of problems of high importance in applications consists of the analysis of the limit behaviour of the solution as one or more parameters vary. This includes problems of *homogenization*, [10] in which some material coefficients oscillate on a small scale; for instance, this occurs in composite materials. Here the problem arises of deriving the macroscopic properties from the *microstructure*. Other examples are offered by *singular perturbations;* [11] here the limit problem is radically different from the perturbed problem; for example it is governed by a P.D.E. of lower order, or the boundary conditions are of different type. It may also happen that the solutions of the perturbed problems do not converge as the perturbation vanishes, but nevertheless one is able to study the limit behaviour of the *nonsingular part* of those solutions.

Space-Distributed Systems. In this book we mainly deal with the analysis of some either stationary or evolutive *space-distributed* systems. By this we mean that the state variables are either stationary or time dependent *fields*. These systems are typically described either via *variational principles,* differential equations, integral equations, or integro-differential equations. Here we confine ourselves to the two former classes.

---

[8] Here the term *model* has the meaning of *prototype* or *example*, not of metaphor of reality.

[9] (Mathematical) models should not be overrated. They only isolate few features of the phenomena, provide simplified and schematic pictures, may be based upon questionable hypotheses, have limited range of validity, or may be in poor agreement with experimental evidence, or may contradict each other.
  But what else do we have at our disposal to describe reality (quantitatively)?

[10] See, e.g., Bensoussan, Lions and Papanicolau [69], Dal Maso [163], and Tartar [524].

[11] See, e.g., Lions [352], Sanchez-Hubert and Sanchez-Palencia [494], and Sanchez-Palencia [495].

For these problems boundary (and initial) conditions are important ingredients, since equations set in the whole space, without boundary or initial conditions, are less frequently encountered in applications. In many cases continuous functions are not suited for these problems, and one is led to deal with functions defined almost everywhere. The necessity then arises of giving a meaning to the *trace* of the solution on the boundary. Here the *functional approach* seems especially convenient.

Although the point of view of space-distributed systems is fairly customary for mathematical analysts, it is not universally regarded as the most convenient for applications. For instance, the constitutive behaviour of materials is typically measured on nondistributed systems, otherwise the design of experimental tests would be much more difficult. The necessity then often arises of extrapolating those properties, to formulate models applicable to distributed systems.

On Linear and Nonlinear P.D.E.s. Infinite dimensional function spaces are the natural environment of P.D.E.s. The infinite number of degrees of freedom is strictly related to the fact that derivatives correlate the behaviour of the solution at different points. To grasp this issue, albeit in a rather indirect way, just remove the derivatives from a differential equation set in some Euclidean domain. The problem then remains infinite dimensional, but it is much more natural to regard it is as a family of independent finite dimensional equations in **R** (or **C**), parameterized by the points of the domain.

For linear P.D.E.s a classification by *types* is well-established, at least for the second order equations that are most frequently encountered in applications. At first one distinguishes the *principal part* of the differential operator; this consists of the terms that contain the *highest order* derivatives. One then speaks of elliptic, parabolic, hyperbolic, or ultrahyperbolic operators (and equations), according to the form of the principal part. [12] This classification reflects the fact that several basic features of the behaviour of solutions, including the well-posedness of corresponding boundary and/or initial value problems, are dictated by the form of the principal part, under appropriate conditions (e.g., on the sign of the terms). [13] This is strictly related to the stability of solutions of linear P.D.E.s with respect to large classes of perturbations.

In nonlinear P.D.E.s, infinite dimension is combined with nonlinearity; this generates a much wider variety of behaviours, which is at the basis of the large range of applications of these equations. For instance, nonlinear lower order terms may com-

---

[12] See, e.g., Baiocchi and Capelo [47; Sect. 7.1], and Renardy and Rogers [464; Sect. 2.1].

[13] This, however, does not apply to parabolic equations. For instance, the principal part of the *heat operator* $a\partial/\partial t - \Delta$ is $-\Delta$; but the well-posedness of the corresponding initial value problem depends on the sign of a.

One might then consider an alternative definition of the principal part, which includes any derivative D^α such that no term of the form $D^{\alpha+\beta}$ with $|\beta| > 0$ occurs (see Sect. XI.1 for the definition of these operators). With this convention the heat operator would coincide with its principal part.

pete with leading order ones, depending on the rate of growth of the nonlinearity. Consequently, nonlinear equations tend to be rather unstable under perturbations.

On Classification of Nonlinear P.D.E.s. As might be expected, the extension of the preceding classification by types to nonlinear P.D.E.s is not satisfactory, since it accounts for infinite dimension but not for nonlinearity.

The nonlinear P.D.E.s that are most usually encountered in applications are usually classified according to the order of the derivatives which occur in the nonlinearities. In this way one accounts for both infinite dimension and nonlinearity. One speaks of *semilinear* operators (or equations) if the principal part is linear; otherwise the operator is said to be *quasilinear (fully nonlinear,* resp.) whenever leading order derivatives occur linearly (nonlinearly, resp.). Here are some typical examples:

$$
\begin{aligned}
\frac{\partial u}{\partial t} - \Delta u &= f & \text{(linear)}, \\
\frac{\partial u}{\partial t} - \Delta u &= g(u) & \text{(semilinear)}, \\
\frac{\partial u}{\partial t} - \nabla \cdot \vec{\alpha}(\nabla u) &= f & \text{(quasilinear)}, \\
\frac{\partial u}{\partial t} - g(\Delta u) &= f & \text{(fully nonlinear)},
\end{aligned}
\qquad (1.1)
$$

(Here f, g, $\vec{\alpha}$ are given functions, $\Delta := \sum_{i=1}^{N} \partial^2/\partial x_i^2$.) [14]

In several cases this classification corresponds to different types and degrees of difficulty in the analysis of these equations; however, it is not optimal either. In several semilinear and quasilinear P.D.E.s, one can then apply the previous *linear* classification (based on the distinction between elliptic, parabolic, hyperbolic, and ultrahyperbolic equations) to the principal part.

Lions [351] suggested classifying nonlinear P.D.E.s by methods of analysis, and studied a large collection of examples accordingly. P.D.E.s are also strictly related to applications, which represent their source, aim, and lifeblood; this offers a way of organizing their analysis and presentation, although similar equations often arise in different physical settings — a fact that confirms their universality.

---

[14] According to this classification an equation of the form $\partial \alpha(u)/\partial t - \Delta u = f$ would be labelled as *semilinear,* since its principal part is $-\Delta u$. Using instead the alternative definition of principal part we previously introduced, this equation would be regarded a *quasilinear.*

In Chaps. II and III we take the freedom of using this latter convention.

I.2 Nonlinear P.D.E.s and Minimization Problems

In this essay we model some physical phenomena either by an initial and boundary value problem for a nonlinear P.D.E., or by a minimization problem, and study their mathematical properties. Here we outline some general features of the analysis of such problems.

Main Questions. What does it mean "to solve a mathematical problem"? [15]

This is not a trivial question. The answer depends not only on the kind of problem, but also on the viewpoint of the researcher. For the mathematical analyst, nowadays proving existence of a solution is the basic requirement, and deriving qualitative properties is part of the standard treatment to which the problem is submitted. On the other hand, the numerical analyst will especially demand an efficient procedure for computing an approximate solution; the pure mathematician will then require some agreement between the (either theoretical or numerical) mathematical results and the experimental evidence.

Let us review some of the issues in which the mathematical analyst is interested.

Existence of a Solution. This is indispensable for the model to make sense. Exceptionally, one is able to prove *nonexistence* of the solution, and so to exclude any validity of the model — which may be a useful piece of information after all.

This point of view required some time to emerge. In the pioneering age of differential equations, much effort was devoted to exhibit solutions in *closed form*, that is, in terms of analytical expressions. This was only achieved in a few cases. Attention was also directed to equations with irregular data, for which derivation of an explicit solution was especially problematic, or simply impossible. Equations without solution were even discovered, and the risk of talking of something that might even not exist was perceived.

All this contributed to drive attention towards the question of existence of a solution. Finally the dramatic improvement of computing power due to electronic computers deeply modified the scene, and influenced the concept of solution. In particular, search for closed form solutions lost most of its previous importance, and more and more attention was devoted to qualitative properties and approximation methods.

In this book we systematically deal with equations with nonsmooth solutions. In particular we study some *free boundary problems*. These problems consist of one or more differential equations set in a domain whose boundary is *free*, that is, unknown. This lack of information is balanced by some extra-conditions, which typically couple the free boundary and the solution of the equation itself. The problem then requires the determination of the free boundary together with the solution of the equation.

A multitude of physical and engineering problems can be formulated in these

---

[15] W. Sawyer acutely suggested that "to solve a problem means to reduce it to a simpler one."

terms. Free boundary problems have been the object of intense research in the last twenty years or so, see the *Book Selection*.

In this volume we are concerned with several initial and boundary value problems for *evolution* P.D.E.s. For those problems different sorts of existence statements may be considered. In some cases one is able to prove existence of a solution in any prescribed time interval. In other cases a solution only exists in a time interval which depends on the specific data. For certain problems one can show existence of a solution for a *bilateral* time interval including the initial instant; this is typical of nondissipative wave propagation phenomena governed by hyperbolic equations. In other cases (e.g., for dissipative phenomena governed by parabolic equations), the solution cannot be extended to times preceding the initial instant.

In this book we mainly deal with dissipative phenomena, consider initial and boundary value problems for parabolic equations or systems, and try to prove existence of a solution for arbitrary unilateral time intervals.

Uniqueness of the Solution. For many problems this is a natural requirement. However, usually one is disposed to accept problems for which uniqueness of the solution is not established. In some cases nonuniqueness is even consistent with the model; it is then of interest to point out properties of the set of solutions, such as existence of maximal or minimal solutions (whenever they are elements of an ordered space).

Qualitative Properties of Equations. This term refers to several questions. Here are some of the main issues.

Continuity. If the data converge, do the corresponding solutions converge? More specifically, let the data and the solutions belong to some metric spaces D and S, respectively, and the solution operator $\mathcal{A} : D \to S$ be well-defined and single-valued. Then does $d_n \to d$ in D imply $\mathcal{A}(d_n) \to \mathcal{A}(d)$ in S? If the mapping $\mathcal{A}$ is continuous the problem is said to be *well-posed in the sense of Hadamard*. [16]

Some classes of problems are *ill-posed:* either the solution does not exist, or it is nonunique, or it depends discontinuously on the data. For instance, this occurs for *final* value problems for parabolic equations; this *pathological* behaviour may be ascribed to the fact that these equations typically represent dissipative phenomena. However, ill-posedness is not confined to *unnatural* problems: *inverse* and *identification problems* are well-motivated by applications, nevertheless they are typically ill-posed. [17]

Compactness and Closure. Let the solution operator $\mathcal{A} : D \to S$ be possibly multi-valued, and $d_n \to d$ in D. Then is it possible to extract a converging subsequence from $\{\mathcal{A}(d_n)\}$ *(compactness)*? If so, does $\mathcal{A}(d_n) \ni s_n \to s$ in S imply $s \in \mathcal{A}(d)$ *(closure)*? These properties are weaker and more fundamental than continuity.

---

[16] See Hadamard [286], of 1923.

[17] See, e.g., Lavrent'ev, Romanov and Shishatskii [344], Lavrent'ev, Romanov and Vasiliev [345], Payne [445], and Romanov [478].

Structural Stability. This concept is even deeper. It means stability of the qualitative behaviour of the system, with respect to perturbations not only of the data but also of other elements of the problem, typically the operator, in a sense to be specified.

Monotonicity. If D and S are ordered spaces, does $d_1 \leq d_2$ imply $\mathcal{A}(d_1) \leq \mathcal{A}(d_2)$? In linear problems this is often related to so-called *maximum and minimum principles*. This leads one to define the concepts of *upper* and *lower solutions*.

Bifurcations. In several cases it is of interest to study how a solution depends on a parameter. This dependence is usually represented by a curve in a function space, that can bifurcate. The analysis of the behaviour of the solution near the branching points is of special interest.

Asymptotic Behaviour. For evolution problems the asymptotic behaviour of a solution (existing for all time) as time tends to $+\infty$ is of paramount importance. In several cases this piece of information is even more important than the analysis of the transient. Several concepts of convergence and stability have been introduced, starting with the pioneering works of Poincaré and Liapounov. Important questions concern the existence of a periodic solution, its asymptotic stability, its attractivity, and so on.

This point of view is at the basis of an important field of research, known as the theory of *dynamical systems,* which uses analytical and topological methods. [18] This subject has been deeply investigated in the case of O.D.E.s for more than a century. More recently this point of view has also been applied to P.D.E.s, that is, to infinite dimensional systems.

Regularity of the Solution. At first one is interested in proving the existence of a solution under minimal assumptions, then in deriving further regularity under additional hypotheses on the data. This is especially relevant when one enlarges the class of admissible solutions by weakening the concept of solution, as one is sometimes induced to do when facing a difficult problem. [19] However, although fishing in a larger sea may help to get the existence of a solution, the exigency then arises of discussing the *quality* of the prey — the solution in this case.

At variance with the linear case, for nonlinear P.D.E.s usually there is a fairly low barrier to the regularity of the solution. For linear P.D.E.s intermediate regularity statements can often be derived via interpolation in function spaces, whereas for nonlinear problems only few results are known in that direction. [20]

One should consider that any applied model has its own (albeit not always precise) limits of validity. This is important for nonlinear problems, which are not *scale invariant* in general, [21] and becomes especially relevant in connection with

---

[18] See, e.g., Hale [288] and Temam [531].

[19] One is tempted to guess that any P.D.E. has a solution, provided that the concept of solution is suitably defined.

[20] See, e.g., Tartar [521].

[21] For instance, the heat equation $\partial \theta / \partial t - \Delta \theta = f$ is invariant under any scaling that preserves

regularity properties. For instance, from the viewpoint of applications, it does not make much sense to investigate very fine regularity properties of crude physical models; this is especially evident when this means considering a fine length scale, at which the model loses its validity.

Qualitative Properties of Minimization Problems. Similar questions can be set for variational problems. Here also, one is interested in the *existence* and (possibly) the *uniqueness* of the minimum point. A natural approach to these problems consists of the study of *minimizing sequences*, which are at the basis of the so-called *direct method* of the calculus of variations. [22]

Here qualitative properties include continuous dependence of the minimum point and of the minimum value on the data *(Hadamard well-posedness)*. Another important property, called *well-posedness in the sense of Tikhonov*, consists of the convergence of any minimizing sequence to a (necessarily unique) minimum point. One speaks of *generalized well-posedness in the sense of Tikhonov*, if from any minimizing sequence a converging subsequence can be extracted, which tends to a (possibly not unique) minimum point. In several cases Hadamard's and Tikhonov's well-posedness are equivalent. [23]

Numerical Approximation. Let us now consider the effective numerical solution of the problem. Even in the few cases in which the solution can be represented by an analytical expression, usually this cannot be computed exactly. Effective solution requires approximation, that is, representing the solution as the limit of solutions of a family of *approximate* problems. Each approximation scheme is a family of *finite dimensional* problems, which can be solved numerically by means of a computer. Typically one considers a sequence of *discretized problems*, which depend on a vanishing discretization parameter, h say. More generally some intermediate (e.g., partially discretized) problems are formulated between the original infinite dimensional problem and the completely discretized one.

The choice of the numerical scheme depends on several features: the properties of the mathematical problem under consideration (e.g., the size of the data), the characteristics of the equipment to be used for numerical computations, the precision of the data, the required accuracy, the budget at disposal, and so on.

There is a wide border region between theoretical and numerical analysis. Let us mention some of the main issues.

Consistency. This means that the exact solution solves the approximate problem, but for an *error* that vanishes as $h \to 0$. This guarantees that one is actually approximating the original problem.

Stability. Since data are inevitably affected by roundoff errors, when using iterative or stepwise schemes there is the risk that errors might grow out of control.

---

the ratio x^2/t; but this property fails for nonlinear equations such as $\partial \alpha(\theta)/\partial t - \Delta\theta = f$.

[22] See Sect. XI.7.

[23] See, e.g., Dontchev and Zolezzi [199; Chap. II].

Stability means that errors remain bounded as $h \to 0$.

Convergence. As $h \to 0$ the approximate solution must converge to a solution of the original problem. This yields existence of a solution, too.

Estimation of the Approximation Error. Evaluation of the convergence rate of the approximate solution in terms of the approximation parameter(s) is of paramount importance in comparing different approximation procedures.

The answer to these questions depends on the specific approximation scheme, on the qualitative properties of the solution of the exact and approximate problems, and on their relations.

Problem Formulation. We do not discuss the advantages offered by the functional approach in analysis. [24] Usually one can choose among several possible formulations of the same problem. At first one must select an appropriate unknown function. The classical transformations of Fourier, Laplace, and others show that in some cases it is convenient to change the unknown; these transformations are most appropriate for *linear* differential equations. In Sect. II.5 we apply transformations by either time or space integration to nonlinear P.D.E.s.

The choice of the specific functional environment is a fundamental issue. For instance, one might deal with P.D.E.s in the classical C^k-spaces, or, as we systematically do in this essay, in Sobolev spaces. In the latter case, derivatives are interpreted in the sense of distributions. [25] Moreover, one might look for a *strong* solution, namely, for one in which the equation is fulfilled almost everywhere, or for a *weak* solution, in which the equation just holds in the sense of distributions. A different (weaker) kind of solution is encountered in the theory of *semigroups of nonlinear contractions;* see Sect. XI.6. In several problems one can use the especially weak concept of *viscosity solution,* [26] or *Young measures.* [27] One may also change the *pivot space;* in Sect. II.6 we see an example in which it is convenient to identify $H^{-1}(\Omega)$ (instead of $L^2(\Omega)$) with its topological dual.

Here we mainly discuss weak formulations, which is the approach we most frequently use in this book. The necessity of proving existence of a solution often leads one to introduce weaker and weaker definitions of solution, to search in larger and larger sets — a method that led to the generalization of many concepts in analysis. But by enlarging the class of admissible solutions there is the risk of including too many objects, at the expense of uniqueness of the solution. Indeed the most successful generalizations are those that provide both existence and uniqueness of the solution (provided that this property is justified by the model).

---

[24] However, there are also other approaches; for instance many results can be derived via real analysis and measure theory.

[25] See Sect. XI.1 for the main definitions.

[26] See, e.g., Barles [59], Crandall, H. Ishii and P.L. Lions [148], Crandall and P.L. Lions [150], Evans [216; Chap. 6], and P.L. Lions [353].

[27] See, e.g., DiPerna [196], Kinderlehrer and Pedregal [323], Tartar [522], and Young [590].

A Class of Nonlinear O.D.E.s. As P.D.E.s generalize O.D.E.s, it is natural to expect that the latter might suggest procedures of analysis for the former. Let us then consider a Cauchy problem of the form $y' = f(y,t)$, $y(0) = y^0$. Here $t \geq 0$, $y = y(t) \in \mathbf{R}^N$ ($N \geq 1$), f is a continuous function $\mathbf{R}^N \times \mathbf{R}^+ \to \mathbf{R}^N$, $y^0 \in \mathbf{R}^N$ is given. For this problem we dispose of three classical results, which are respectively based on *compactness, contraction, and monotonicity.*

Peano's theorem states existence of a (nonnecessarily unique) solution in a small (bilateral) time interval. The simple argument is based on application of the *Schauder fixed-point theorem* to the mapping $\Phi : y \mapsto y^0 + \int_0^t f(y(\tau), \tau) d\tau$ in the space of continuous functions $[0,T] \to \mathbf{R}^N$ (for a sufficiently small T). Here only the vectorial and topological structures are used.

On the other hand the *Cauchy-Lipschitz-Picard theorem* assumes that f is Lipschitz continuous with respect to the first argument, uniformly with respect to the second one, and states Hadamard's well-posedness for any time. In this case existence and uniqueness of a fixed point for the mapping Φ is derived via the classical *Banach contraction mapping principle.* Here the metric structure is used.

Another result assumes $-f$ to be monotone with respect to the first argument, and provides Hadamard's well-posedness for positive times.

Many evolution P.D.E.s are regarded as *abstract O.D.E.s,* representing evolution in function spaces. It is then of little surprise that the preceding results can be extended in a number of ways to nonlinear P.D.E.s, and are the basis of important theories. For instance, several semilinear P.D.E.s of evolution can be represented as fixed point problems, and well-posedness can be proved by means of the contraction mapping principle. Later we see that compactness and monotonicity are also fairly universal tools. [28]

Methods for Proving Existence of a Solution. Here we discuss two fundamental classes of techniques for the proof of existence of a solution for nonlinear equations: iterative methods and the "approximation — a priori estimate — passage to the limit" procedure.

Iterative Methods. These techniques play a central role in numerical analysis, and are especially suitable for effective computation. Important examples are the Newton method with its generalizations, and the argument of the Cauchy-Lipschitz-Picard theorem, which we briefly illustrate here.

The fixed point for the mapping Φ (previously defined) can be determined by means of an iterative scheme: choose any continuous function y_0, and define the sequence $\{y_n\}$ via *successive approximations,* $y_n := \Phi(y_{n-1})$. If $T > 0$ is sufficiently small, the operator Φ is a (strict) contraction in the space $C^0([0,T])$; hence y_n converges to the unique fixed point.

---

[28] See Sect. XI.3 for a brief review of compactness results, and Sects. XI.5 and XI.6 for results based on monotonicity methods for Cauchy problems in Banach spaces.

This iteration procedure is especially satisfactory, also for numerical purposes: the *initial guess* y_0 is arbitrary, the whole sequence converges, the procedure is stable with respect to a large class of perturbations, the approximation error can be estimated, and so on. The procedure can be repeated in the interval $[T, 2T]$, and so on.

Approximation Methods. As we saw, these methods consist of replacing the original problem by a family, typically a sequence, of *simpler* problems, for which one can prove existence of a solution, and which *formally* converge to the original one. Usually one can choose among several approximation procedures; some of them may also be used for numerical computations. Here are some examples.

Finite Dimensional Approximation, for instance, via discretization, or by projection on a subspace of the function space to which the solution is expected to belong. These techniques include the Galerkin method for stationary space-distributed P.D.E.s, the Faedo-Galerkin method for evolution space-distributed P.D.E.s, finite element approximation, finite difference approximation, spectral methods, and so on.

For evolution P.D.E.s one can choose among (at least) three sorts of discretization procedures:

(i) Time discretization yields a stationary P.D.E.. For nonlinear problems *implicit* time discretized schemes usually behave better than explicit procedures, in that they are more stable in the sense we previously mentioned. In this book we often use this technique.

(ii) Space discretization yields a system of O.D.E.s.

(iii) Coupled space and time discretization yields a finite dimensional problem. This method may be especially useful in view of the effective numerical solution of the problem.

Time Delay Approximation. This method consists of approximating an evolution equation by a delayed equation; this is especially convenient for semilinear equations. For instance, a term $t \mapsto f(u(t))$ might be replaced by $t \mapsto f(u(t - \tau))$ for $\tau > 0$. For an initial value problem, the evolution of u in $[0, \tau]$ must be prescribed. The behaviour of $f(u(t - \tau))$ for $\tau \le t \le 2\tau$ is then known from that of u in $[0, \tau]$; the approximate problem can then be solved stepwise in successive intervals of length τ. One then passes to the limit as $\tau \to 0$.

Regularization. This can be obtained via *perturbation*. For instance, by adding the term $-\varepsilon \partial^2/\partial t^2$ ($\varepsilon > 0$), a second order parabolic operator is transformed in an elliptic one *(elliptic regularization)*. Similarly a parabolic operator is obtained by adding $\varepsilon \partial/\partial t$ to a second order elliptic operator *(parabolic regularization)*. One then passes to the limit as $\varepsilon \to 0$.

Regularization can also be obtained via *smoothing* procedures. For instance, if f is an irregular function, a term $f(u)$ may be replaced by $(f * \varphi_\varepsilon)(u)$, where ε is a positive parameter, φ_ε is a so-called *mollifier,* and $*$ represents convolution.

Penalization. This is especially convenient for minimization problems. For instance, a convex constraint $u \in K$ may be replaced by adding a contribution such as $\mathrm{dist}(u, K)/n$ to the functional ($\mathrm{dist}(u, K)$:= distance of u from K). In the limit as $n \to \infty$, the constraint should then be recovered.

Cut-Off. For instance, an unbounded function $f(u)$ might be replaced by $f_n(u) := \min\{\max\{f(u), -n\}, n\}$. One must then pass to the limit as $n \to \infty$.

Usually existence of a solution for any of these approximate problems relies (more or less directly) upon a fixed-point theorem.

In many problems there is no preferable approximation technique, although one method may be more natural than another, or may simplify the analysis, or may suggest further results. Then the choice often depends on the taste and the background of the analyst.

A Priori Estimates. [29] Let us come to the second step of our proof of existence of a solution. After *solving* the approximate problems (or rather proving that each of these problems has a solution), one shows that these approximate solutions are uniformly bounded in some function space. Then the same bound also holds for the limit solution, if it exists and belongs to the same space. [30] Uniform estimates may also provide regularity results. In several cases these estimates have a meaning in applications, a typical example is offered by *energy estimates.* [31]

For minimization problems a priori estimates typically follow from *coerciveness* assumptions; this refers to the fact that the functional to be minimized is larger than certain quantities, which are then necessarily bounded for any minimizing sequence. For P.D.E.s there is no general rule on how to prove a priori estimates; however, they are usually derived by multiplying the approximate equation by suitable functions of the approximate solution. Other tools may also be used, for example, the maximum principle, and comparison among the terms of the approximate equation.

As we said, one can often choose among several approximation procedures. Moreover, even when there is a most convenient one, this may not be evident a priori. It is then expedient at first to proceed *formally* in the derivation of the estimates, [32] then on the basis of these results to choose the functional framework

---

[29] The name is due to the fact that they are established *a priori* of (i.e., before) the determination of the solution.

[30] This only applies to estimates in spaces that have a *predual,* cf. Theorem XI.3.1. For instance, this excludes spaces such as $C^0([0, T])$: this sort of regularity may be lost in the limit (however, as $C^0([0, T]) \subset L^\infty(0, T) = L^1(0, T)'$ with continuous injection, the limit function belongs to the latter space).

[31] The term *energy estimate* tends to be used rather extensively. For instance, in heat diffusion problems the heat equation represents energy conservation; instead the estimate obtained multiplying that equation by the temperature accounts for the second principle of thermodynamics, although some authors call it an energy estimate.

[32] In this volume we often speak of *formal procedures* and the like. These are heuristic arguments in which analytical calculations are accomplished without caring about their admissibility (e.g.,

and an appropriate approximation procedure, and to reproduce those estimates rigourously. This final step is absolutely necessary for the validity of the result.

Limit Procedure. A priori estimates enable one to apply the *weak compactness theorem* (see Theorem XI.3.1), which provides existence of weakly (or weakly star) converging subsequences. [33]

At this point one faces the crucial task: *to pass to the limit* in the approximate problem. Linear operators acting in Banach spaces are weakly continuous if and only if they are strongly continuous, [34] but in general this property fails for nonlinear operators. Therefore under weak convergence, linear relations are often preserved in the limit, whereas difficulties may arise in passing to the limit in the nonlinear terms.

For instance, $u_n \to u$ and $f(u_n) \to \xi$ weakly in $L^2(\Omega)$ do not imply $\xi = f(u)$ in Ω, whenever f is a nonlinear continuous real function. As a counterexample, consider $u_n(x) := \text{sign}(\sin(nx))$ for any $x \in]0, 2\pi[$, and $f(\xi) := \xi^2$ for any $\xi \in \mathbf{R}$. Then $u_n \to 0$ and $f(u_n) = 1 \to 1$ weakly in $L^2(0, 2\pi)$, but $f(0) \neq 1$.

In nonlinear problems passage to the limit is usually the hardest step. Much of the analysis of nonlinear P.D.E.s turns around the necessity of deriving sufficiently precise a priori estimates, such that they allow passing to the limit in the nonlinear terms of the equation.

Several techniques have been devised to accomplish that task; most of them stem from two fundamental tools: *monotonicity* and *compactness.* One can also consider methods of *compactness and monotonicity.* these include *pseudo-monotonicity*, [35] *compensated compactness*, [36] and so on. *Semicontinuity* often plays an important role, as we see in Chaps. II and III.

It must be noted that in the last years a considerable amount of research in the analysis of P.D.E.s has been devoted to tools beyond compactness and convexity, such as *viscosity solutions*, [37] *concentrated compactness*, [38] *quasi-convexity*, [39] *Young measures*, and others.

---

without checking regularity). Of course such an *optimistic* attitude must then be justified by a rigourous derivation.

Formal developments typically intervene in the derivation of the model, too. Here the applied researcher is assumed to be assisted by some insight into the phenomenon, to avoid fallacies.

[33] In nonmetrizable spaces one must deal with *nets* in place of sequences; see, e.g., Brézis [85], Ekeland and Temam [207; Sect. II.3]. However, here we confine ourselves to metrizable spaces.

[34] See, e.g., Brézis [92; p. 39].

[35] See, e.g., Brézis [85], Browder [99], and Lions [351; Sect. 2.2].

[36] These methods include the classical *div-curl lemma*, Theorem XI.3.13. See, e.g., Dacorogna [160], Evans [216; Chap. 5], Murat [409], Struwe [517; Chap. I.3], and Tartar [522, 523].

[37] See a previous footnote for some references.

[38] See, e.g., Evans [216; Chap. 4], P.L. Lions [354, 355], and Struwe [517; Chap. I], and references therein.

[39] See, e.g., Acerbi and Fusco [2], Ball [51], Dacorogna [160, 161; Chap. 5], and Evans [216; Chap. 3].

Analysis of Minimization Problems. Here the main property is lower semicontinuity of the functional, which may be compared with continuity of the operator in P.D.E.s. The most favourable setting is that in which the functional is *convex*, since in that case strong and weak lower semicontinuity are equivalent, cf. Proposition XI.4.2. For nonconvex functionals, compactness methods are in order.

For nonlinear P.D.E.s, several approximation procedures are at disposal, whereas for the minimization of functionals there are two fundamental techniques to prove the existence of a solution. The so-called *direct method* consists of the following steps, in its most simple form:

(i) consider a minimizing sequence;
(ii) prove that it is uniformly bounded in some reflexive Banach space B;
(iii) select a weakly converging subsequence, via some compactness argument, cf. Theorem XI.3.1;
(iv) show that the limit minimizes the functional via some lower semicontinuity property.

Alternatively, if B is the dual of a separable space, a weakly star converging subsequence is extracted.

The reader will notice that the basic features of this method are analogous to those we met in proving existence for P.D.E.s: approximation, derivation of a priori estimates, extraction of a converging subsequence, and passage to the limit.

Another technique consists of showing that the minimizer of the functional fulfills either an equation or a *variational inequality* (see later), and then proving existence of a solution for the latter problem. This is known as the *Euler-Lagrange method*.

Compactness. We already mentioned *weak compactness*, cf. Theorem XI.3.1, which plays a key role in the methods we are illustrating. In that case, by compactness one means existence of a converging subsequence. The same term may also indicate some *strengthening of the convergence*, that is, passage from weak to strong convergence; for instance, this is illustrated by the classical results of Ascoli-Arzelà, Rellich-Kondrachov, and Lions-Aubin (see Sect. XI.3).

Let us go back to the previous example in which $u_n \to u$ and $f(u_n) \to \xi$ weakly in $L^2(\Omega)$, with f continuous and nonlinear. Assume that Ω is bounded and has a *regular* boundary. If one can prove that $\{u_n\}$ is uniformly bounded in the Sobolev space $W^{1,1}(\Omega)$ (say), by the Rellich-Kondrachov theorem, then $u_n \to u$ strongly in $L^1(\Omega)$, hence in measure. Therefore the continuity of f yields $f(u_n) \to f(u)$ in measure, and we conclude that $\xi = f(u)$ a.e. in Ω.

Compactness is also a fundamental tool for *nonconvex* minimization problems. For instance, let $\varphi : \mathbf{R} \to \mathbf{R}$ be lower semicontinuous and bounded from below, and Ω have finite measure. If $\{v_n\}$ is uniformly bounded in $W^{1,1}(\Omega)$, then there exists $v \in L^1(\Omega)$ such that, possibly extracting a subsequence, $v_n \to v$ strongly in $L^1(\Omega)$, hence in measure. By the Fatou lemma, one can easily show that this entails $\liminf_{n \to \infty} \int_\Omega \varphi(v_n) dx \geq \int_\Omega \varphi(v) dx$.

Monotonicity. In elementary analysis and in linear functional analysis, monotonicity entails several properties. For instance, it leads to the concepts of *maximal monotonicity* and *m-accretiveness*, which are at the basis of the theory of *semi-groups of (nonlinear) contractions* in Hilbert and Banach spaces; see Sects. XI.5 and XI.6.

Convexity may be regarded as a variational counterpart of monotonicity: for example, the derivative of any differentiable convex function $\mathbf{R} \to \mathbf{R}$ is nondecreasing. In *convex analysis* the concept of differential is extended to that of *subdifferential* operator, denoted by ∂, cf. Sect. XI.4. Subdifferentials of lower semicontinuous convex functionals are maximal monotone. The converse does not hold in general: not all maximal monotone operators are subdifferentials (consider, e.g., a rotation by an angle $\varphi \in [-\pi/2, \pi/2]$ in $\mathbf{R}^2$). [40]

Any solution of a minimization problem also solves the corresponding *Euler equation*; actually, $J(x_0) = \inf J(\in \mathbf{R})$ if and only if $\partial J(x_0) \ni 0$. In several cases one can then apply the theory of equations to the study of minimization problems, and conversely for the equations that can be represented as the Euler equation of some functional. The latter are called *variational equations*. [41]

Maximal monotone operators and in particular subdifferentials may be multi-valued, and are usually identified with their graphs. This leads to the formulation of *inclusions* rather than equations. Maximal monotonicity and subdifferential relations can also be expressed through *variational inequalities*. In this book we often encounter objects of this sort, and we devote Chaps. II and III to the analysis of some classes of nonlinear parabolic variational inequalities.

Uniqueness. Here is a simple but important technique. Assuming that u_1 and u_2 are two solutions of some equation, take the difference of the corresponding equations and multiply it by a suitable function of $u_1 - u_2$, provided that this makes sense. Finally exploit some *strong monotonicity* [42] and/or Lipschitz property to conclude that $u_1 = u_2$.

Such a procedure may be used in problems that have a favourable monotone structure, and/or where nonlinearities act on lower order terms and are expressed through Lipschitz continuous functions. Difficulties may arise if the solution is not regular enough, a fact that may be either intrinsic to the problem, or due to an

---

[40] Convexity plays an important role in *linear* functional analysis as well, through the concept of *locally convex topological space*. For instance, in Banach spaces strongly closed convex sets are weakly closed; hence in some cases convexity may surrogate compactness.

Strict convexity may have a compactness-like effect occasionally. For instance, in locally uniformly convex Banach space, a sequence is strongly convergent whenever it is weakly convergent and the sequence of the norms converges. See Chap. X for other results in this direction and for some applications.

[41] The term *variational* tends to be used rather extensively. For instance, weak formulations are often called variational problems, even if they are not derived from any minimization problem, as it typically occurs for evolution equations.

[42] Cf., e.g., (II.2.13).

excessive extension of the concept of solution. In minimization problems, the use of strict convexity can be regarded as a counterpart of this technique.

In some linear problems one can prove uniqueness of the solution via the *Holmgren method,* based on transposition of the differential operator. [43] This can be extended to several nonlinear problems. In many cases establishing uniqueness may be a rather difficult task, although one may often expect that property to hold because of the phenomenon that is described by the model. [44] In the most favourable cases, one is able to prove existence and uniqueness of the solution in the same class of functions. In other cases, the set in which one gets uniqueness is smaller than that in which existence was derived. Such a result cannot be regarded as conclusive. An answer to this difficulty might reside in the analysis, but a revision of the formulation of the problem might also be required.

In several cases nonuniqueness is intrinsic, and the model itself points to a multiplicity of solutions. Existence of more than one stationary solution *(multi-stability)* is typical of problems that lack a convex (or monotone) structure. In those cases the evolution may exhibit *hysteresis,* which is representative of the tendency of a system to stay close to its previous state — a sort of *homeostasis.* When strong deviations of the state variables cannot be delayed any longer, discontinuities may occur, and the system may lose some symmetry properties *(symmetry breaking).* In Chap. VIII, dealing with nucleation in phase transitions, we encounter an example of such a setting.

Finally, we notice that typically uniqueness of the solution implies convergence of the *whole* approximating sequence, whereas the converse does not necessarily hold.

Comparison Between P.D.E.s and Minimization Problems. As we saw, several problems can be equivalently formulated in either form. Let us briefly and schematically review these two classes.

P.D.E.s. One may distinguish the following cases, for which we indicate the corresponding main technique for passage to the limit, when an approximation procedure is used.

(i) Linear P.D.E.s, for example, $(1.1)_1$.

(ii) Semilinear P.D.E.s, for example, $(1.1)_2$. Here one typically uses compactness methods, or monotonicity techniques in some cases.

(iii) Quasilinear P.D.E.s with monotone terms, for example, $(1.1)_3$. Here monotonicity procedures are in order.

---

[43] See Theorem XI.9.5. Here the basic statement is that the solution is unique if and only if the transposed (or adjoint) problem has a solution. Examples occur even in finite systems of linear algebraic equations.

[44] We stress that this cannot be used as an argument to support uniqueness of the solution of the mathematical problem, whose adequacy to represent the phenomenon under study is to be established. Rather the opposite holds: the validity of the model can only be corroborated by correspondence between the mathematical results and the properties of the phenomenon — and this is one of the main purposes of the analysis we are illustrating here.

(iv) Quasilinear P.D.E.s with nonmonotone nonlinearities in lower order terms, for example, $(1.1)_3$ with $g(u)$ in place of f. Here both monotonicity and compactness methods are typically used.

Functional Minimization. As we saw, the direct method is typically used to prove existence of a solution. Let us distinguish two cases:

(v) Convex functionals. Here convexity techniques are in order.

(vi) Nonconvex functionals that are convex with respect to leading order terms. Here one uses both convexity and compactness methods.

Analogies and Differences. Generally speaking, for equations continuity properties are crucial for existence of a solution, and so is strict monotonicity for uniqueness. On the other hand, lower semicontinuity properties play a key role for existence of a solution of minimization problems, and so does strict convexity for uniqueness.

The solutions of variational problems of the forms (v) and (vi) also fulfill the corresponding Euler equations, which, respectively, are of the classes (iii) and (iv). In case of cyclic monotonicity the converse also holds for (iii) and (v), but may fail for (iv) and (vi): stationary points of a nonconvex functional are not necessarily minimizers. Schematically: (v) $\Leftrightarrow$ (iii), and (vi) $\Rightarrow$ (iv), but (iv) $\not\Rightarrow$ (vi).

There is another important issue where the analogy breaks. For equations, linearity is a very strong property; but (if coupled with self-adjointness) it corresponds to quadraticity for functionals, which does not seem to play a relevant role in minimization. In fact for functionals convexity is the key property.

Here we have not considered *variational inequalities,* which partecipate partly of the properties of equations, and partly of those of variational problems.

Gradient Flows. Many phenomena are at first studied under stationary conditions. The question then arises of describing evolution consistently with the stationary model. Let us assume that the latter consists of the minimization of a differentiable functional $\Psi : B \to \mathbf{R}$, where B is a Banach space with dual B'. For instance, Ψ might represent the total *free energy* of a thermodynamical system, as we see in Sect. VII.6.

In many cases, evolution is driven by *steepest descent* of the point $(u, \Psi(u))$ along the graph of the potential Ψ in $B \times \mathbf{R}$. In first approximation, this can be described by a *gradient dynamics:* [45]

$$a\frac{du}{dt} + \nabla\Psi(u) = 0 \qquad \text{in } B'\,(a > 0). \tag{2.1}$$

This setting can be extended to nondifferentiable functionals. If the potential F is convex, one can replace the gradient ∇ by the subdifferential; this yields an inclusion in place of the equality. This setting has been extensively studied. In

---

[45] In Sect. VII.6 we derive the so-called *mean curvature flow* as a gradient dynamics.

many cases the functional Ψ is nonconvex, and the subdifferential operator must be properly extended. [46]

I.3 Examples of Nonlinear P.D.E.s

In this section we briefly illustrate how the preceding techniques can be applied to two classical physical problems.

Physical Principles and Constitutive Behaviour. Several physical phenomena lead to problems of the following sort. A *basic physical principle,* such as mass or moment conservation, or the Newton law, or the Maxwell laws, yields one or more linear P.D.E.s, which apply to a broad class of phenomena.

The behaviour of specific classes of materials must be then described through additional (either linear or nonlinear) relations, which are called *constitutive laws;* these may be represented either in the form of equations or through variational principles. The formulation of these laws is one of the main issues of modelling. Here experimental tests play an important role, especially in determining material coefficients. However, the identification of these laws cannot be reduced to a matter of measurement: no physical law is derived just by brutal interpolation of experimental data, not to mention that certain quantities (e.g., stresses in continuum mechanics, entropy in thermodynamics) are hardly accessible to direct measurement.

It is especially difficult to use measurements accomplished on space-distributed systems to identify constitutive laws; actually, such a procedure is more appropriate for testing the overall validity of a model after numerical computation. For instance, in problems of continuum mechanics, stress-strain relations are identified through experiments made on specimens that are (supposed to be) uniform in space, and are then extrapolated to space-distributed systems. On the other hand, time-dependence is often included; for instance, this allows the identification of viscosity effects.

Extrapolation from space-uniform to space-distributed systems may have some drawbacks. For instance, a relation derived in this way hardly includes space derivatives, which may be then introduced only on a theoretic basis, if specific clues point to them.

Surface tension effects in multi-phase systems are an example of this setting. As we illustrate in Chap. VII, surface tension is a *mesoscopic* (i.e., small-scale) phenomenon. Nevertheless, it provides a convincing explanation of undercooling prior to phase nucleation. This leads one to account for surface tension even in *macroscopic* systems, although the law that describes this phenomenon (the *Gibbs-Thomson law,* see Sect. IV.2) is hardly accessible to direct laboratory measurement.

---

[46] See, e.g., the operator ∂^- of De Giorgi, Marino and Tosques [181]; cf. Sect. XI.4.

Heat Diffusion. Let us consider a (bounded) space domain Ω occupied by a heat conducting material, fix a $T > 0$ and set $Q := \Omega \times]0, T[$. Let us denote the energy density by u, the heat flux by $\vec{q}$, the temperature by θ, the intensity of a given distributed heat source or sink by f. Neglecting convection, the *energy balance* reads

$$\frac{\partial u}{\partial t} + \nabla \cdot \vec{q} = f \qquad \text{in } Q \ (\nabla \cdot := \text{div}), \tag{3.1}$$

in a sense to be specified. This law must be coupled with two constitutive relations. $\vec{q}$ is correlated with the thermal gradient; the nonlinear *Fourier law* $\vec{q} = -k(\theta)\nabla\theta$ with $k > 0$ is consistent with the *second principle of thermodynamics*, and in most cases is fairly satisfactory.

Another constitutive law must relate u with θ. In a single-phase system, one can assume that u is a differentiable function of θ, $u = \alpha(\theta)$; $\alpha'(\theta)$ (> 0) then represents the *specific heat*. The nonlinear parabolic P.D.E. that is obtained by coupling equation (3.1) with these laws can be studied along the lines which we indicated in the previous section.

Multi-phase systems are the focus of this book. A typical example is the evolution of solid–liquid systems, which can be described by the classical *Stefan model;* see Chap. IV. This is an example of *free boundary problem,* since the evolution of the phase interface is a priori unknown. [47] On those surfaces θ is continuous, whereas u has a jump; hence there the specific heat diverges. This setting can be described by means of a relation of the form

$$u \in \alpha(\theta) \qquad \text{in } Q, \tag{3.2}$$

where α is a maximal monotone graph. The equation (3.1) must be then interpreted in the sense of distributions.

Boundary and initial value problems for the system (3.1) and (3.2) are studied in Chap. II according to the lines we previously indicated: approximation by implicit time discretization, derivation of a priori estimates by L^2- and L^1-techniques, and passage to the limit via compactness and monotonicity methods. Regularity results are derived by means of further a priori estimates.

The discussion of this macroscopic model is the starting point for proposing a *mesoscopic* model of phase transitions which includes surface tension effects (see Chaps. VI, VII, VIII). From the analytical point of view, that model is characterized by lack of convexity; this is compensated by further compactness, due to the presence of space derivatives in the constitutive law.

Maxwell's Equations. Now let us briefly consider electromagnetism. With standard notation, we denote the density of electric current by $\vec{J}$, the electric field by $\vec{E}$, the electric displacement by $\vec{D}$, the magnetic field by $\vec{H}$, and the magnetic induction

---

[47] In this case one also speaks of a *moving* boundary problem.

by $\vec{B}$. Using *Gauss units*, the *Ampère* and *Faraday laws*, respectively, read

$$c\nabla \times \vec{H} = 4\pi \vec{J} + \frac{\partial \vec{D}}{\partial t}, \qquad c\nabla \times \vec{E} = \frac{\partial \vec{B}}{\partial t} \qquad \text{in } Q \, (\nabla \times := \text{curl}), \qquad (3.3)$$

where c is the speed of the light in vacuum. Moreover, we have the laws

$$\nabla \cdot \vec{D} = 4\pi \rho, \qquad \nabla \cdot \vec{B} = 0 \qquad \text{in } Q \, (\nabla \cdot := \text{div}), \qquad (3.4)$$

where ρ is the electric charge density. These laws must be coupled with constitutive laws, which respectively relate $\vec{J}$ with $\vec{E}$, $\vec{D}$ with $\vec{E}$, $\vec{B}$ with $\vec{H}$. Let us assume that the first two of these relations are linear:

$$\vec{D} = \epsilon \vec{E}, \qquad \vec{J} = \sigma(\vec{E} + \vec{g}) \qquad \text{in } Q; \qquad (3.5)$$

here ϵ is the *dielectric permeability*, σ is the *electric conductivity* (both assumed to be constant), and $\vec{g}$ represents a given electric source (the second one is the classical Ohm law). By coupling these laws with $(3.3)_1$, we get the following P.D.E.:

$$\epsilon \frac{\partial^2 \vec{B}}{\partial t^2} + 4\pi \sigma \frac{\partial \vec{B}}{\partial t} + c^2 \nabla \times \nabla \times \vec{H} = 4\pi c \sigma \nabla \times \vec{g} \qquad \text{in } Q. \qquad (3.6)$$

Let us now assume that the fields $\vec{B}$ and $\vec{H}$ are related by a nonlinear law of the form

$$\vec{B} \in \mathcal{F}(\vec{H}) \qquad \text{in } Q, \qquad (3.7)$$

where $\mathcal{F} : \mathbf{R}^3 \to \mathbf{R}^3$ is a maximal monotone graph. For instance, this may represent the behaviour of soft iron for high field saturation. If $\mathcal{F}$ is multi-valued, this corresponds to a free boundary problem, see Sect. IV.8.

By coupling equations $(3.4)_2$ and (3.6) with (3.7), we get a quasilinear hyperbolic vectorial P.D.E.. The analysis of systems of this sort is at the forefront of research.

Now let us consider a different (and simpler) setting. In metals, at usual frequencies, the ϵ-term is much smaller than the σ-term; then the *displacement current* term $\partial \vec{D}/\partial t$ can be neglected. Accordingly, in place of (3.6), we get the equation

$$4\pi \sigma \frac{\partial \vec{B}}{\partial t} + c^2 \nabla \times \nabla \times \vec{H} = 4\pi c \sigma \nabla \times \vec{g} \qquad \text{in } Q. \qquad (3.8)$$

The analysis of the vectorial system $(3.4)_2$, (3.7), (3.8) is easier than that of $(3.4)_2$, (3.6), (3.7). Its structure is similar to that of the scalar system (3.1), (3.2), since $\nabla \times \nabla \times$ is an elliptic operator. Most of the analysis that we develop in Chap II also applies to this vectorial problem.

A Simple P.D.E. Let us briefly see on an example how the tool which we illustrated in the previous section can be applied to the analysis of a quasilinear P.D.E.s. Let us consider the equation

$$\frac{\partial u}{\partial t} - \nabla \cdot \vec{\gamma}(\nabla u) = f(u) \qquad \text{in } Q := \Omega \times]0, T[, \qquad (3.9)$$

coupled with initial and boundary data. Let $\vec{\gamma}$ and f be continuous, with linear growth at ∞, and $\vec{\gamma}$ monotone (hence maximal monotone).

Assume that one can efficiently approximate this problem, and also derive a priori estimates in some L^p-space ($1 \leq p \leq +\infty$). Let us denote the approximation parameter by $n \in \mathbf{N}$. Multiplying the approximate equation by u_n, by the classical *Gronwall lemma* (see Theorem XI.9.1.), one easily gets that u_n and ∇u_n are uniformly bounded in $L^2(Q)$ and $L^2(Q; \mathbf{R}^N)$, respectively. Comparing the terms of the approximate equation, one sees that u_n is uniformly bounded in $H^1\left(0, T; H^{-1}(\Omega)\right)$. [48] Therefore there exists

$$u \in L^2\left(0, T; H^1(\Omega)\right) \cap H^1\left(0, T; H^{-1}(\Omega)\right) (=: X)$$

such that, possibly extracting a subsequence, $u_n \to u$ weakly in X. Then this sequence also converges strongly in $L^2(Q)$, by Lions-Aubin's Theorem XI.3.5 (ii). This yields $f(u_n) \to f(u)$ strongly in $L^2(Q)$.

Finally, one has to take the limit in the leading order term, $\nabla \cdot \vec{\gamma}(\nabla u)$. Because of the maximal monotonicity of $\vec{\gamma}$, one can express this nonlinearity in terms of a variational inequality, and then pass to the limit via a lower semicontinuity property. In fact we have $\vec{g} = \vec{\gamma}(\nabla u)$ a.e. in Q if and only if

$$\iint_Q [\vec{g} - \vec{\gamma}(\nabla v)] \cdot \nabla(u - v) dx dt \geq 0 \quad \forall v \in H^1(Q);$$

hence (3.9) is equivalent to

$$\iint_Q \vec{\gamma}(\nabla v) \cdot \nabla(u - v) dx dt + \frac{1}{2} \int_\Omega \left[u(x,T)^2 - u(x,0)^2 \right] dx$$

$$+ \iint_Q \left(f(u)(v-u) + u \frac{\partial v}{\partial t} \right) dx dt \leq \int_\Omega \left[(uv)(x,T) - (uv)(x,0) \right] dx \quad (3.10)$$

$$\forall v \in H^1(Q).$$

Once this variational inequality has been derived for u_n, it is easy to pass to the limit, because of the lower semicontinuity of the norm.

Equations like (3.9) are systematically studied in Chap. II.

A Simple Variational Problem. Let us now see an example of minimization problem. Let us consider the functional $J := J_1 + J_2$, where

$$J_1(u) := \int_\Omega |\nabla u|^2 dx, \quad J_2(u) := \int_\Omega \left[(u^2 - 1)^2 - gu \right] dx \quad (3.11)$$

$$\forall u \in H^1(\Omega),$$

---

[48] See Sect. XI.2 for the definition of Sobolev spaces.

for a given $g \in L^{4/3}(\Omega)$. Although J is not convex, its principal part J_1 has that property. One can prove the existence of a minimizer, since J is coercive and lower semicontinuous in appropriate function spaces.

In fact any minimizing sequence $\{u_n\}$ is obviously bounded in $Y := H^1(\Omega) \cap L^4(\Omega)$. Hence there exists $u \in Y$ such that, possibly extracting a subsequence, $u_n \to u$ weakly in Y. As the H^1-seminorm is lower semicontinuous, then we have $\liminf J_1(u_n) \geq J_1(u)$. By the Rellich-Kondrachov Theorem XI.3.4, $u_n \to u$ in measure in Ω, provided that Ω is regular enough; hence by the Fatou lemma $\liminf J_2(u_n) \geq J_2(u)$. Therefore $\liminf J(u_n) \geq J(u)$, whence $J(u) = \inf J$ as $\{u_n\}$ is a minimizing sequence.

In general, uniqueness of the solution is excluded, since the functional is not convex.

Another Variational Problem. We now introduce another example, which illustrates some of the main issues of this volume [49] in a simplified way. The reader will notice the interplay of analytical and modelling features.

Let $\theta \in L^1(\Omega)$ be a given function, and consider the problem of minimizing the functional $J_3(u) := -\int_\Omega \theta u \, dx$ in the set of characteristic functions $X := \{u \in L^\infty(\Omega) : |u| = 1 \text{ a.e. in } \Omega\}$. One might interpret θ as the relative temperature (up to a multiplicative constant), and the domains respectively characterized by $u = -1$ and $u = 1$ as different *phases* (e.g., solid and liquid) of a material.

Although the constraint $|u| = 1$ is nonconvex, it is obvious that this problem has at least one solution: it suffices to minimize the integrand pointwise. It is easy to see that there exists a measurable function u such that $u = -1$ where $\theta < 0$, $u = 1$ where $\theta > 0$, either $u = -1$ or $u = 1$ where $\theta = 0$.

One may also *convexify* the problem, and replace X by its *closed convex hull* [50] $\overline{\text{co}}(X) = \{u \in L^\infty(\Omega) : |u| \leq 1 \text{ a.e. in } \Omega\}$. In this way one allows for $|u| \leq 1$ where $\theta = 0$; intermediate values of u between -1 and 1 can be interpreted as space averages of characteristic functions (or *fuzzy sets*). This corresponds to a two-scale model: on the finer length scale one only deals with pure phases in the framework of a nonconvex model; on the other hand, on the coarser scale one also allows for phase mixtures, and uses a nonconvex model.

Let us now account for *surface tension*, by inserting the contribution $J_4(u) := (\sigma/2) \int_\Omega |\nabla u|$ into the potential. [51] Here σ is a positive parameter, and determines the ratio between the two length scales. By the direct method of the calculus of variations, it is easy to see that there exists at least one minimizer of the potential $J_3 + J_4$ in X. Notice that any minimizing sequence is bounded in $BV(\Omega)$, and this space is included in $L^1(\Omega)$ with continuous and compact injection. Hence one can

---

[49] See Chaps. VI through VIII.
[50] See Sect. XI.4.
[51] See Sect. XI.1 and for the definition of the *total variation* $\int_\Omega |\nabla u|$ and of the associated space $BV(\Omega)$.

extract a subsequence which converges a.e. in Ω; therefore the constraint $u \in X$ is preserved in the limit.

I.4 Comments

The preceding discussion was intended as a presentation of some general methods of analysis of P.D.E.s and variational problems for didactic purposes, and has gone beyond the necessities of this essay.

Methodological digressions are seldom developed in mathematical monographs, [52] with remarkable and authoritative exceptions, for example, Ladyženskaja [333], Lin and Segel [347], Lions [351], Thom [532], and Zeidler [591, 592, 593]. Nevertheless, reflections of that sort underlie several decisions not only in *macro research* (namely, strategic choices involving a large number of researchers), but also in *micro research* (namely, choices of the single researcher about her/his own activity). Of course the purpose of the preceding discussion was more modest.

We stress that the panorama of analytical techniques at our disposal for dealing with nonlinear P.D.E.s and minimization problems is much richer than it might appear from this schematic overview. Actually, any presentation of this field will hardly deal with *the* methods, but just with *some* methods, consistently with the title and program of Lions [351], which also inspired part of Sect. I.2.

---

[52] Is this due to the fact that such questions lack the objectivity to which we suppose mathematics should strive? However, inevitably mathematics (especially applied mathematics) is more *biased* than we would like it to be.

Chapter II. A Class of Quasilinear Parabolic P.D.E.s

Outline

This chapter deals with (possibly degenerate) quasilinear [1] parabolic inclusions of the form

$$\frac{\partial}{\partial t}\alpha(u) - \Delta u \ni f \qquad \left(\Delta := \sum_{i=1}^{N} \frac{\partial^2}{\partial x_i^2}\right),$$

where α is a maximal monotone graph. Existence of a solution is proved via approximation by time-discretization, derivation of a priori estimates, and passage to the limit by compactness and monotonicity procedures. Uniqueness and regularity results are derived via L^2- and L^1-techniques.

Integral transformations and semigroup methods are also outlined.

Prerequisites. Some acquaintance with the methods of analysis of linear and nonlinear P.D.E.s in Sobolev spaces is assumed, although the presentation is rather didactical. Variational inequalities are used in Sect. II.5.

Some definitions and standard results are recalled in Chap. XI.

II.1 Variational Techniques of L^2-Type

In this section we deal with an initial and boundary value problem for a class of quasilinear parabolic P.D.E.s, in the framework of Sobolev spaces of Hilbert type. In Sect. IV.4 we see that this setting includes the weak formulation of the multi-dimensional Stefan and Hele-Shaw problems.

These developments are mainly standard; our aim is just to introduce some basic results and techniques, and we do not strive for generality.

A Model Problem. Let Ω be a bounded domain of $\mathbf{R}^N$ ($N \geq 1$) of Lipschitz class. We denote by Γ its boundary, fix any $T > 0$ and set $Q := \Omega \times]0, T[$, $\Sigma := \Gamma \times]0, T[$. We assume that $\hat{f}_1 : Q \to \mathbf{R}$ is a given function and [2]

$$\alpha : \text{Dom}(\alpha) (\subset \mathbf{R}) \to 2^{\mathbf{R}} \text{ is a maximal monotone graph.} \tag{1.1}$$

---

[1] We use the term *quasilinear* in the sense specified in one of the footnotes of Sect. I.1.

[2] See Sect. XI.5 for the definition of *maximal monotone graph*.

We deal with the system

$$\begin{cases} \dfrac{\partial w}{\partial t} - \Delta u = \hat{f}_1 & \text{in } \mathcal{D}'(Q), \\ w \in \alpha(u) & \text{in } Q, \end{cases} \qquad (1.2)$$

coupled with appropriate initial and boundary conditions. We set $\beta := \alpha^{-1}$, and

$$a(\xi) := \int_{\xi_0}^{\xi} \tilde{\alpha}(\eta) d\eta \qquad \forall \xi \in \text{Dom}(\alpha) \ (\xi_0 \in \text{Dom}(\alpha) \text{ fixed}),$$

$$b(\xi) := \int_{\xi_1}^{\xi} \tilde{\beta}(\eta) d\eta \qquad \forall \xi \in \text{Dom}(\beta) \ (\xi_1 \in \text{Dom}(\beta) \text{ fixed}), \qquad (1.3)$$

for any section $\tilde{\alpha}$ of α and $\tilde{\beta}$ of β; these integrals do not depend on the choice of the section. Then [3]

$$\alpha = \partial a, \qquad \beta = \partial b, \qquad b = a^*. \qquad (1.4)$$

Later we deal with a graph α defined on the whole **R**. We can assume that

$$\alpha(0) \ni 0, \qquad (1.5)$$

with no loss of generality since in our equation the α-term is differentiated. This allows us to take $\xi_0 = \xi_1 = 0$ in (1.3).

Weak Formulation. Let us fix a (possibly empty) relatively open subset Γ_1 of Γ, and set

$$V := H^1_{\Gamma_1}(\Omega) := \{ v \in H^1(\Omega) : \gamma_0 v = 0 \text{ on } \Gamma_1 \}, \qquad (1.6)$$

where γ_0 denotes the *trace* operator. [4]

We identify the space $L^2(\Omega)$ with its topological dual $L^2(\Omega)'$. As the injection of V into $L^2(\Omega)$ is continuous and *dense* (i.e., V equipped with the strong topology of $L^2(\Omega)$ is a dense subspace of $L^2(\Omega)$ itself), $L^2(\Omega)'$ can be identified with a subspace of V'. This yields the *Hilbert triplet* [5]

$$V \subset L^2(\Omega) = L^2(\Omega)' \subset V', \qquad (1.7)$$

with dense and compact injections.

---

[3] See Sect. XI.4 for the definition of the *subdifferential* operator ∂ and of the *convex conjugate* function a^*.

[4] See Sect. XI.2.

[5] See Sect. XI.9.

Let us denote by $_{V'}\langle\cdot,\cdot\rangle_V$ the duality pairing between V' and V. We define the linear and continuous operator $A : V \to V'$ as follows

$$_{V'}\langle Au, v\rangle_V := \int_\Omega \nabla u \cdot \nabla v \, dx \qquad \forall u, v \in V.$$

Hence $Av = -\Delta v$ in $\mathcal{D}'(\Omega)$, and, after an obvious identification, $A : L^2(0,T;V) \to L^2(0,T;V')$ is linear and continuous. We assume that

$$w^0 \in L^2(\Omega), \qquad f \in L^2(0,T;V'), \tag{1.8}$$

and introduce a *weak formulation*.

Problem 1.1 *To find* $u \in L^2(0,T;V)$ *and* $w \in L^2(Q)$ *such that*

$$\iint_Q \left(-w\frac{\partial v}{\partial t} + \nabla u \cdot \nabla v\right) dx dt = \int_0^T {}_{V'}\langle f, v\rangle_V \, dt + \int_\Omega w^0 v(\cdot, 0) dx \tag{1.9}$$

$$\forall v \in L^2(0,T;V) \cap H^1\left(0,T;L^2(\Omega)\right), v(\cdot, T) = 0,$$

$$w \in \alpha(u) \qquad \text{a.e. in } Q. \tag{1.10}$$

Interpretation. By (1.1), (1.10) is equivalent to a *variational inequality*:

$$(w - \eta)(u - \xi) \geq 0 \qquad \text{a.e. in } Q, \forall (\xi, \eta) \in \text{graph}(\alpha). \tag{1.11}$$

(1.9) yields

$$\frac{\partial w}{\partial t} + Au = f \qquad \text{in } H^{-1}(0,T;V'), \tag{1.12}$$

whence $\partial w/\partial t = f - Au \in L^2(0,T;V')$. Therefore $w \in H^1(0,T;V')$ and (1.12) is satisfied in the sense of $L^2(0,T;V')$. Hence, by integrating (1.9) by parts in time, we get

$$w|_{t=0} = w^0 \qquad \text{in } V' \text{ (in the sense of the traces of } H^1(0,T;V')). \tag{1.13}$$

In turn (1.12) and (1.13) yield (1.9).

Let us now interpret (1.12). For instance, let us take

$$\hat{f}_1 \in L^2(Q), \qquad \hat{f}_2 \in L^2(\Gamma_2 \times]0,T[) \quad (\Gamma_2 := \Gamma \setminus \Gamma_1), \tag{1.14}$$

and define $f \in L^2(0,T;V')$ by

$$_{V'}\langle f(t), v\rangle_V := \int_\Omega \hat{f}_1(x,t)v(x)dx + \int_{\Gamma_2} \hat{f}_2(\sigma,t)\gamma_0 v(\sigma)d\sigma \tag{1.15}$$

$$\forall v \in V, \text{ for a.a. } t \in]0,T[.$$

34   II. A Class of Quasilinear Parabolic P.D.E.s

Then (1.9) yields the differential equation

$$\frac{\partial w}{\partial t} - \Delta u = \hat{f}_1 \qquad \text{in } \mathcal{D}'(Q), \tag{1.16}$$

and by integration by parts one *formally* gets the boundary condition

$$\frac{\partial u}{\partial \nu} = \hat{f}_2 \qquad \text{on } \Gamma_2 \times]0, T[, \tag{1.17}$$

where $\partial/\partial \nu$ denotes the external normal derivative (this also is a trace operator). [6] Moreover by definition (1.6) of V we have

$$\gamma_0 u = 0 \qquad \text{on } \Gamma_1 \times]0, T[. \tag{1.18}$$

In conclusion, if (1.14) is satisfied and f is as in (1.15), then (1.12) corresponds to the system (1.16), (1.17), (1.18). □

Lemma 1.1 *After an obvious identification, one has*

$$L^2(Q) \cap H^1(0, T; V') \subset C^0\left([0, T]; V'\right) \tag{1.19}$$

with compact injection.

Proof. Let us set $V_r := \{v \in H^r(\Omega) : \gamma_0 v = 0 \text{ on } \Gamma_1\}$ for any $r > \frac{1}{2}$. We claim that

$$L^2\left(0, T; L^2(\Omega)\right) \cap H^1(0, T; V') \subset H^r(0, T; V'_r)$$
$$\subset H^s(0, T; V') \subset C^0\left([0, T]; V'\right) \qquad \forall r, s, \tfrac{1}{2} < s < r < 1,$$

with continuous injections. The first inclusion follows from interpolation between Sobolev spaces, see Theorem XI.2.7. [7] The other inclusions are obvious. The second one is compact by a simple generalization of the Rellich-Kondrachov Theorem XI.3.4(i). □

Theorem 1.2 *(Existence) Assume that (1.1), (1.5), (1.8) are satisfied, and that*

$$\exists L, M > 0 : \forall (u, w) \in \text{graph}(\alpha), |w| \leq L|u| + M, \tag{1.20}$$

---

[6] More precisely, (1.17) holds in $H^{-1}\left(0, T; H_{00}^{1/2}(\Gamma_2)'\right) = H_0^1\left(0, T; H_{00}^{1/2}(\Gamma_2)\right)'$. In fact a comparison in (1.16) yields $\Delta u \in H^{-1}\left(0, T; L^2(\Omega)\right)$, and in general $v \in H^1(\Omega)$, $\Delta v \in L^2(\Omega)$ imply $\partial v/\partial \nu \in H_{00}^{1/2}(\Gamma_2)'$; see, e.g., Lions and Magenes [356; vol. I, Chap. 2].

$H_{00}^{1/2}(\Gamma_2)$ is the closure in the topology of $H^{1/2}(\Gamma_2)$ of the smooth functions $\Gamma_2 \to \mathbf{R}$ having compact support. Hence if $\Gamma_2 = \Gamma$ then $H_{00}^{1/2}(\Gamma_2) = H^{1/2}(\Gamma)$.

[7] See, e.g., Lions and Magenes [356; vol. II, Chap. 4] for more general results, and Bergh and Löfström [72], and Triebel [536] for the theory of interpolation between function spaces.

$$b(w^0) \in L^1(\Omega), \tag{1.21}$$

$$0 < |\Gamma_1| := (N-1)\text{-dimensional Hausdorff measure of } \Gamma_1. \tag{1.22}$$

Then Problem 1.1 has a solution such that $w \in L^\infty\left(0, T; L^2(\Omega)\right)$.

Proof. By the generalized *Poincaré inequality*, cf. Proposition XI.9.3, (1.22) allows us to equip the space V with the equivalent norm $\|v\|_V := \left(\int_\Omega |\nabla v|^2 dx\right)^{1/2}$.

We prove existence of a solution via the *approximation — a priori estimate — limit* procedure, cf. Sect. I.2.

(i) *Approximation.* Let us fix any $m \in \mathbf{N}$, set $k := T/m$, $w_m^0 := w^0$ and

$$f_m^n := \frac{1}{k} \int_{(n-1)k}^{nk} f(\tau) d\tau \qquad \text{in } V', \text{ for } n = 1, \ldots, m. \tag{1.23}$$

We approximate our problem by *implicit time-discretization*.

Problem 1.1$_m$ *To find $u_m^n \in V$ and $w_m^n \in L^2(\Omega)$ for $n = 1, \ldots, m$, such that*

$$\frac{w_m^n - w_m^{n-1}}{k} + A u_m^n = f_m^n \qquad \text{in } V', \text{ for } n = 1, \ldots, m, \tag{1.24}$$

$$w_m^n \in \alpha(u_m^n) \qquad \text{a.e. in } \Omega, \text{ for } n = 1, \ldots, m. \tag{1.25}$$

We claim that this problem can be solved step by step. Let us fix any $n \in \{1, \ldots, m\}$, assume that $w_m^{n-1} \in L^2(\Omega)$ is known, and consider the problem of determining u_m^n and w_m^n. We define the convex functional

$$J_m^n(v) := \int_\Omega \left(a(v) + \frac{k}{2}|\nabla v|^2 - w_m^{n-1} v\right) dx - k\,{}_{V'}\langle f_m^n, v\rangle_V \qquad \forall v \in V.$$

By the *direct method* of the calculus of variations (see Theorem XI.7.3), there exists $u_m^n \in V$ such that $J_m^n(u_m^n) = \inf J_m^n \in \mathbf{R}$. Hence by Theorem XI.4.6(iv), $0 \in \partial J_m^n(u_m^n)$ in V'. Moreover, by Theorem XI.4.7 (or by direct computation), $\partial J_m^n(u_m^n) = \alpha(u_m^n) + kAu_m^n - w_m^{n-1} - kf_m^n$ in V'. This yields

$$0 \in \alpha(u_m^n) + kAu_m^n - w_m^{n-1} - kf_m^n \qquad \text{in } V', \forall n \in \{1, \ldots, m\},$$

which is equivalent to the system (1.24) and (1.25).

(ii) *A Priori Estimates.* Let us multiply (1.24) by ku_m^n and sum for $n = 1, \ldots, \ell$, for any $\ell \in \{1, \ldots, m\}$. By (1.4) and (1.25), we have

$$(w_m^n - w_m^{n-1})u_m^n \geq b(w_m^n) - b(w_m^{n-1}) \qquad \text{a.e. in } \Omega, \text{ for } n = 1, \ldots, m.$$

Thus we get

$$\int_\Omega \left[b(w_m^\ell) - b(w^0)\right] dx + k \sum_{n=1}^\ell \int_\Omega |\nabla u_m^n|^2 dx$$
$$\leq k \sum_{n=1}^\ell \|f_m^n\|_{V'} \|u_m^n\|_V \leq \|f\|_{L^2(0,T;V')} \left(k \sum_{n=1}^\ell \|u_m^n\|_V^2\right)^{1/2} \quad (1.26)$$

for any $\ell \in \{1, \ldots, m\}$. We claim that

$$b(\eta) \geq \frac{1}{2L}(\eta - M)^2 \quad \forall \eta \in \mathbf{R}. \quad (1.27)$$

The equations (1.5) and (1.20) yield $\xi \geq (\eta - M)/L$ for any $\eta \geq 0$ and $\xi \in \beta(\eta)$; by integrating in $[0, \eta]$, we then get (1.27). Any $\eta \leq 0$ can be treated similarly.

The equations (1.26) and (1.27) then yield

$$\max_{n=1,\ldots,m} \|w_m^n\|_{L^2(\Omega)}, k \sum_{n=1}^m \|u_m^n\|_V^2 \leq \text{Constant}. \quad (1.28)$$

Let us set

$$w_m(x, \cdot) := \text{linear time interpolate of } \{w_m(x, nk) := w_m^n(x)\}_{n=0,\ldots,m}, \quad (1.29)$$

$$\bar{w}_m(x, t) := w_m^n(x) \quad \text{if } (n-1)k < t \leq nk, \text{ for } n = 1, \ldots, m, \quad (1.30)$$

for almost any $x \in \Omega$, and define $\bar{u}_m, \bar{f}_m$ similarly. [8] The equations (1.24) and (1.28) then read

$$\frac{\partial w_m}{\partial t} + A\bar{u}_m = \bar{f}_m \quad \text{in } V', \text{ a.e. in }]0, T[, \quad (1.31)$$

$$\|w_m\|_{L^\infty(0,T;L^2(\Omega))}, \|\bar{u}_m\|_{L^2(0,T;V)} \leq \text{Constant (independent of } m\text{)}. \quad (1.32)$$

Hence $A\bar{u}_m$ is uniformly bounded in $L^2(0, T; V')$, and by comparing the terms of (1.31) we get

$$\|w_m\|_{H^1(0,T;V')} \leq \text{Constant}. \quad (1.33)$$

(iii) *Limit Procedure.* By the preceding estimates and by classical compactness results, cf. Theorem XI.3.1, there exist u, w such that, possibly taking $m \to \infty$ along a subsequence,

$$\bar{u}_m \to u \quad \text{weakly in } L^2(0, T; V), \quad (1.34)$$

$$w_m \to w \quad \text{weakly star in } L^\infty\left(0, T; L^2(\Omega)\right) \cap H^1(0, T; V'). \quad (1.35)$$

---

[8] This sort of notation is used systematically in the following, to transform finite difference schemes into approximate continuous equations.

Hence by taking the limit in (1.31) we get (1.12).

We are left with the proof of (1.10). By Lemma 1.1, we have

$$w_m \to w \quad \text{strongly in } C^0([0,T];V'). \tag{1.36}$$

Therefore $\|\bar{w}_m - w_m\|_{V'} \to 0$ uniformly in $]0,T[$, and we get

$$\iint_Q \bar{w}_m \bar{u}_m \, dx dt = \int_0^T {}_{V'}\langle \bar{w}_m - w_m, \bar{u}_m \rangle_V \, dt + \iint_Q w_m \bar{u}_m \, dx dt \tag{1.37}$$
$$\to \iint_Q wu \, dx dt.$$

By (1.25), (1.34), (1.35), and Lemma XI.5.1, then we get (1.10).

Finally, by (1.36), the initial condition (1.13) is preserved in the limit. □

Let us now replace the assumption $f \in L^2(0,T;V')$ by

$$f = f_1 + f_2, \quad f_1 \in L^1\left(0,T;L^2(\Omega)\right), \quad f_2 \in L^2(0,T;V'). \tag{1.38}$$

In this case the formulation of Problem 1.1 must be slightly modified: in (1.9), $\int_0^T {}_{V'}\langle f,v\rangle_V \, dt$ is replaced by $\iint_Q f_1 v \, dx dt + \int_0^T {}_{V'}\langle f_2,v\rangle_V \, dt$.

Proposition 1.3 *Assume that (1.1), (1.5), (1.20), (1.21), and (1.38) are satisfied, and*

$$\exists \hat{L}, \hat{M} > 0 : \forall (u,w) \in \text{graph}(\alpha), |u| \leq \hat{L}|w| + \hat{M}. \tag{1.39}$$

Then Problem 1.1 has a solution such that $u,w \in L^\infty\left(0,T;L^2(\Omega)\right)$.

Outline of the Proof. The argument is similar to that of Theorem 1.2. Here, by (1.27) and (1.39), a uniform estimate on u_m in $L^\infty\left(0,T;L^2(\Omega)\right)$ can be derived from (1.26), via the classical *Gronwall lemma*, cf. Theorem XI.9.1. Then the analogous estimate for w follows by (1.20). □

Theorem 1.4 [9] *Assume that (1.1), (1.5), (1.21), and (1.22) are satisfied, and*

$$w^0 \in L^2(\Omega), \quad f \in L^1\left(0,T;L^2(\Omega)\right) \cap L^2(0,T;V'). \tag{1.40}$$

Then Problem 1.1 has a solution such that $w \in L^\infty\left(0,T;L^2(\Omega)\right)$.

Outline of the Proof. The problem may be approximated by time discretization as previously. Here the first estimate procedure used for Theorem 1.2 only yields a uniform bound for u_m in $L^2(0,T;V)$, as $b \geq 0$.

---

[9] See Colli and Savaré [140].

A uniform estimate for w_m in $L^\infty(0,T;L^2(\Omega))$ is then *formally* obtained multiplying the approximate equation by w_m^n. Here a space regularization is in order, for w_m cannot be expected to be an element of $L^2(0,T;V)$. □

Notice that in the last result no assumption is required on the order of growth of α and β, as in Theorem 2.1.

Remarks. (i) Problem 1.1 is just a model problem, and the preceding results can be extended in several ways. For instance, one can deal with the abstract setting, in which $V \subset H^1(\Omega)$ and $L^2(\Omega)$ are replaced by generic Hilbert spaces W, H such that $W \subset H = H' \subset W'$ with continuous and dense inclusions, and $A : W \to W'$. Actually, W may be replaced by a Banach space, and more general settings may also be treated as well.

(ii) If in equation (1.16) Δu is replaced by $\nabla \cdot (k(u)\nabla u)$, with k continuous positive function, the transformation $K : u \mapsto \tilde{u} := \int_0^u k(\xi)d\xi$ allows us to rewrite that nonlinear term as $\Delta \tilde{u}$. The condition $(1.2)_2$ must be then replaced by "$w \in \tilde{\alpha}(\tilde{u}) := \alpha\left(K^{-1}(\tilde{u})\right)$". As $k > 0$, $\tilde{\alpha}$ is also a maximal monotone graph. Thus a problem like 1.1 is retrieved. □

Exercises.

1.1 Give examples of $f \in L^2(0,T;V')$ alternative to (1.14) and (1.15).

1.2 Does a statement analogous to Lemma 1.1 hold with $H^1(0,T;V')$ replaced by $H^s(0,T;V')$? for which values of s?

1.3 Detail the derivation of (1.28).

1.4 Prove Theorem 1.2 by using the *Faedo-Galerkin method* to approximate Problem 1.1, under the simplifying assumption that $\alpha \in C^1(\mathbf{R})$ and $\alpha' > 0$. That is, approximate V by a sequence $\{V_n\}$ of finite-dimensional subspaces of V, which fill up V itself; for any $n \in \mathbf{N}$ then solve the Cauchy problem obtained by restricting (1.12) and (1.13) to V_n. Note that for any n this is equivalent to a system of O.D.E.s. The argument then goes on as previously, by deriving uniform estimates on the approximate solutions and passing to the limit along a subsequence.

1.5 Prove Proposition 1.3 and Theorem 1.4.

1.6 In Theorem 1.2, can the assumption (1.20) be weakened, by allowing $|\alpha(v)|$ to grow more than linearly as $v \to \pm\infty$?

Same question for $|\alpha|$ and $|\beta|$, respectively, in (1.20) and (1.39), for Proposition 1.3.

1.7 Let us fix any $g \in L^2\left(0,T;H^1(\Omega)\right)$, and replace (1.18) by the nonhomogeneous Dirichlet condition $\gamma_0 u = \gamma_0 g$ on $\Gamma_1 \times]0,T[$ in Problem 1.1. This yields Problem 1.1_g. Assume that $g \in W^{1,1}\left(0,T;L^2(\Omega)\right) \cap L^2\left(0,T;H^1(\Omega)\right)$, and derive a priori estimates of the form (1.28) and (1.32) for the time discretization of this problem.

Hint. For any n, set $g_m^n := \frac{1}{k}\int_{(n-1)k}^{nk} g(\cdot,\tau)d\tau$ a.e. in Ω, and multiply the time-discretized equation by $u_m^n - g_m^n (\in V)$.

II.2 Further Results via L^2-Techniques

In this section we continue our analysis of Problem 1.1. In particular, further a priori estimates are derived multiplying the approximate equation by the time derivative of the approximate solution.

Some Generalizations. Although in the proof of Theorem 1.2 we used the compactness of the injection $V \to L^2(\Omega)$, cf. (1.36), this property is not really needed. To show this, here we derive (1.10) by means of a different technique, which is based on the use of the structure of the equation in the limit procedure. [10] Let us set

$$z(\cdot, t) := \int_0^t u(\cdot, \tau)d\tau \qquad \text{a.e. in } \Omega, \text{ in } [0, T],$$

$$F(t) := \int_0^t f(\tau)d\tau + w^0 \qquad \text{in } V', \text{ in } [0, T], \tag{2.1}$$

$$z_m^n := k\sum_{j=0}^n u_m^j \quad \text{a.e. in } \Omega, \qquad F_m^n := k\sum_{j=0}^n f_m^j \quad \text{in } V',$$

for $n = 1, \ldots, m$, and define $z_m, \bar{z}_m, F_m, \bar{F}_m$ as in (1.29) and (1.30). By summing (1.24) with respect to n, we get

$$\bar{w}_m + A\bar{z}_m = \bar{F}_m \qquad \text{in } V', \text{ in } [0, T], \tag{2.2}$$

whence taking $m \to \infty$

$$w + Az = F \qquad \text{in } V', \text{ in } [0, T]. \tag{2.3}$$

Let us set $Q_{\tilde{t}} := \Omega \times]0, \tilde{t}[$ for any $\tilde{t} \in]0, T]$. By the two latter equations, Proposition XI.4.11, and the weak lower semicontinuity of the norm, we have

$$\limsup_{m\to\infty} \iint_{Q_{\tilde{t}}} \bar{w}_m \bar{u}_m dx dt = \limsup_{m\to\infty} \int_0^{\tilde{t}} {}_{V'}\langle \bar{F}_m - A\bar{z}_m, \bar{u}_m\rangle_V dt$$

$$\leq \lim_{m\to\infty} \int_0^{\tilde{t}} {}_{V'}\langle \bar{F}_m, \bar{u}_m\rangle_V dt - \frac{1}{2}\liminf_{m\to\infty}\int_\Omega |\nabla z_m(\cdot, \tilde{t})|^2 dx \tag{2.4}$$

$$\leq \int_0^{\tilde{t}} {}_{V'}\langle F, u\rangle_V dt - \frac{1}{2}\int_\Omega |\nabla z(\cdot, \tilde{t})|^2 dx$$

$$= \int_0^{\tilde{t}} {}_{V'}\langle F - Az, u\rangle_V dt = \iint_{Q_{\tilde{t}}} wu\, dx dt \qquad \forall \tilde{t} \in]0, T].$$

---

[10] In the following we use this procedure again, and refer to it as the *equation + lower semicontinuity* technique.

Then we get (1.10) by Lemma XI.5.1. □

A Weaker Formulation. The previous limit procedure allows us to prove existence of a solution for a weaker formulation of Problem 1.1. Assume that

$$w^0 \in V', \qquad f \in L^2(0, T; V'). \tag{2.5}$$

Problem 2.1 *To find $u \in L^2(0, T; V)$ and $w \in L^2(0, T; V')$ such that*

$$\int_0^T {}_{V'}\left\langle w^0 - w, \frac{\partial v}{\partial t} \right\rangle_V dt + \iint_Q \nabla u \cdot \nabla v \, dx dt = \int_0^T {}_{V'}\langle f, v \rangle_V dt \tag{2.6}$$

$$\forall v \in H^1(0, T; V), v(\cdot, T) = 0,$$

$$\int_0^T {}_{V'}\langle w, u - v \rangle_V \, dt \geq \iint_Q \bigl[a(u) - a(v)\bigr] dx dt \qquad \forall v \in L^2(0, T; V). \tag{2.7}$$

Interpretation. This problem can be interpreted as Problem 1.1. In particular, (2.6) yields (1.12) in $L^2(0, T; V')$, whence $w \in H^1(0, T; V')$. The equation (2.7) is a weak formulation of (1.10). Note that $a \geq 0$, by (1.5) and our choice of $\xi_0 = \xi_1 = 0$ in (1.3); therefore $\iint_Q a(v) dx dt\, (\leq +\infty)$ is defined for any measurable function $v : Q \to \mathbf{R}$.

Theorem 2.1 *Assume that (1.1), (1.5), (1.21), (1.22), and (2.5) are satisfied. Then Problem 2.1 has a solution.*

Outline of the Proof. It suffices to approximate the problem by implicit time discretization, and then derive a uniform estimate for u_m in $L^2(0, T; V)$, multiplying the approximate equation by u_m, as in the proof of Theorem 1.2. The limit procedure can be performed as in (2.1) through (2.4), replacing the integral over Q by the duality pairing between $L^2(0, T; V')$ and $L^2(0, T; V)$. □

Remark. The latter result can also be applied to other equations of the form (1.12), with A replaced by a different (coercive) *self-adjoint* operator. Note that we used the latter property to get the equality

$$_{V'}\left\langle Az, \frac{\partial z}{\partial t} \right\rangle_V = \frac{1}{2} \frac{d}{dt} {}_{V'}\langle Az, z \rangle_V \qquad \text{a.e. in }]0, T[,$$

cf. Proposition XI.4.11 (and in a similar inequality for $\bar{u}_m$ and $\bar{z}_m$). For instance, if $\Omega \subset \mathbf{R}^3$, we can replace the scalar functions u, w, and f by vectors of $\mathbf{R}^3$, and A by the operator $\tilde{A}$ associated with the *double curl* operator:

$$\tilde{A} : \tilde{V} := \bigl\{ v \in L^2\left(\Omega; \mathbf{R}^3\right) : \nabla \times v \in L^2\left(\Omega; \mathbf{R}^3\right) \bigr\} \to \tilde{V}',$$

$$_{\tilde{V}'}\langle \tilde{A} v, w \rangle_{\tilde{V}} := \int_\Omega \nabla \times v \cdot \nabla \times w \, dx \qquad \forall \vec{v}, \vec{w} \in \tilde{V} \ (\nabla \times := \text{curl}).$$

Thus $\tilde{A}v := \nabla \times \nabla \times v$ in $\mathcal{D}'(\Omega; \mathbf{R}^3)$. Notice that the injection of $\tilde{V}$ into $L^2(\Omega; \mathbf{R}^3)$ is not compact. [11] □

Theorem 2.2 (L^2-*Lipschitz Continuous Dependence on the Data and Uniqueness*) *Let (1.1) and (1.5) be satisfied. Then there exists a constant C such that, if (for $i = 1, 2$) f_i and w_i^0 fulfill (1.8) and (u_i, w_i) solves the corresponding Problem 1.1, then*

$$\|w_1 - w_2\|_{L^\infty(0,T;V')} + \left\| \int_0^t (u_1 - u_2)(\cdot, \tau) d\tau \right\|_{L^\infty(0,T;V)} \quad (2.8)$$
$$\leq C \left(\|w_1^0 - w_2^0\|_{V'} + \|f_1 - f_2\|_{L^1(0,T;V')} \right).$$

Proof. Let us integrate the equation (1.12) with respect to t. Defining z and F as in (2.1), we have (2.3). Let us set $\tilde{w} := w_1 - w_2$, and similarly use the *tilde* to denote the difference of other quantities indexed by $i = 1, 2$: $\tilde{z}, \tilde{w}^0, \tilde{u}, \tilde{F}$. So we have

$$\tilde{w} + A\tilde{z} = \tilde{F} \qquad \text{in } V', \text{ in } [0, T]. \quad (2.9)$$

Let us multiply this equation by $\tilde{u} = \partial \tilde{z}/\partial t$. The equation (1.1) yields $\tilde{w}\tilde{u} \geq 0$ a.e. in Q, and we get

$$\frac{1}{2} \frac{d}{dt} \int_\Omega |\nabla \tilde{z}|^2 dx \leq {}_{V'}\left\langle \tilde{F}, \frac{\partial \tilde{z}}{\partial t} \right\rangle_V \qquad \text{a.e. in }]0, T[,$$

whence

$$\frac{1}{2} \int_\Omega |\nabla \tilde{z}(x,t)|^2 dx \leq {}_{V'}\langle \tilde{F}(t), \tilde{z}(\cdot, t) \rangle_V - \int_0^t {}_{V'}\left\langle \frac{\partial \tilde{F}}{\partial \tau}, \tilde{z} \right\rangle_V d\tau$$
$$\forall t \in [0, T].$$

This yields the estimate (2.8) for $\tilde{z}$. The estimate for $\tilde{w}$ then follows by comparison in (2.9). □

Corollary 2.3 *Under the assumptions (1.1), (1.5), (1.8), and (1.22), Problem 1.1 has at most one solution.*

Proof. By (2.8), w is unique; then a simple comparison in (1.12) yields the uniqueness of Au. Hence u is also unique, by (1.22). □

Theorem 2.4 (*Weakly Continuous Dependence on the Data*) *Assume that the sequences $\{\alpha_n\}, \{f_n\}, \{w_n^0\}$ fulfill the assumptions of Theorem 1.2, that (1.20)*

---

[11] In this case Problem 2.1 may represent the system of *Maxwell equations* for a nonlinear conducting magnetic material, once the displacement current has been neglected; see Sects. I.3 and IV.8.

holds for L and M independent of n, and that (defining b_n as in (1.3) for any n) the sequence $\{b_n(w_n^0)\}$ is bounded in $L^1(\Omega)$. Let us define the corresponding functions $\{a_n\}$ as in (1.4), and assume that

$$\begin{cases} a_n \to a & \text{uniformly in } \mathbf{R}, \\ f_n \to f & \text{weakly in } L^2(0, T; V'), \\ w_n^0 \to w^0 & \text{weakly in } L^2(\Omega). \end{cases} \qquad (2.10)$$

Let (u_n, w_n) be the solution of the corresponding Problem 1.1_n, [12] and (u, w) be the solution of Problem 1.1. Then

$$u_n \to u \quad \text{weakly in } L^2(0, T; V), \qquad (2.11)$$
$$w_n \to w \quad \text{weakly star in } L^\infty\left(0, T; L^2(\Omega)\right) \cap H^1(0, T; V'). \qquad (2.12)$$

Outline of the Proof. One can multiply the equation $(1.12)_n$ by u_n, then use the procedure of Theorem 1.2 to derive uniform estimates and to pass to the limit on a subsequence. As the solution is unique, the whole sequence converges. □

Theorem 2.5 *(First Regularity Result) Assume that (1.1), (1.5), (1.8), and (1.20) are satisfied, and that* [13]

$$\exists c > 0 : \forall (u_i, w_i) \in \operatorname{graph}(\alpha) \, (i = 1, 2), \\ (w_1 - w_2)(u_1 - u_2) \geq c(u_1 - u_2)^2, \qquad (2.13)$$

$$u^0 := \beta(w^0) \in V, \qquad (2.14)$$

$$f = f_1 + f_2, \qquad f_1 \in L^2(Q), \qquad f_2 \in W^{1,1}(0, T; V'). \qquad (2.15)$$

Then the solution of Problem 1.1 has the following further regularity:

$$u \in H^1\left(0, T; L^2(\Omega)\right) \cap L^\infty(0, T; V), \qquad w \in L^\infty\left(0, T; L^2(\Omega)\right). \qquad (2.16)$$

Proof. Let us multiply (1.24) by $u_m^n - u_m^{n-1}$ and sum for $n = 1, \ldots, \ell$, for any $\ell \in \{1, \ldots, m\}$. By (2.13), we have

$$\int_\Omega (w_m^n - w_m^{n-1})(u_m^n - u_m^{n-1}) dx \geq c \int_\Omega (u_m^n - u_m^{n-1})^2 dx \qquad (2.17)$$
$$\text{for } n = 1, \ldots, \ell.$$

---

[12] In the following we often use this sort of notation, with an index representing dependence on a parameter.

[13] A graph fulfilling (2.13) is said to be *strongly monotone*.

Moreover,

$$\sum_{n=1}^{\ell} {}_{V'}\langle Au_m^n, u_m^n - u_m^{n-1}\rangle_V \geq \frac{1}{2}\sum_{n=1}^{\ell}\int_\Omega \left(|\nabla u_m^n|^2 - |\nabla u_m^{n-1}|^2\right) dx$$
$$= \frac{1}{2}\int_\Omega \left(|\nabla u_m^\ell|^2 - |\nabla u^0|^2\right) dx.$$

Denoting by C_1 a constant independent of m, we get

$$ck\sum_{n=1}^{\ell}\int_\Omega \left(\frac{u_m^n - u_m^{n-1}}{k}\right)^2 dx + \frac{1}{2}\int_\Omega \left(|\nabla u_m^\ell|^2 - |\nabla u^0|^2\right) dx$$
$$\leq \sum_{n=1}^{\ell} {}_{V'}\langle f_m^n, u_m^n - u_m^{n-1}\rangle_V = \sum_{n=1}^{\ell}\int_\Omega f_{1m}^n(u_m^n - u_m^{n-1})dx$$
$$+ {}_{V'}\langle f_{2m}^\ell, u_m^\ell\rangle_V - {}_{V'}\langle f_{2m}^1, u^0\rangle_V - \sum_{n=2}^{\ell} {}_{V'}\langle f_{2m}^n - f_{2m}^{n-1}, u_m^{n-1}\rangle_V$$
$$\leq \left(k\sum_{n=1}^{\ell}\int_\Omega (f_{1m}^n)^2 dx\right)^{1/2} \cdot \left[k\sum_{n=1}^{\ell}\int_\Omega \left(\frac{u_m^n - u_m^{n-1}}{k}\right)^2 dx\right]^{1/2} \quad (2.18)$$
$$+ \left(2\max_{n=1,\ldots,\ell}\|f_{2m}^n\|_{V'} + k\sum_{n=2}^{\ell}\left\|\frac{f_{2m}^n - f_{2m}^{n-1}}{k}\right\|_{V'}\right)\max_{n=0,\ldots,\ell}\|u_m^n\|_V$$
$$\leq \frac{1}{2c}\|f_1\|_{L^2(Q)}^2 + \frac{kc}{2}\sum_{n=1}^{\ell}\int_\Omega\left(\frac{u_m^n - u_m^{n-1}}{k}\right)^2 dx$$
$$+ C_1\|f_2\|_{W^{1,1}(0,T;V')}^2 + \frac{1}{4}\max_{n=0,\ldots,\ell}\|u_m^n\|_V^2.$$

This yields

$$k\sum_{n=1}^{m}\left\|\frac{u_m^n - u_m^{n-1}}{k}\right\|_{L^2(\Omega)}^2, \max_{n=1,\ldots,m}\|u_m^n\|_V \quad (2.19)$$
$$\leq \text{Constant (independent of } m\text{)},$$

that is,

$$\|u_m\|_{H^1(0,T;L^2(\Omega))\cap L^\infty(0,T;V)} \leq \text{Constant.} \quad (2.20)$$

Then, by (1.20), w_m is uniformly bounded in $L^\infty\left(0,T;L^2(\Omega)\right)$. The rest of the argument follows as in the proof of Theorem 1.2. □

Theorem 2.6 *(Second Regularity Result) Assume that the assumptions of Theorem 2.5 are satisfied and that α is Lipschitz continuous. Then $w \in H^1\left(0,T;L^2(\Omega)\right)$.*

If moreover f is as in (1.14) and (1.15), then:
(i) $\Delta u \in L^2(Q)$;
(ii) the equation (1.16) is satisfied a.e. in Q;
(iii) if Ω is of class $C^{1,1}$,

$$u \in L^2\left(0, T; H^2_{\text{loc}}(\Omega) \cap H^r(\Omega)\right) \qquad \forall r < \tfrac{3}{2}. \tag{2.21}$$

(iv) if $\hat{f}_2 \in L^2(0, T; H^s(\Gamma_2))$ for some $s \in \left[0, \tfrac{1}{2}\right]$, then for any neighbourhood N of $\bar{\Gamma}_1 \cap \bar{\Gamma}_2$, $u \in L^2\left(0, T; H^{s+3/2}(\Omega \setminus N)\right)$. If $\bar{\Gamma}_1 \cap \bar{\Gamma}_2 = \emptyset$, then (2.21) holds for $r = 3/2$ as well.

Part (iv) implies (1.17) a.e. on $\Gamma_2 \times \,]0, T[$, so in this case (u, w) is a *strong solution*.

Proof. If α is Lipschitz continuous, then the regularity of w is a direct consequence of (2.16). Then $\Delta u \in L^2(Q)$ by comparison in (1.16), and the rest follows from standard regularity results for elliptic problems. [14] □

Theorem 2.7 *(Third Regularity Result) Assume that (1.1), (1.5), (1.8), (1.20), and (2.13) are satisfied. Then the solution of Problem 1.1 has the following further regularity:*

$$u \in H^r\left(0, T; L^2(\Omega)\right) \qquad \forall r < \tfrac{1}{2}. \tag{2.22}$$

Proof. The first part of the proof of Theorem 1.2 yields (1.32). We derive a further estimate for the solution of Problem 1.1_m. At first let us fix any $h \in \,]0, T[$ and set

$$(\tau_h \varphi)(t) := \varphi(t + h) \qquad \forall t \in \mathbf{R}, \forall \varphi : \mathbf{R} \to \mathbf{R}. \tag{2.23}$$

Let us set $u_m(\cdot, t) := u_m(\cdot, 0)$ for any $t < 0$, multiply (1.31) by $u_m - \tau_{-h} u_m$, and integrate with respect to t. By (1.32) and (1.33), we have

$$\int_h^T dt \int_\Omega \frac{\partial w_m}{\partial t}(u_m - \tau_{-h} u_m) dx \leq 2 \left\| \frac{\partial w_m}{\partial t} \right\|_{L^2(0,T;V')} \|u_m\|_{L^2(0,T;V)} \tag{2.24}$$
$$\leq \text{Constant (independent of } m, h).$$

By (1.35), $\sigma_{m,h} := \partial w_m / \partial t - (w_m - \tau_{-h} w_m)/h$ is bounded in $L^2(0, T; V')$, uniformly with respect to h, m. Moreover, (1.25) and (2.13) yield

$$\int_\Omega (w_m - \tau_{-h} w_m)(u_m - \tau_{-h} u_m) dx \geq c \int_\Omega (u_m - \tau_{-h} u_m)^2 dx \qquad \forall t_1, t_2 \in [0, T].$$

---

[14] See, e.g., Lions and Magenes [356; vol. I, Chap. 2].

Hence we get

$$c \int_h^T dt \int_\Omega \frac{(u_m - \tau_{-h} u_m)^2}{h} dx \leq \int_h^T dt \int_\Omega \frac{w_m - \tau_{-h} w_m}{h} (u_m - \tau_{-h} u_m) dx$$

$$\leq \int_h^T dt \int_\Omega \frac{\partial w_m}{\partial t} (u_m - \tau_{-h} u_m) dx + \|\sigma_{m,h}\|_{L^2(0,T;V')} \|u_m - \tau_{-h} u_m\|_{L^2(0,T;V)}$$

$$\leq \text{Constant (independent of } m, h\text{)}.$$

Therefore for any $\varepsilon \in]0, \frac{1}{2}[$ we have

$$\|u_m\|_{H^{1/2-\varepsilon}(0,T;L^2(\Omega))}$$

$$= \|u_m\|_{L^2(Q)} + \left\{ \iint_{]0,T[^2} \frac{\|u_m(\cdot,t') - u_m(\cdot,t'')\|^2_{L^2(\Omega)}}{|t'-t''|^{2-2\varepsilon}} dt' dt'' \right\}^{1/2}$$

$$= \|u_m\|_{L^2(Q)} + \left\{ 2 \int_0^T dt \int_0^t \frac{\|u_m(\cdot,t) - u_m(\cdot,t-h)\|^2_{L^2(\Omega)}}{h^{2-2\varepsilon}} dh \right\}^{1/2}$$

$$= \|u_m\|_{L^2(Q)} + \left\{ 2 \int_0^T dh \, h^{2\varepsilon-1} \int_h^T \frac{\|u_m(\cdot,t) - \tau_{-h} u_m(\cdot,t)\|^2_{L^2(\Omega)}}{h} dt \right\}^{1/2}$$

$$\leq \text{Constant (independent of } m, h\text{)}.$$

This yields (2.22) by letting $m \to \infty$. □

Theorem 2.8 *(Local Time Regularization) Assume that (1.1), (1.5), (1.8), (1.20), (2.13) are satisfied, and that*

$$f = f_1 + f_2, \qquad \sqrt{t} f_1 \in L^2(Q), \qquad \sqrt{t} \frac{\partial f_2}{\partial t} \in L^1(0,T;V'). \tag{2.25}$$

Then the solution of Problem 1.1 has the following regularity:

$$\sqrt{t} \frac{\partial u}{\partial t} \in L^2(Q), \qquad \sqrt{t} u \in L^\infty(0,T;V). \tag{2.26}$$

Hence, even if $u^0 \notin V$ for any function such that $u^0 \in \beta(w^0)$ a.e. in Ω,

$$u \in H^1\left(\delta, T; L^2(\Omega)\right) \cap L^\infty(\delta, T; V) \qquad \forall \delta > 0. \tag{2.27}$$

Outline of the Proof. This can be compared with that of Theorem 2.5; the main difference is that here one multiplies (1.24) by $nk(u_m^n - u_m^{n-1})$. Note that

$$\sum_{n=1}^\ell {}_{V'}\langle A u_m^n, nk(u_m^n - u_m^{n-1}) \rangle_V \geq \frac{\ell k}{2} \int_\Omega |\nabla u_m^\ell|^2 dx - \frac{k}{2} \sum_{n=0}^{\ell-1} \int_\Omega |\nabla u_m^\ell|^2 dx.$$

A calculation similar to (2.18) yields a uniform estimate corresponding to (2.26). □

On Some Quasilinear Systems. So far we dealt with equations for *scalar* variables u, w. Now we want to discuss the case in which u, w are *vectors*; that is, we deal with a class of quasilinear systems of P.D.E.s. So let $M \in \mathbf{N}$ and assume that

$$\alpha : \text{Dom}(\alpha) \,(\subset \mathbf{R}^M) \to 2^{(\mathbf{R}^M)} \text{ is a maximal monotone graph.} \qquad (2.28)$$

Defining V as in (1.6), $\mathcal{V} := V^M$ is a Hilbert space; note that $\mathcal{V}' = (V')^M$. Let $\mathcal{A} : \mathcal{V} \to \mathcal{V}'$ be a *vectorial* linear, bounded, second order, elliptic operator:

$$_{\mathcal{V}'}\langle \mathcal{A}v, v\rangle_{\mathcal{V}} \geq C\|v\|_{\mathcal{V}}^2 \qquad \forall v \in \mathcal{V} \text{ (with } C\text{: constant} > 0\text{)}. \qquad (2.29)$$

Let also $F \in L^2(0, T; \mathcal{V}')$ and consider a vectorial system analogous to (1.10), (1.12):

$$\begin{cases} \dfrac{\partial w}{\partial t} + \mathcal{A}u = F & \text{in } H^{-1}(0, T; \mathcal{V}'), \\ w \in \alpha(u) & \text{a.e. in } Q. \end{cases} \qquad (2.30)$$

In this framework, two properties play an important role: the self-adjointness of $\mathcal{A}$, and the fact that α is the subdifferential of some lower semicontinuous functional $\mathbf{R}^M \to \mathbf{R} \cup \{+\infty\}$. [15] Let us distinguish the following cases.

(i) If α is cyclically monotone and $\mathcal{A}$ is self-adjoint, the results of Sects. II.1 and II.2 can be extended, with essentially unchanged arguments.

(ii) If α is cyclically monotone but $\mathcal{A}$ is not self-adjoint, then Theorem 1.2 can be extended, since it is possible to multiply the vectorial equation by u. On the other hand, as we pointed out in the first remark of this section, difficulties arise in multiplying the time integral of the equation by u to prove uniqueness of the solution, and also to extend Theorems 2.5 and 2.7.

(iii) If $\mathcal{A}$ is self-adjoint but α is not cyclically monotone, difficulties are met in multiplying the equation by u to prove Theorem 1.2. However, existence of a solution in $H^1(Q)$ can be obtained multiplying the equation by $\partial u/\partial t$, and Theorems 2.5 and 2.7 can be extended. Here it is also possible to multiply the time integral of the equation by u, hence to prove uniqueness of the solution and Theorem 2.2.

(iv) If neither α is cyclically monotone nor $\mathcal{A}$ is self-adjoint, then even in the linear case the initial boundary value problem may be ill-posed, no matter how smooth are the data. To see this, let $M = 2$ and set

$$B_\varphi = \begin{pmatrix} \cos\varphi & \sin\varphi \\ -\sin\varphi & \cos\varphi \end{pmatrix} \qquad \forall \varphi \in \mathbf{R},$$

---

[15] The latter condition is equivalent to the *cyclical monotonicity* of α (see (XI.5.4)), and is stronger than the monotonicity.

which represents rotation by the angle φ. Let $-\pi/2 < \varphi < \pi/2$, so that the matrix B_φ is positive definite, as well as $B_\varphi^{-1} = B_{-\varphi}$. Hence the operator $A := -B_\varphi \Delta$ is elliptic but not self-adjoint, and $\alpha : v \mapsto B_{-\varphi} v$ is maximal monotone but not a subdifferential. After multiplication by B_φ, (2.30) is equivalent to $\partial u/\partial t - B_{2\varphi} \Delta u = F$ in $\mathcal{V}'$. If $\pi/2 < 2\varphi < \pi$, $B_{2\varphi}$ is negative definite and this system is *backward* parabolic.

Exercises.

2.1 Justify the first inequality of (2.4).

2.2 Detail the derivation of (2.19).

2.3 Detail the proof of Theorem 2.4.

2.4 Let $g \in W^{1,1}(0,T; H^1(\Omega))$, and consider the nonhomogeneous Dirichlet Problem 1.1$_g$ (i.e., with $\gamma_0 u = \gamma_0 g$ on $\Gamma_1 \times]0, T[$ in place of (1.18)). Then derive estimates of the forms (1.28) and (2.20) for the discrete approximation of Problem 1.1$_g$.

2.5 Detail the proof of Theorem 2.8. Then derive a priori estimates, multiplying the approximate equation (1.31) by $t^\lambda \partial u_m / \partial t$ for a fixed $\lambda > 1$, and state a corresponding result analogous to Theorem 2.8.

2.6 Formulate a boundary and initial value problem associated with the following *pseudo-parabolic* inclusion

$$\frac{\partial}{\partial t} \alpha(u) - \Delta u + A \frac{\partial u}{\partial t} \ni f, \qquad (2.31)$$

where $A : V \to V'$ is a linear, bounded, second order, elliptic operator. Then discuss existence and uniqueness of the solution.

2.7 Use the strong monotonicity condition (2.13) to improve the stability condition (2.8).

II.3 Techniques of L^1- and L^∞-Type

In this section we study Problem 1.1 in L^1-spaces, and show a maximum principle.

An L^1-Technique. The following result can be compared with Theorem 2.2.

Theorem 3.1 (*Monotone and L^1-Lipschitz Continuous Dependence on the Data*) *Assume that the assumptions of Theorem 1.2 are satisfied. For $i = 1, 2$, let* [16]

$$w_i^0 \in L^2(\Omega), \quad f_i \in L^2(0,T; V'), \quad f_1 - f_2 \in L^1(Q), \qquad (3.1)$$

---

[16] In this section we denote by f_1, f_2 different choices of the datum f.

and (w_i, u_i) be any solution of the corresponding Problem 1.1. Then

$$\int_\Omega (w_1 - w_2)^+(x, t)dx \leq \int_\Omega (w_1^0 - w_2^0)^+ dx + \int_0^t d\tau \int_\Omega (f_1 - f_2)^+(x, \tau)dx \qquad (3.2)$$

for a.a. $t \in]0, T[$.

Hence $w_1 \leq w_2$ a.e. in Q, whenever $w_1^0 \leq w_2^0$ a.e. in Ω and $f_1 \leq f_2$ a.e. in Q.
Moreover, (3.2) is also satisfied if the positive part is replaced by the absolute value.

Proof. At first we assume that

$$\begin{aligned} &\alpha \text{ is Lipschitz continuous and strongly monotone, cf. (2.13),} \\ &\text{and } w_i^0, f_i \ (i = 1, 2) \text{ fulfill (2.14), (2.15);} \end{aligned} \qquad (3.3)$$

later on we drop these restrictions. By Theorem 2.5 and by the assumption on α, we have $w_i, u_i \in H^1(0, T; L^2(\Omega))$ for $i = 1, 2$. Let us set $\tilde{u} := u_1 - u_2$, $\tilde{w} := w_1 - w_2$, $\tilde{f} := f_1 - f_2$, and

$$H(\eta) := \begin{cases} \{0\} & \text{if } \eta < 0, \\ [0, 1] & \text{if } \eta = 0, \\ \{1\} & \text{if } \eta > 0, \end{cases} \qquad H_j(\eta) := \begin{cases} 0 & \text{if } \eta \leq 0, \\ j\eta & \text{if } 0 \leq \eta \leq \frac{1}{j}, \\ 1 & \text{if } \eta \geq \frac{1}{j}, \end{cases} \qquad (3.4)$$

for any $j \in \mathbf{N}$. Let us take the difference between (1.12) written for $i = 1, 2$, multiply it by $H_j(\tilde{u})$, and integrate in Ω. Since

$$\int_\Omega \nabla \tilde{u} \cdot \nabla H_j(\tilde{u}) dx = \int_\Omega H_j'(\tilde{u}) |\nabla \tilde{u}|^2 dx \geq 0 \qquad \text{a.e. in }]0, T[,$$

we get

$$\int_\Omega \frac{\partial \tilde{w}}{\partial t} H_j(\tilde{u}) dx \leq \int_\Omega \tilde{f} H_j(\tilde{u}) dx \leq \int_\Omega \tilde{f}^+ dx \qquad \text{a.e. in }]0, T[.$$

Let us now pass to the limit as $j \to \infty$. Note that

$$H_j(\tilde{u}) \to \psi := \begin{cases} 0 & \text{where } \tilde{u} \leq 0, \\ 1 & \text{where } \tilde{u} > 0, \end{cases} \qquad \text{a.e. in } Q.$$

As α and β are single-valued, we have $H(\tilde{u}) = H(\tilde{w})$ a.e. in Q; hence $\psi \in H(\tilde{w})$ a.e. in Q. So we get

$$\frac{d}{dt} \int_\Omega \tilde{w}^+(x, t) dx = \int_\Omega \frac{\partial \tilde{w}}{\partial t} \psi dx \leq \int_\Omega \tilde{f}^+ dx \qquad \text{a.e. in }]0, T[,$$

which yields (3.2) after time integration.

In the general setting, we can approximate α, w_i^0, f_i ($i = 1, 2$) by means of sequences fulfilling (3.3) (indexed by $n \in \mathbf{N}$, say); this yields $(3.2)_n$ for any n.

By applying the following Lemma 3.2 with $H := L^2(\Omega)$ and $W := V'$, by (2.12) we have

$$w_{1n} - w_{2n} \to w_1 - w_2 \qquad \text{weakly in } L^2(\Omega), \forall t \in [0, T],$$

whence

$$\liminf_{n \to \infty} \int_\Omega (w_{1n} - w_{2n})^+(x, t)dx \geq \int_\Omega (w_1 - w_2)^+(x, t)dx \qquad \forall t \in [0, T]. \quad (3.5)$$

By passing to the inferior limit in $(3.2)_n$, then we get (3.2).

The final statement of the theorem is simply obtained by exchanging w_1 and w_2 in (3.2), and then adding the two inequalities. □

Lemma 3.2 *Let H, W be Banach spaces, H be reflexive, and $H \subset W$ with continuous injection. Let the sequence $\{v_n\}$ be bounded in $L^\infty(0, T; H)$, and*

$$v_n \to v \qquad \text{strongly in } C^0([0, T]; W). \quad (3.6)$$

Then

$$v_n(t) \to v(t) \qquad \text{weakly in } H, \forall t \in [0, T]. \quad (3.7)$$

Proof. Let us fix any $t \in [0, T]$. By the reflexivity of H, there exists $z_t \in H$ such that, possibly extracting a subsequence, $v_n(t) \to z_t$ weakly in H. By the continuity of the injection of H into W, we have $z_t = v(t)$. As this limit does not depend on the subsequence, we conclude that the whole sequence converges weakly in H. □

The technique of Theorem 3.1 also allows to derive some time regularity.

Theorem 3.3 *(Time Regularity) Let the hypotheses of Theorem 1.2 hold and* [17]

$$f \in W^{1,1}\left(0, T; C_c^0(\Omega)'\right). \quad (3.8)$$

Moreover, assume that there exist sequences $\{w_n^0\} \subset L^2(\Omega)$, $\{u_n^0\} \subset V$, $\{a_n : \mathbf{R} \to \mathbf{R}\}$, such that for any n, a_n is convex, $\alpha_n := \partial a_n$ is Lipschitz continuous, and, setting $w_n^0 := \alpha_n(u_n^0)$ a.e. in Ω,

$$\Delta u_n^0 \in L^1(\Omega), \qquad \|\Delta u_n^0\|_{L^1(\Omega)} \leq \text{Constant (independent of } n\text{)}, \quad (3.9)$$

---

[17] $C_c^0(\Omega)'$ is the space of *Radon measures*, cf. Sect. XI.1.

$$\begin{cases} a_n \to a & \text{uniformly in } \mathbf{R}, \\ w_n^0 \to w^0 & \text{weakly in } L^2(\Omega), \\ u_n^0 \to u^0 & \text{strongly in } L^1(\Omega), \\ \Delta u_n^0 \to \Delta u^0 & \text{weakly star in } C_c^0(\Omega)'. \end{cases} \quad (3.10)$$

Then for the solution of Problem 1.1 we have [18]

$$w, \frac{\partial w}{\partial t}, \Delta u \in L^\infty_{w*}\left(0, T; C_c^0(\Omega)'\right). \quad (3.11)$$

(In general, the existence of sequences $\{w_n^0\}$, $\{u_n^0\}$, $\{a_n\}$ as in the preceding does not seem granted a priori.)

Proof. At first we assume that w^0, u^0, α fulfill the conditions that we required for the approximating sequences, and that $f \in W^{1,1}\left(0, T; L^1(\Omega)\right) \cap L^2(0, T; V')$. Later on we drop these restrictions.

Let us fix any $h \in]0, T[$, define τ_h as in (2.23), and apply Theorem 3.1 with $f_1 = \tau_h f$, $f_2 = f$, $w_1^0 = w(\cdot, h)$, $w_2^0 = w(\cdot, 0) = w^0$. This yields

$$\int_\Omega |\tau_h w - w|(x, t) dx \leq \int_\Omega |w(\cdot, h) - w^0| dx \\ + \int_0^t d\tau \int_\Omega |\tau_h f - f|(x, \tau) dx \quad \forall t \in [0, T - h]. \quad (3.12)$$

Let us define H and the sequence $\{H_j\}$ as in (3.4), and approximate the multi-valued function sign $:= 2H - 1$, cf. (XI.5.3), by the sequence

$$\text{sign}_j(\xi) := H_j(\xi) - H_j(-\xi) \quad \forall \xi \in \mathbf{R}, \forall j \in \mathbf{N}. \quad (3.13)$$

Now let us multiply (1.12) by $\text{sign}_j(u - u^0)$ ($\in L^2(0, T; V)$), and integrate in space and time for any $j \in \mathbf{N}$. This yields

$$\begin{aligned} {}_{V'}\left\langle \frac{\partial w}{\partial t}, \text{sign}_j(u - u^0)\right\rangle_V &= {}_{V'}\langle f - Au, \text{sign}_j(u - u^0)\rangle_V \\ &= {}_{V'}\langle f - Au^0, \text{sign}_j(u - u^0)\rangle_V - {}_{V'}\langle A(u - u^0), \text{sign}_j(u - u^0)\rangle_V \\ &\leq {}_{V'}\langle f - Au^0, \text{sign}_j(u - u^0)\rangle_V = \int_\Omega (f + \Delta u^0) \text{sign}_j(u - u^0) dx \\ &\leq \int_\Omega |f + \Delta u^0| dx \quad \text{a.e. in }]0, T[. \end{aligned} \quad (3.14)$$

---

[18] Here $L^\infty_{w*}\left(0, T; C_c^0(\Omega)'\right)$ is the space of essentially bounded *weakly star measurable* functions $[0, T] \to C_c^0(\Omega)'$; see, e.g., Diestel and Uhl [195; Chap. IV], and Kufner, John and Fučik [330; Sect. 2.22]. This space can be identified with $L^1\left(0, T; C_c^0(\Omega)\right)'$, but is different from $L^\infty\left(0, T; C_c^0(\Omega)'\right)$ since $C_c^0(\Omega)'$ is not separable.

As we assumed that α is single-valued, we have

$$\operatorname{sign}_j(u - u^0) \to \psi \in \operatorname{sign}(u - u^0) \subset \operatorname{sign}(w - w^0) \quad \text{a.e. in } Q.$$

By Theorem 2.6, we have $\partial w / \partial t \in L^2(Q)$; hence, by taking $j \to \infty$ in (3.14), we get

$$\frac{d}{dt} \int_\Omega |w(\cdot, t) - w^0| dx = \int_\Omega \frac{\partial w}{\partial t} \psi dx \leq \int_\Omega |f + \Delta u^0| dx \quad \text{a.e. in }]0, T[.$$

By integrating in $]0, h[$, we have

$$\int_\Omega |w(\cdot, h) - w^0| dx \leq \int_0^h d\tau \int_\Omega |f(\cdot, \tau) + \Delta u^0| dx.$$

Therefore, by dividing (3.12) by h and passing to the limit as $h \to 0$, we get

$$\int_\Omega \left|\frac{\partial w}{\partial t}\right|(x, t) dx \leq \int_\Omega |f(\cdot, 0) + \Delta u^0| dx + \int_0^t d\tau \int_\Omega \left|\frac{\partial f}{\partial \tau}\right|(x, \tau) dx \quad (3.15)$$

$$\text{for a.a. } t \in]0, T[.$$

The analogous estimate for Δu then follows by comparison in (1.12).

We now drop the restrictions on the data. Let us approximate α, w^0 and u^0 by three sequences $\{\alpha_n\}$, $\{w_n^0\}$, $\{u_n^0\}$ fulfilling the assumptions of the theorem, set $w_n^0 := \alpha_n(u_n^0)$ a.e. in Ω, and consider the corresponding Problem 1.1$_n$. The latter has one and only one solution (u_n, w_n), which also fulfills the regularity properties stated in Theorem 2.5.

As we saw, $\partial w_n / \partial t$ is uniformly bounded in $L^\infty(0, T; L^1(\Omega))$. By comparison in (1.12), Δu_n is then also uniformly bounded in the same space. By Theorem 2.4, we know that (u_n, w_n) converges to the solution (u, w) of Problem 1.1. Hence (3.11) is satisfied for (u, w), too. □

The following statement does not yield much extra-regularity, on the other hand it does not require (3.9) and (3.10). It can also be compared with Theorem 2.8.

Proposition 3.4 *(Time Regularization) Assume that the hypotheses of Theorem 2.5 hold, that α is Lipschitz continuous, and*

$$tf \in W^{1,1}\left(0, T; L^1(\Omega)\right). \quad (3.16)$$

Then the solution of Problem 1.1 is such that

$$t\frac{\partial w}{\partial t}, t\Delta u \in L^\infty_{w*}\left(0, T; L^1(\Omega)\right). \quad (3.17)$$

Hence
$$\frac{\partial w}{\partial t}, \Delta u \in L^\infty_{w*}(\delta, T; L^1(\Omega)) \qquad \forall \delta > 0. \tag{3.18}$$

Outline of the Proof. Let us apply the time increment operator δ_k to the approximate equation (1.31), multiply it by $tH_j(\delta_k u_m)$, and take $j \to +\infty$. As α is single-valued, we have $H(\delta_k u_m) \subset H(\delta_k w_m)$ and then get

$$\int_\Omega t \frac{\partial}{\partial t} |\delta_k w_m| dx \leq \int_\Omega t |\delta_k f_m| dx \qquad \text{for a.a. } t \in \,]0, T[.$$

By time integration we have

$$\int_\Omega t |\delta_k w_m| dx \leq \int_0^t d\tau \int_\Omega |\delta_k w_m| dx + \iint_Q \tau |\delta_k f_m| dx d\tau \qquad \forall t \in \,]0,T].$$

This yields a uniform estimate for $t\partial w_m/\partial t$ in $L^\infty_{w*}(a, T; L^1(\Omega))$, as $\partial w_m/\partial t$ is uniformly bounded in $L^2(Q)$ by Theorem 2.6. An analogous estimate for $t\Delta u_m$ then follows by comparison in (1.31). Therefore, possibly extracting subsequences,

$$t \frac{\partial w_m}{\partial t} \to t \frac{\partial w}{\partial t}, \quad t\Delta u_m \to t\Delta u \qquad \text{weakly star in } L^\infty_{w*}(0, T; C^0_c(\Omega)');$$

these sequences also converge weakly star in $L^\infty_{w*}(0, T; L^1(\Omega))$, as the limit functions belong to $L^2(Q)$. □

L^∞-Results. Results of boundedness are here derived in two ways, multiplying the equation either by a *cut-off* of the solution, or by a power of the solution and than taking this power to infinity.

Proposition 3.5 *(Maximum and Minimum Principles) Assume that (1.1), (1.5), (1.8), (1.21), and (1.22) hold, and that there exists a constant $M > 0$ such that*

$$\exists u^0 \in L^1(\Omega) : u^0 \in \beta(w^0), \ u^0 \leq M \ (u^0 \geq -M, \text{ resp.}) \ \text{a.e. in } \Omega, \tag{3.19}$$

$$f \leq 0 \quad (f \geq 0, \text{ resp.}) \qquad \text{in the sense of } \mathcal{D}'(Q). \tag{3.20}$$

Then the solution of Problem 1.1 is such that

$$u \leq M \quad (u \geq -M, \text{ resp.}) \qquad \text{a.e. in } Q. \tag{3.21}$$

Proof. By Theorem 1.4 and Corollary 2.3, Problem 1.1 has one and only one solution. At first, let us assume that $u^0 \leq M$ and $f \leq 0$.
Let us set $\Phi(v) := \int_0^v (\tilde{\beta}(\xi) - M)^+ d\xi \ (\geq 0)$, for any $v \in \text{Dom}(\beta)$ and any section $\tilde{\beta}$ of β (the integral is independent from this choice).

Then let us multiply (1.12) by $(u-M)^+ \left(\in L^2(0,T;V)\right)$, and integrate in time. Note that (3.19) and (3.20), respectively, yield

$$\Phi(w^0) = 0 \quad \text{a.e. in } \Omega, \qquad {}_{V'}\langle f, (u-M)^+\rangle_V d\tau \leq 0 \quad \text{a.e. in }]0,T[.$$

Moreover, as $(u-M)^+ \in \left(\tilde{\beta}(w)-M\right)^+$ a.e. in Q, by Proposition XI.4.11 we have $\int_\Omega \Phi(w)dx \in W^{1,1}(0,T)$ and ${}_{V'}\langle \frac{\partial w}{\partial t}, (u-M)^+\rangle_V = \frac{d}{dt}\int_\Omega \Phi(w)dx$. So we get

$$\int_\Omega \Phi(w(x,T))dx + \int_0^T dt \int_\Omega |\nabla(u-M)^+|^2 dx \leq 0.$$

Hence $\nabla(u-M)^+ = 0$ a.e. in Q, which yields $u \leq M$ by (1.22).

The case of $u^0 \geq M$ and $f \geq 0$ can be treated by the same argument, using $-(u-M)^-$ in place of $(u-M)^+$. □

Proposition 3.6 *Assume that (1.1), (1.5), (1.21), (1.22) hold, and that for some $q > 2$*

$$w^0 \in L^q(\Omega), \qquad f \in L^q(Q) \cap L^2(0,T;V'). \tag{3.22}$$

Then the solution of Problem 1.1 is such that $w \in L^\infty(0,T;L^q(\Omega))$, and

$$\|w(\cdot,t)\|_{L^q(\Omega)} \leq \left[1 + \left(q\|f\|_{L^q(Q)}\right)^{1/q} + \|w^0\|_{L^q(\Omega)}\right] \exp\left(\|f\|_{L^q(Q)}t\right) \tag{3.23}$$

$$\forall t \in]0,T].$$

(This estimate is not optimal, but it suffices for our purposes.)

Proof. By Theorem 1.4 and Corollary 2.3, Problem 1.1 has one and only one solution. At first we assume that α is Lipschitz continuous.

Fix any $M > 0$, and set $\varphi_q(v) := |v|^{q-2}v$, $\psi_M(v) := \min\{\max\{v,-M\},M\}$ for any $v \in \mathbf{R}$, $w_M := \psi_M(w)$ a.e. in Q. Note that the function $\varphi_q \circ \psi_M \circ \alpha$ is Lipschitz continuous and nondecreasing, and

$$\frac{1}{q}|(\psi_M(v)|^q \leq \int_0^{|v|} |\psi_M(\xi)|^{q-1}d\xi = \int_0^v \varphi_q(\psi_M(\xi))d\xi \leq \frac{1}{q}|v|^q \qquad \forall v \in \mathbf{R}.$$

Let us multiply the equation (1.12) by $\varphi_q(w_M) = \varphi_q(\psi_M(\alpha(u))) \in L^2(0,T;V)$ and integrate in time. Note that ${}_{V'}\langle Au, \varphi_q(w_M)\rangle_V \geq 0$. By the Schwarz-Hölder inequality, then we get

$$\frac{1}{q}\int_\Omega |w_M(x,t)|^q dx - \frac{1}{q}\int_\Omega |w^0(x)|^q dx$$

$$\leq \|f\|_{L^q(\Omega\times]0,t[)} \||w_M|^{q-1}\|_{L^{q/(q-1)}(\Omega\times]0,t[)}$$

$$= \|f\|_{L^q(Q)}\left(\int_0^t d\tau \int_\Omega |w_M(x,\tau)|^q dx\right)^{(q-1)/q}$$

$$\leq \|f\|_{L^q(Q)}\left(1 + \int_0^t d\tau \int_\Omega |w_M(x,\tau)|^q dx\right) \qquad \forall t \in]0,T]$$

By the Gronwall lemma, cf. Theorem XI.9.1, we get

$$\int_\Omega |w_M(x,t)|^q dx \le \left(q\|f\|_{L^q(Q)} + \int_\Omega |w^0(x)|^q dx \right) \exp\left(q\|f\|_{L^q(Q)} t\right),$$

for any $t \in]0, T]$. By passing to the limit as $M \to +\infty$, we get (3.23).

If α is not Lipschitz continuous, one can approximate it by a sequence $\{\alpha_n\}$ of regular monotone functions, and then apply Theorem 2.4. □

Corollary 3.7 *Assume that (1.1), (1.5), (1.21), (1.22) hold, and*

$$w^0 \in L^\infty(\Omega), \qquad f \in L^\infty(Q) \cap L^2(0, T; V'). \tag{3.24}$$

Then the solution of Problem 1.1 is such that $w \in L^\infty(Q)$.

Proof. It suffices to apply the previous Proposition for any $q > 2$, and then to pass to the limit as $q \to +\infty$ in (3.23). □

Exercises.

3.1 Under the assumptions of Theorem 3.3, is it possible to replace the space $C_c^0(\Omega)'$ by $L^1(\Omega)$?

Hint. What are the respective compactness properties of bounded subsets of these spaces?

3.2 In the proof of Proposition 3.6, may equation (1.12) be multiplied by $\varphi_q(w)$ in place of $\varphi_q(w_M)$?

3.3 Under the assumptions of Theorem 3.3, assume that $\partial f/\partial t \le 0$ in $\mathcal{D}'(Q)$ and $f(\cdot, 0) + \Delta u_n^0 \le 0$ in $\mathcal{D}'(\Omega)$ for any $n \in \mathbf{N}$. Then prove that $\partial w/\partial t \le 0$ in $\mathcal{D}'(Q)$.

Hint. Derive a formula analogous to (3.12), with positive parts in place of absolute values, then use an argument similar to that of Theorem 3.3.

II.4 Local Regularity Results

In this section we prove two results of local regularity, by means of L^2- and L^1- techniques, respectively.

At first we show that if the datum f fulfills suitable integrability properties in a *parabolic cylinder* $Q_2 \subset Q$, then the solution of Problem 1.1 has some extra-regularity in any parabolic cylinder $Q_1 \subset\subset Q_2$. (This means that the closure of Q_1 is a compact subset of Q_2.)

Theorem 4.1 *(Local H^1-Regularity in Space and Time) Assume that the hypotheses of Theorem 1.2 hold. Let*

$$\Omega_1 \subset\subset \Omega_2 \subset \Omega, \qquad 0 \le T_2 < T_1 < T. \tag{4.1}$$

If α is strongly monotone, cf. (2.13), and $f \in L^2(\Omega_2 \times]T_2, T[)$, then

$$u \in H^1\left(T_1, T; L^2(\Omega_1)\right) \cap L^\infty\left(T_1, T; H^1(\Omega_1)\right). \tag{4.2}$$

Proof. By Theorems 1.2 and 2.2, Problem 1.1 has a unique solution (u, w). We use a simple *cut-off* technique and then derive a priori estimates by an L^2-procedure. Let $g : \Omega \to [0, 1]$ be of class C^2 and such that $g = 1$ in Ω_1, $g = 0$ in $\bar{\Omega} \setminus \Omega_2$. Set

$$\varphi_1(x) := g(x)^2 \qquad \forall x \in \Omega, \tag{4.3}$$

$$\varphi_2(t) := \min\left\{\max\left\{\frac{t - T_2}{T_1 - T_2}, 0\right\}, 1\right\} \qquad \forall t \in]0, T[,$$

$$\varphi(x, t) = \varphi_1(x)\varphi_2(t) \qquad \forall (x, t) \in Q.$$

Denoting the Lipschitz constant of g by L, we have

$$|\nabla \varphi| = |\nabla \varphi_1|\varphi_2 \leq 2L\sqrt{\varphi_1}\varphi_2 \leq 2L\sqrt{\varphi} \qquad \text{a.e. in } Q. \tag{4.4}$$

Now we multiply (1.31) by $(\partial u_m/\partial t)\varphi$ and integrate in time. We have

$$\int_\Omega \frac{\partial w_m}{\partial t} \frac{\partial u_m}{\partial t} \varphi \, dx \geq c \int_\Omega \left(\frac{\partial u_m}{\partial t}\right)^2 \varphi \, dx, \tag{4.5}$$

$$\int_\Omega \nabla \bar{u}_m \cdot \nabla \left(\frac{\partial u_m}{\partial t} \varphi\right) dx$$
$$= \int_\Omega \nabla \bar{u}_m \cdot \frac{\partial \nabla u_m}{\partial t} \varphi \, dx + \int_\Omega \nabla \bar{u}_m \cdot \nabla \varphi \frac{\partial u_m}{\partial t} dx$$
$$\geq \frac{1}{2} \int_\Omega \left(\frac{\partial}{\partial t} |\nabla u_m|^2\right) \varphi \, dx + \int_\Omega \nabla \bar{u}_m \cdot \nabla \varphi \frac{\partial u_m}{\partial t} dx \tag{4.6}$$
$$\geq \frac{1}{2} \frac{d}{dt} \int_\Omega |\nabla u_m|^2 \varphi \, dx - \frac{1}{2} \int_\Omega |\nabla u_m|^2 \frac{\partial \varphi}{\partial t} dx$$
$$- 2L \int_\Omega |\nabla \bar{u}_m| \sqrt{\varphi} \left|\frac{\partial u_m}{\partial t}\right| dx \qquad \text{a.e. in }]0, T[.$$

Hence we get

$$c \int_0^t d\tau \int_\Omega \left(\frac{\partial u_m}{\partial t}\right)^2 \varphi \, dx dt + \frac{1}{2} \int_\Omega |\nabla u_m|^2(x, t) \varphi(x, t) dx$$
$$\leq \frac{1}{2(T_1 - T_2)} \|u_m\|^2_{L^2(0,T;V)} + (2L\|\bar{u}_m\|_{L^2(0,T;V)} +$$
$$+ \|\bar{f}_m\|_{L^2(\Omega_2 \times]T_2, T[)}) \left[\int_0^t d\tau \int_\Omega \left(\frac{\partial u_m}{\partial t}\right)^2 \varphi \, dx dt\right]^{1/2}.$$

56   II. A Class of Quasilinear Parabolic P.D.E.s

By (1.32), then we have

$$\left\|\frac{\partial u_m}{\partial t}\sqrt{\varphi}\right\|_{L^2(Q)}, \|\nabla u_m \sqrt{\varphi}\|_{L^\infty(0,T;L^2(\Omega;\mathbf{R}^N))} \qquad (4.7)$$
$$\leq \text{Constant (independent of } m),$$

which yields a uniform estimate corresponding to the regularity (4.2).

Remarks. (i) A result of local regularity in space in the whole $]0,T[$ can be similarly proved. If $f \in L^2(\Omega_2 \times]0,T[)$ and $u^0 \in H^1(\Omega_2)$, then, by using $\varphi = \varphi_1$ as a cut-off function, one gets (4.2) with $T_1 = 0$.

(ii) Similarly a result of local regularity in time holds in the whole Ω. If $f \in L^2(\Omega \times]T_2,T[)$, then, by using $\varphi = \varphi_2$ as a cut-off function, one gets (4.2) with $\Omega_1 = \Omega$.

(iii) Theorem 4.1 can be extended to systems of P.D.E.s of the form (2.30), provided that the maximal monotone graph α is *cyclically monotone* and the elliptic operator A is self-adjoint. In fact under these assumptions an existence result like Theorem 1.2 holds, and one can multiply the approximate vectorial equation by $(\partial u_m/\partial t)\varphi$. □

A Space Regularity Result for w. At first for any $D \subset\subset \Omega$ let us set

$$\xi := \tfrac{1}{2}\min\left\{|x-y| : x \in \bar{D}, y \in \mathbf{R}^N \setminus \Omega\right\},$$
$$D_\xi := \{x \in \Omega : \exists y \in D : |x-y| < \xi\},$$
$$\delta_h v(x) := v(x+h) - v(x) \qquad \forall v \in L^1(\Omega), \forall x \in D, \forall h \in \mathbf{R}^N : |h| \leq \xi.$$

The main interest of the following result stays in the technique, which yields some space regularity for w even if α is multi-valued.

Theorem 4.2 *(Local BV-Regularity in Space) Assume that the hypotheses of Theorem 2.5 hold, and that* [19]

$$\Omega_1 \subset\subset \Omega_2 \subset \Omega, \qquad (4.8)$$

$$w^0 \in BV(\Omega_2), \qquad f \in L^1_{w*}(0,T;BV(\Omega_2)). \qquad (4.9)$$

Then

$$w \in L^\infty_{w*}(0,T;BV(\Omega_1)). \qquad (4.10)$$

---

[19] $L^1_{w*}(0,T;BV(\Omega))$ is the space of integrable *weakly star measurable* functions $[0,T] \to BV(\Omega)$; see, e.g., Diestel and Uhl [195; Chap. IV], and Kufner, John and Fučik [330; Sect. 2.22]. This space is larger than $L^1\big(0,T;BV(\Omega)\big)$ since $BV(\Omega)$ is not separable.

Similar properties hold for $L^\infty_{w*}(0,T;BV(\Omega))$. Since $BV(\Omega)$ is the dual of a separable Banach space, X say, we have $L^\infty_{w*}(0,T;BV(\Omega)) = L^1(0,T;X)'$.

II.4 Local Regularity Results

Proof. Here also we use a simple *cut-off* technique, and then derive a priori estimates by an L^1 method. At first let us assume that α is single-valued.

Let us define φ_1 as in (4.3), the sequence $\{\text{sign}_j\}$ as in (3.13), and approximate the absolute value function by $m_j(\zeta) := \int_0^\zeta \text{sign}_j(\eta) d\eta \ (\leq |\zeta|)$ for any $\zeta \in \mathbf{R}$. Let us fix any $j \in \mathbf{N}$ and any $h \in \mathbf{R}^N$ with $|h| \leq \text{dist}\,(\Omega_1, \mathbf{R}^N \setminus \Omega)$. Then apply the operator δ_h to the restriction of (1.16) to $\Omega_1 \times\,]0, T[$, and multiply it by $\varphi_1 \text{sign}_j(\delta_h w) \in L^2(0, T; V)$. We denote by C_i suitable positive constants.

By Theorem 2.5, $u \in L^\infty(0, T; V)$; hence part (i) of Proposition XI.2.2 yields $\|\delta_h u(\cdot, t)\|_{L^2(\Omega)} \leq h \|u\|_{L^\infty(0,T;V)}$ a.e. in $]0, T[$. Therefore

$$-\int_\Omega (\Delta \delta_h u)\,\text{sign}_j(\delta_h u)\varphi_1 dx = \int_\Omega \nabla \delta_h u \cdot \nabla \left[\text{sign}_j(\delta_h u)\varphi_1\right] dx$$

$$= \int_\Omega |\nabla \delta_h u|^2 \, \text{sign}'_j(\delta_h u)\varphi_1 dx + \int_\Omega \text{sign}_j(\delta_h u) \nabla \delta_h u \cdot \nabla \varphi_1 dx$$

$$\geq \int_\Omega \nabla m_j(\delta_h u) \cdot \nabla \varphi_1 dx = -\int_\Omega m_j(\delta_h u) \Delta \varphi_1 dx$$

$$\geq -\|\delta_h u(\cdot, t)\|_{L^2(\Omega)} \|\Delta \varphi_1\|_{L^2(\Omega)} \geq -C_1 h \qquad \text{a.e. in }]0, T[.$$

Hence we get

$$\int_\Omega \frac{\partial \delta_h w}{\partial t} \text{sign}_j(\delta_h u) \varphi_1 dx \leq \int_\Omega |\delta_h f| \varphi_1 dx + C_1 h \qquad \text{a.e. in }]0, T[.$$

Now we follow a procedure analogous to that of the proof of Theorem 3.1. At first in the latter inequality we pass to the limit as $j \to \infty$, and get the same inequality with $\text{sign}_j(\delta_h u)$ replaced by a measurable function $\psi \in \text{sign}(\delta_h u)$ a.e. in Q. As here we assumed α to be single-valued, we have $\text{sign}(\delta_h u) \subset \text{sign}(\delta_h w)$, whence $(\partial \delta_h w/\partial t)\psi = \partial/\partial t |\delta_h w|$. After time integration, we get

$$\int_\Omega \left[|\delta_h w(\cdot, t)| - |\delta_h w^0|\right] \varphi_1 dx \leq \int_0^t d\tau \int_\Omega |\delta_h f|\varphi_1 dx + C_1 t h \qquad (4.11)$$

for a.a. $t \in]0, T[$.

So, by (4.11) and part (ii) of Proposition XI.2.2, we have

$$\int_{\Omega_1} |\delta_h w(\cdot, t)| dx \leq \int_{\Omega_2} |\delta_h w^0| dx + \int_0^T d\tau \int_{\Omega_2} |\delta_h f| dx + C_1 T h$$

$$\leq h \|\nabla w^0\|_{C_c^0(\Omega_2; \mathbf{R}^N)'} + h \|\nabla f\|_{L^1_{w*}(0,T;BV(\Omega_2))} + C_1 T h =: C_2 h$$

for a.a. $t \in]0, T[$;

here C_2 is a constant independent of h and t (but not of Ω_1). This yields (4.10) by Proposition XI.2.2.

If α is multi-valued we approximate it by a sequence of single-valued functions $\alpha_n = a'_n$, such that $a_n \to a$ (defined in (1.3)) uniformly. We then notice that the estimate is independent of this approximation, and apply Theorem 2.4. □

Exercises.

4.1 Justify the first inequality of (4.6). (Note that the property $\varphi \geq 0$ is essential.)

4.2 Prove the results stated in the two remarks after Theorem 4.1.

4.3 May Theorems 4.1 and 4.3 be extended to systems of the form (2.30)? If so, under which assumptions?

II.5 Integral Transformations

In this section we discuss two simple and natural transformations of Problem 1.1. By integrating the equation (1.12) in time, we get a differential inclusion, that contains the maximal monotone graph α. *Dually*, by applying the operator A^{-1} to (1.12), we obtain an inclusion that contains $\beta \left(:= \alpha^{-1}\right)$. Either problem can be formulated as a variational inequality, and well-posedness can easily be proved. Actually, the two transformations turn out to be equivalent, and indeed they yield the same regularity result.

Time Integral Transformation. As in (2.1) let us set

$$z(\cdot,t) := \int_0^t u(\cdot,\tau)d\tau \qquad \text{a.e. in } \Omega, \forall t \in [0,T], \tag{5.1}$$

$$F(t) := \int_0^t f(\tau)d\tau + w^0 \qquad \text{in } V', \forall t \in [0,T]. \tag{5.2}$$

By integrating (1.12) in time and coupling it with (1.10), we get

$$\alpha\left(\frac{\partial z}{\partial t}\right) + Az \ni F \qquad \text{in } V', \forall t \in [0,T]. \tag{5.3}$$

If $\partial z/\partial t \in V$ a.e. in $]0,T[$, (5.3) is equivalent to the following variational inequality, where a is defined as in (1.3):

$$_{V'}\left\langle Az - F, \frac{\partial z}{\partial t} - v \right\rangle_V + \int_\Omega \left[a\left(\frac{\partial z}{\partial t}\right) - a(v)\right] dx \leq 0 \qquad \forall v \in V, \tag{5.4}$$

$$\text{a.e. in }]0,T[.$$

Let us assume that

$$F \in W^{1,1}(0,T;V'); \tag{5.5}$$

obviously this condition is fulfilled if (1.8) is satisfied. We can now formulate (5.3) requiring less regularity for z.

Problem 5.1 *To find $z \in L^\infty(0,T;V) \cap H^1\left(0,T;L^2(\Omega)\right)$ such that*

$$\frac{1}{2}\int_\Omega |\nabla z(\cdot,\tilde{t})|^2 dx + \int_0^{\tilde{t}} {}_{V'}\!\left\langle \frac{\partial F}{\partial t}, z\right\rangle_V dt - {}_{V'}\!\left\langle F(\tilde{t}), z(\cdot,\tilde{t})\right\rangle_V$$
$$+ \int_0^{\tilde{t}} dt \int_\Omega \left[a\left(\frac{\partial z}{\partial t}\right) - a(v)\right] dx \leq \int_0^{\tilde{t}} {}_{V'}\!\left\langle Az - F, v\right\rangle_V dt \qquad (5.6)$$
$$\forall v \in V, \text{ for a.a. } \tilde{t} \in]0,T[,$$

$$z(\cdot,0) = 0 \qquad \text{a.e. in } \Omega. \qquad (5.7)$$

Theorem 5.1 *(Existence and Uniqueness) Assume that (1.1), (1.5), (1.20), (1.39) are satisfied. Then Problem 5.1 has one and only one solution.*

Outline of the Proof. Let us set $z_m^0 := 0$, $F_m^n := F(nk)$ for any $n \in 1, \ldots, m$, and approximate (5.3) by implicit time discretization:

$$\alpha\left(\frac{z_m^n - z_m^{n-1}}{k}\right) + Az_m^n \ni F_m^n \qquad \text{in } V', \text{ for } n = 1, \ldots, m. \qquad (5.8)$$

This inclusion is also equivalent to the implicit time-discretization of the variational inequality (5.4), as well as to the minimization of the functional

$$\hat{J}_m^n(v) := \int_\Omega \left[k\, a\left(\frac{v - z_m^{n-1}}{k}\right) + \frac{1}{2}|\nabla v|^2\right] dx - {}_{V'}\!\left\langle F_m^n, v\right\rangle_V \qquad \forall v \in V.$$

By the direct method of the calculus of variations, cf. Theorem XI.7.3, there exists a (unique) minimum point of this functional.

Let us define the time interpolate functions z_m as in (1.29). Multiplying (5.8) by $z_m^n - z_m^{n-1}$, one easily derives a uniform estimate for z_m in $L^\infty(0,T;V) \cap H^1\left(0,T;L^2(\Omega)\right)$. Therefore there exists z such that, possibly extracting a subsequence, $z_m \to z$ weakly star in the latter space. By taking the limit in the approximate variational inequality, then one gets (5.6) by lower semicontinuity.

The proof of uniqueness is straightforward. □

Inversion of the Laplace Operator. In view of our second integral transformation, let us revisit the functional setting of Problem 1.1.

If $\Gamma_1 = \Gamma$ (i.e. $V = H_0^1(\Omega)$), A^{-1} can be interpreted as the inverse of the operator $-\Delta$ associated with the homogeneous Dirichlet boundary condition. That is, for any $v \in H^{-1}(\Omega)$,

$$u = A^{-1}v \quad \text{iff} \quad \begin{cases} u \in H^1(\Omega), \\ -\Delta u = v & \text{in } \mathcal{D}'(\Omega), \\ \gamma_0 u = 0 & \text{on } \partial\Omega. \end{cases} \qquad (5.9)$$

More generally, we assume that the $(N-1)$-dimensional Hausdorff measure of Γ_1 does not vanish, so that the operator $A : V \to V'$ is an isomorphism. Here the interpretation of A^{-1} is less obvious: *formally* one would replace the condition on $\partial\Omega$ by

$$\gamma_0 u = 0 \quad \text{on } \Gamma_1, \qquad \frac{\partial u}{\partial \nu} = 0 \quad \text{on } \Gamma_2,$$

However, in general the latter condition has no precise meaning for $v \in H^{-1}(\Omega)$ (although it holds in $H_{00}^{1/2}(\Gamma_2)'$ whenever $v \in L^2(\Omega)$).

By applying the operator A^{-1} to (1.12), we have

$$A^{-1}\frac{\partial w}{\partial t} + u = A^{-1}f =: g \qquad \text{in } V, \text{ a.e. in }]0, T[. \tag{5.10}$$

By coupling this equation with (1.2)$_2$, we get the inclusion

$$A^{-1}\frac{\partial w}{\partial t} + \beta(w) \ni g \qquad \text{in } V, \text{ a.e. in }]0, T[, \tag{5.11}$$

which is equivalent to the following variational inequality, where b is defined as in (1.3):

$$_V\left\langle A^{-1}\frac{\partial w}{\partial t} - g, w - v \right\rangle_{V'} + \int_\Omega [b(w) - b(v)]dx \leq 0 \tag{5.12}$$

$$\forall v \in L^2(\Omega), \text{ a.e. in }]0, T[.$$

Let us assume that

$$w^0 \in V', \qquad g \in L^2(0, T; V). \tag{5.13}$$

Problem 5.2 *To find $w \in L^2(Q) \cap H^1(0, T; V')$ such that (5.12) is satisfied, and*

$$w|_{t=0} = w^0 \qquad \text{in } V' \text{ (in the sense of the traces of } H^1(0, T; V')\text{)}. \tag{5.14}$$

Theorem 5.2 *(Existence and Uniqueness) Assume that (1.1), (1.5), (1.20), (1.22) and (5.13) are satisfied. Then Problem 3.2 has one and only one solution.*

Outline of the Proof. One can approximate (5.11) via implicit time-discretization, and then derive a priori estimates multiplying the approximate equation by the approximate solution w_m^n. This yields a uniform estimate for the linear interpolate function w_m in $L^2(Q) \cap L^\infty(0, T; V')$. Hence a suitable subsequence of $\{w_m\}$ converges weakly star in the latter space. The equation (5.12) is then derived via lower semicontinuity, by passing to the limit in the approximate variational inequality.

The proof of uniqueness is straightforward. □

Remarks. (i) Other equations can also be transformed by inversion of an elliptic operator, which does not need to be self-adjoint, cf. the Lax-Milgram Theorem XI.9.2.

(ii) The transformations that we discussed in this section can easily be extended to systems of P.D.E.s of the form (2.30). □

Exercises.

5.1 Detail the proof of Theorem 5.1.

5.2 Retrieve Theorem 1.2, multiplying (5.8) by $A(z_m^n - z_m^{n-1})$, or equivalently multiplying the time increment of (5.8) by $z_m^n - z_m^{n-1}$.

II.6 Semigroup Techniques

In this section we show that both L^2- and L^1-semigroup techniques (see Sect. XI.6) can be used to deal with Problem 1.1.

Change of Pivot Space. Here we continue our discussion on the procedure based on the inversion of the operator A, under the assumption that the $(N-1)$-dimensional Hausdorff measure of Γ_1 does not vanish.

The bilinear forms

$$(u,v)_V := \int_\Omega \nabla u \cdot \nabla v \, dx \qquad \forall u,v \in V, \tag{6.1}$$

$$(u,v)_{V'} := \left(A^{-1}u, A^{-1}v\right)_V = {}_V\langle A^{-1}u, v\rangle_{V'} \qquad \forall u,v \in V' \tag{6.2}$$

are scalar products in the Hilbert spaces V and V', respectively. Note that

$$(Au,v)_{V'} := {}_V\langle u,v\rangle_{V'} = \int_\Omega uv \, dx \qquad \forall u \in V, \forall v \in L^2(\Omega). \tag{6.3}$$

Here we denote the space V' by $\mathcal{H}$, to avoid any possible confusion with the dual spaces that now we introduce.

Let us consider the *Riesz operator* $\mathcal{R}: \mathcal{H} \to \mathcal{H}'$, defined by ${}_{\mathcal{H}'}\langle \mathcal{R}u,v\rangle_{\mathcal{H}} = (u,v)_{\mathcal{H}}$ for any $u,v \in \mathcal{H}$. As $L^2(\Omega) \subset \mathcal{H}$ with continuous and dense injection, we can identify $\mathcal{H}'$ with a subspace of $L^2(\Omega)'$. In this way we get [20]

$$\mathcal{R}L^2(\Omega) \subset \mathcal{R}\mathcal{H} = \mathcal{H}' \subset L^2(\Omega)' \tag{6.4}$$

with dense and compact injections.

---

[20] Cf. also Sect. XI.9.

Henceforth we omit the operator $\mathcal{R}$. Then the space $\mathcal{H}$ is identified with its topological dual, and accordingly plays the role of *pivot space*. This approach can be compared with the more usual procedure of identifying $L^2(\Omega)$ with its dual, cf. (1.7).

By (6.3), the equation (1.12) in $\mathcal{H}$ reads

$$\left(\frac{\partial w}{\partial t}, v\right)_{\mathcal{H}} + \int_{\Omega} uv dx = (f,v)_{\mathcal{H}} \qquad \forall v \in \mathcal{H}, \tag{6.5}$$

which is equivalent to (5.10). So this is just another way of deriving Problem 5.2.

L^2-Semigroups. Let us assume (1.20), so that $b(w) \in L^1(\Omega)$ for any $w \in L^2(\Omega)$, cf. (1.27).

The operator $w \mapsto -\Delta\beta(w)$ is nonmonotone in $L^2(\Omega)$ whenever β is nonlinear. On the other hand this operator is monotone in $\mathcal{H}$, and this is the reason why we chose the latter as a pivot space. More specifically, we claim that the (possibly multi-valued) operator

$$\begin{cases} B : \text{Dom}(B) := L^2(\Omega) \subset \mathcal{H} \to 2^{\mathcal{H}} : \\ w \mapsto A\beta(w) := \{Au : u \in V, u \in \beta(w) \text{ a.e. in } \Omega\} \end{cases} \tag{6.6}$$

coincides with the subdifferential of the proper, convex, and lower semicontinuous functional

$$\mathcal{H} \to \mathbf{R} : w \mapsto \begin{cases} \int_{\Omega} b(w) dx & \text{if } w \in L^2(\Omega), \\ +\infty & \text{otherwise.} \end{cases} \tag{6.7}$$

In fact, for any $u \in L^2(\Omega)$ and any $w \in V$, one has $u \in \beta(w)$ a.e. in Ω iff

$$(Au, w-v)_{\mathcal{H}} = \int_{\Omega} u(w-v) dx \geq \int_{\Omega} b(w) dx - \int_{\Omega} b(v) dx \qquad \forall v \in L^2(\Omega).$$

This allows the application of the classical theory of *semigroups of nonlinear contractions* in Hilbert spaces [21] to the equation

$$\frac{\partial w}{\partial t} + B(w) = f \qquad \text{in } \mathcal{H}, \text{ a.e. in }]0,T[. \tag{6.8}$$

In this way one retrieves Theorem 5.2, among other results. In fact, this semigroup approach is essentially equivalent to the inversion of the Laplace operator.

This discussion can easily be extended to systems of P.D.E.s of the form (2.30).

L^1-Semigroups. Theorem 2.2 suggests investigating the accretiveness properties of the (possibly multi-valued) operator $-\Delta\beta(w)$ in $L^1(\Omega)$. Let us set

$$\begin{cases} \hat{B} : \text{Dom}(\hat{B}) \subset L^1(\Omega) \to L^1(\Omega) : w \mapsto -\Delta\beta(w) := \\ \{\Delta u \in L^1(\Omega) : u \in L^1(\Omega), \gamma_0 u = 0 \text{ on } \Gamma_1, u \in \beta(w) \text{ a.e. in } \Omega\} \end{cases} \tag{6.9}$$

---

[21] See Sect. XI.5.

(here $\Gamma_1 \subset \Gamma$ is of positive $(N-1)$-Hausdorff measure, as previously). Notice that the trace $\gamma_0 u$ is meaningful on Γ_1, as $\Delta u \in L^1(\Omega)$. [22] The operators $\hat{B}$ and B are obviously different.

The next result allows us to apply the theory of L^1-*contraction semigroups* [23] to the following Cauchy problem, for any prescribed

$$f \in L^1\left(0,T;L^1(\Omega)\right), \qquad w^0 \in L^1(\Omega). \tag{6.10}$$

Problem 6.1 *To find $w : [0,T] \to L^1(\Omega)$ such that (in a sense to be specified)*

$$\frac{\partial w}{\partial t} + \hat{B}(w) \ni f \qquad \text{in } L^1(\Omega), \text{ in }]0,T[, \tag{6.11}$$

$$w(0) = w^0. \tag{6.12}$$

In general this problem has no *strong solution*, [24] hence we are satisfied with the *integral solution*. [25]

We assume the Dirichlet condition on the whole Ω, and use the following lemma, which is interesting in itself.

Lemma 6.1 [26] *Assume that γ is a maximal monotone graph in $\mathbf{R}^2$ that contains the origin. Let $p \in [1,+\infty[$ and set $p' := p/(p-1)$ ($p' := +\infty$ if $p = 1$). Let $u \in L^p(\Omega)$ be such that $\Delta u \in L^p(\Omega)$ and $\gamma_0 u = 0$ a.e. on $\partial\Omega$, and $g \in L^{p'}(\Omega)$ be such that $g \in \gamma(u)$ a.e. in Ω. Then $-\int_\Omega g\Delta u\,dx \geq 0$.*

For the proof we refer to Brézis and Strauss [95].

Theorem 6.2 *Assume that (1.1) and (1.20) hold, and set $\beta := \alpha^{-1}$. Let $\Gamma_1 = \partial\Omega$. Then the operator $\hat{B}$ is T- and m-accretive in $L^1(\Omega)$.*

Proof. (i) At first we show that $\hat{B}$ is T-accretive.

For $i = 1, 2$, for any $w_i \in \text{Dom}(\hat{B})$, let $u_i \in \beta(w_i)$ a.e. in Ω be such that $u_i, \Delta u_i \in L^1(\Omega)$. Let us set

$$h(x) := \begin{cases} 1 & \text{if either } w_1(x) > w_2(x) \text{ or } u_1(x) > u_2(x) \\ 0 & \text{otherwise} \end{cases} \quad \text{for a.a. } x \in \Omega.$$

---

[22] Cf. Theorem XI.2.3.

[23] See Sect. XI.6 for definitions and results on accretive operators.

[24] See, for instance, the weak formulation of the classical Stefan problem, which we introduce in Sect. IV.4. Indeed the presence of phase interfaces is not consistent with the regularity $\Delta\beta(w) \in L^1(\Omega)$, where $\beta(w)$ represents the temperature.

[25] See Sect. XI.6 for these concepts of solution.

[26] See Brézis and Strauss [95; p. 566], where this result is stated in more general form, for a class of unbounded m-accretive operators in $L^1(\Omega)$ that fulfill a maximum principle.

Note that h is defined a.e. in Ω and measurable, and $h \in H(w_1 - w_2) \cap H(u_1 - u_2)$ a.e. in Ω. By Lemma 6.1, we have $-\int_\Omega h\Delta(u_1 - u_2)\,dx \geq 0$. Hence

$$\int_\Omega [w_1 - w_2 - \lambda\Delta(u_1 - u_2)]^+ dx \geq \int_\Omega [w_1 - w_2 - \lambda\Delta(u_1 - u_2)]h\,dx$$
$$\geq \int_\Omega (w_1 - w_2)h\,dx = \int_\Omega (w_1 - w_2)^+ dx.$$

Thus $\hat{B}$ is T-accretive, hence it is also accretive since $L^1(\Omega)$ fulfills (XI.6.13).

(ii) Now we prove that $\hat{B}$ is m-accretive in $L^1(\Omega)$, that is,

$$\forall \lambda > 0, \forall f \in L^1(\Omega), \exists w \in \text{Dom}(\hat{B}) : w - \lambda\Delta\beta(w) \ni f \text{ a.e. in } \Omega. \qquad (6.13)$$

At first, let us fix any $f \in L^2(\Omega)$. The functional

$$J : V \to \mathbf{R} \cup \{+\infty\} : v \mapsto \int_\Omega \left(a(v) + \frac{\lambda}{2}|\nabla v|^2 - fv \right) dx$$

(with a as in (1.3)) is convex, lower semicontinuous, and coercive; hence it has a minimum point u. Therefore $\partial J(u) \ni 0$ in V' and $w := f + \lambda\Delta u \in \alpha(u)$ in V'; hence $w \in L^2(\Omega)$, by (1.20). We conclude that for any $f \in L^2(\Omega)$ there exists a pair $(u, w) \in V \times L^2(\Omega)$ such that $\Delta u \in L^2(\Omega)$ and

$$u \in \beta(w), \quad w - \lambda\Delta u = f \qquad \text{a.e. in } \Omega. \qquad (6.14)$$

Now let us consider any $f \in L^1(\Omega)$, any sequence $\{f_n\} \subset L^2(\Omega)$ that converges to f strongly in $L^1(\Omega)$, and for any n let (u_n, w_n) solve (6.14), with f_n in place of f. By the accretiveness of $\hat{B}$, $\{w_n\}$ is a Cauchy sequence in $L^1(\Omega)$; hence it converges to some w strongly in this space.

Then, by comparison in (6.14)$_n$, Δu_n is uniformly bounded in $L^1(\Omega)$. Hence, possibly extracting a subsequence, u_n converges to some u strongly in $L^1(\Omega)$. Therefore, possibly extracting further subsequences, w_n and u_n converge a.e.. Hence $u \in \beta(w)$ a.e. in Ω, and by taking the limit in the approximate equation we get $w - \lambda\Delta u = f$ a.e. in Ω. □

By Theorems XI.6.1 and XI.6.3, the latter theorem yields the following result.

Theorem 6.3 *Assume that (1.1) and (1.20) hold, and that $\Gamma_1 = \partial\Omega$. Then, for any f and w^0 as in (6.10), Problem 6.1 has one and only one integral solution $w \in C^0\left([0, T]; L^1(\Omega)\right)$.*

This solution depends Lipschitz continuously and monotonically on the data.

Moreover, if $f \in BV\left(0, T; L^1(\Omega)\right)$ and $w^0 \in \text{Dom}(\hat{B})$, then $w : [0, T] \to L^1(\Omega)$ is Lipschitz continuous.

Remarks. (i) The space $L^1(\Omega)$ does not fulfill the *Radon-Nikodým property* (cf. Sect. XI.6). Hence w may not be a *strong solution*, consistently with our previous remark.

(ii) If the growth assumption (1.20) is dropped, a more complex argument is needed; see Brézis and Strauss [95].

(iii) More general boundary conditions can also be considered; see, for example, Bénilan [65], and Magenes, Verdi and V. [377]. □

II.7 Comments

Here are the main techniques that we used in this chapter (and use in the next one):

(i) approximation — derivation of a priori estimates — passage to the limit;

(ii) derivation of a priori estimates in L^p-spaces, for $p = 1, 2, \infty$, by suitable choices of test functions;

(iii) compactness results, in particular, compactness of Sobolev inclusions, cf. Theorem XI.3.4;

(iv) what we called the *equation + lower semicontinuity* technique, which is essentially a convexity method;

(v) compactness and monotonicity procedures, such as Lemma XI.5.1;

(vi) cut-off techniques, to derive local regularity results;

(vii) integral transformations;

(viii) L^2- and L^1- semigroups of nonlinear contractions.

The system (1.1), (1.2) is an example of a wide class of quasilinear parabolic equations, which have been intensively and extensively studied. It represents a typical *gradient flow* (in the sense of Sect. VII.6), and includes the weak formulation of the multi-dimensional *Stefan* and *Hele-Shaw problems,* which we introduce in Chap. IV, a model of gas filtration through porous media, [27] and several others.

In this chapter we just discussed a model problem; these developments might be extended in several ways, for instance by including lower order terms or different boundary conditions. Problems of this sort have also been treated in abstract form as equations in Banach spaces; in Sects. XI.5 and XI.6 we review some fundamental results of that theory.

The L^p-techniques that we used in Sects. II.1 through II.3 are quite classical; see, for example, Lions [351], and Brézis [87]. The local regularity procedure of Theorem 4.2 seems to be less standard; the author learned it by Huang [303] and Verdi [545].

---

[27] See, e.g., Aronson [26] and references therein.

Variational inequalities are at the basis of the problems we dealt with in this chapter. They were introduced in the 1960s, mainly in connection with problems that issued from applications, for example, the Signorini problem, the obstacle problem, elasto-plasticity; see, for instance, Fichera [232, 233], Lions and Stampacchia [357, 358], Stampacchia [514, 515], and Moreau [400, 401, 402]. They were then intensively studied, see the *Book Selection*.

The transformation (5.1) was independently introduced by Duvaut [203, 204] and Frémond [244] for the *Stefan problem*. This technique was inspired by a similar integral transformation, which was successfully used by Baiocchi [44, 45] to solve a free boundary problem representing porous medium filtration (the so-called *dam problem*). Similar transformations were then also used for other free boundary problems; see, for example, Baiocchi [46], and Baiocchi and Capelo [47; Chap. 13]. The change of pivot space was studied by Lions [351; Sect. 2.3].

Statement	Estimate Procedure	Derived Regularity
Theor. 1.2	eq. $\times u$	$w \in L^\infty\left(0,T; L^2(\Omega)\right), u \in L^2(0,T;V)$
Theor. 1.4	eq. $\times u$, eq. $\times w$	$w \in L^\infty\left(0,T; L^2(\Omega)\right), u \in L^2(0,T;V)$
Theor. 2.5	eq. $\times \frac{\partial u}{\partial t}$	$u \in H^1\left(0,T; L^2(\Omega)\right) \cap L^\infty(0,T;V)$
Theor. 2.7	eq. $\times (u - \tau_{-h}u)$	$u \in H^r\left(0,T; L^2(\Omega)\right) \forall r < \frac{1}{2}$
Theor. 2.8	eq. $\times t\frac{\partial u}{\partial t}$	$\sqrt{t}\frac{\partial u}{\partial t} \in L^2(Q)$, $\sqrt{t}u \in L^\infty(0,T;V)$
Theor. 3.3	$\frac{\partial}{\partial t}$eq. $\times \operatorname{sign}\left(\frac{\partial u}{\partial t}\right)$	$\frac{\partial w}{\partial t}, \Delta u \in L^\infty_{w*}\left(0,T; C^0_c(\Omega)'\right)$
Prop. 3.5	eq. $\times (u - M)^+$	$u \leq M \ (M \in \mathbf{R}^+)$
Prop. 3.6	eq. $\times |w|^{q-2}w \ (q > 2)$	$w \in L^\infty\left(0,T; L^q(\Omega)\right)$

Table 1. Main estimate procedures, and related regularity.

Some shortwriting is used. For instance, $\frac{\partial}{\partial t}$eq. $\times \frac{\partial u}{\partial t}$ stays for the procedure of multiplying the time incremental ratio of the discretized equation by the time incremental ratio of the approximate u, and then to integrate in space and time.

The *semigroup approach* to equations like (1.2) was investigated, by Brézis [86, 87], Brézis and Pazy [94], Bénilan [65], Crandall and Pierre [153], Bénilan and Crandall [66], and Bénilan, Crandall and Sacks [67] among others. Berger, Brézis and Rogers [70] and others applied the L^1-semigroup method to the numerical solution of the Stefan problem.

As we saw, several results, in particular those based on L^2-techniques, can be extended to quasilinear parabolic systems such as (2.30). In the next chapter, we deal with a more general class of quasilinear and fully nonlinear parabolic equations.

An Overview. The main results and techniques of this chapter are displayed in Table 1.

Chapter III. Doubly Nonlinear Parabolic P.D.E.s

Outline

Existence results are derived for (possibly degenerate) doubly nonlinear parabolic equations, in particular

$$\frac{\partial}{\partial t}\alpha(u) - \nabla \cdot \vec{\gamma}(\nabla u) \ni f, \qquad \alpha\left(\frac{\partial u}{\partial t}\right) - \nabla \cdot \vec{\gamma}(\nabla u) \ni f,$$

$$\frac{\partial}{\partial t}\alpha(u) - \Delta\tilde{\gamma}(u) \ni f, \qquad \alpha\left(\frac{\partial u}{\partial t}\right) - \Delta\tilde{\gamma}(u) \ni f;$$

here $\alpha, \tilde{\gamma} : \mathbf{R} \to 2^{\mathbf{R}}$ and $\vec{\gamma} : \mathbf{R}^N \to 2^{(\mathbf{R}^N)}$ are maximal monotone operators.

The arguments are based on implicit time-discretization, derivation of a priori estimates, and passage to the limit via compactness and monotonicity techniques. Part of these developments can also be extended to systems of P.D.E.s.

Prerequisites. Same as for Chap. II. Here also the presentation is rather didactical. *Compactness by strict monotonicity* (see Sect. X.3) is used in Sect. III.4. Definitions and standard results are recalled in Chap. XI.

III.1 Doubly Nonlinear Parabolic Equations of First Type

So far we studied quasilinear [1] parabolic P.D.E.s containing a single nonlinear term; in this chapter we deal with equations [2] containing two (or more) monotone functions. In Sects. III.1 and III.2 we study equations of the form

$$\frac{\partial}{\partial t}\alpha(u) - \nabla \cdot \vec{\gamma}(\nabla u) \ni f \qquad \text{in } Q, \tag{1.1}$$

$$\alpha\left(\frac{\partial u}{\partial t}\right) - \nabla \cdot \vec{\gamma}(\nabla u) \ni f \qquad \text{in } Q, \tag{1.2}$$

respectively, each coupled with suitable initial and boundary conditions.

---

[1] We still use the term *quasilinear* in the sense specified in one of the footnotes of Sect. I.1.
[2] Or rather *inclusions;* we often use these terms equivalently.

We still assume that Ω is a bounded domain of $\mathbf{R}^N$ ($N \geq 1$) of Lipschitz class, denote its boundary by Γ, fix any $T > 0$, and set $Q := \Omega \times {]0, T[}$, $\Sigma := \Gamma \times {]0, T[}$. f is a given function $Q \to \mathbf{R}$,

$$\alpha : \mathbf{R} \to \mathbf{R} \text{ and } \vec{\gamma} : \mathbf{R}^N \to \mathbf{R}^N \tag{1.3}$$

are maximal monotone (possibly multi-valued) functions.

The inclusions (1.1) and (1.2) stay for the following systems:

$$\begin{cases} \dfrac{\partial w}{\partial t} - \nabla \cdot \vec{z} = f & \text{in } Q, \\ w \in \alpha(u) & \text{in } Q, \\ \vec{z} \in \vec{\gamma}(\nabla u) & \text{in } Q; \end{cases} \tag{1.4}$$

$$\begin{cases} w - \nabla \cdot \vec{z} = f & \text{in } Q, \\ w \in \alpha\left(\dfrac{\partial u}{\partial t}\right) & \text{in } Q, \\ \vec{z} \in \vec{\gamma}(\nabla u) & \text{in } Q. \end{cases} \tag{1.5}$$

Each of these inclusions is equivalent to a variational inequality.

We say that (1.4) and (1.5) are *doubly nonlinear parabolic equations* of first and second type, respectively. [3] In general, (1.4) and (1.5) are not equivalent; however, if α is linear they obviously coincide. If $\vec{\gamma}$ is linear, a simple transformation relates the two classes: $(u, w, \vec{z})$ is a solution of (1.2) if and only if $(\partial u/\partial t, w, \vec{z})$ solves the inclusion obtained by replacing f with $\partial f/\partial t$ in (1.1).

In this and the next section, we prove existence results for these systems, allowing α and $\vec{\gamma}$ to be multi-valued and nonstrictly monotone. This includes several degenerate parabolic and elliptic-parabolic equations.

For the sake of simplicity, we consider the homogeneous Dirichlet boundary condition for u, and set $V := H_0^1(\Omega)$. However, our results can easily be extended to other boundary conditions, for example, (II.1.17) and (II.1.18).

Doubly Nonlinear Parabolic Equations of First Type. Let us assume that (1.3) is satisfied, and that

$$w^0 \in V', \qquad f \in L^2(0, T; V'). \tag{1.6}$$

Problem 1.1 *To find*

$$u \in L^2(0, T; V), \quad w \in L^2(Q), \quad \vec{z} \in L^2\left(Q; \mathbf{R}^N\right) \tag{1.7}$$

---

[3] This terminology has some analogy with that used by Duvaut and Lions [205; Chaps. I, II] for simply nonlinear parabolic equations (which correspond to single variational inequalities). Note that the equation (1.4) is quasilinear, whereas (1.5) is fully nonlinear.

such that

$$\iint_Q \left(-w\frac{\partial v}{\partial t} + \vec{z}\cdot\nabla v\right)dxdt = \int_0^T {}_{V'}\langle f,v\rangle_V\,dt + {}_{V'}\langle w^0, v(\cdot,0)\rangle_V \tag{1.8}$$
$$\forall v \in H^1(0,T;V) : v(\cdot,T) = 0,$$

$$\iint_Q (w-\xi)(u-v)dxdt \geq 0 \qquad \forall v \in L^2(Q), \forall \xi \in \alpha(v), \tag{1.9}$$

$$\iint_Q (\vec{z}-\vec{\eta})\cdot(\nabla u - \vec{v})dxdt \geq 0 \qquad \forall \vec{v} \in L^2\left(\Omega;\mathbf{R}^N\right), \forall \vec{\eta} \in \vec{\gamma}(\vec{v}). \tag{1.10}$$

The equation (1.8) yields the differential equation $(1.4)_1$ in V' a.e. in $]0,T[$, whence $w \in H^1(0,T;V')$ and the initial condition

$$w|_{t=0} = w^0 \quad \text{in } V' \text{ (in the sense of the traces of } H^1(0,T;V')). \tag{1.11}$$

The equations (1.9) and (1.10) are respectively equivalent to $(1.4)_2$ and $(1.4)_3$. By definition of V, we also have $\gamma_0 u = 0$ on $\partial\Omega\times]0,T[$ (in the sense of the traces).

Theorem 1.1 *(Existence) Assume that (1.3) is satisfied, and*

$$\exists L, M \in \mathbf{R}, (L > 0) : \forall (u,w) \in \text{graph}(\alpha), |w| \leq L|u| + M, \tag{1.12}$$

$$\exists \tilde{L}, \tilde{M} \in \mathbf{R}(\tilde{L} > 0) : \forall (\vec{v}, \vec{z}) \in \text{graph}(\vec{\gamma}), \quad \vec{z}\cdot\vec{v} \geq \tilde{L}|\vec{v}|^2 - \tilde{M}|\vec{v}|, \tag{1.13}$$

$$\exists \hat{L}, \hat{M}(\hat{L} > 0) : \forall (\vec{v}, \vec{z}) \in \text{graph}(\vec{\gamma}), \quad |\vec{z}| \leq \hat{L}|\vec{v}| + \hat{M}, \tag{1.14}$$

$$w^0 \in L^2(\Omega). \tag{1.15}$$

Then Problem 1.1 has a solution such that

$$w \in L^\infty\left(0,T;L^2(\Omega)\right) \cap H^1(0,T;V'). \tag{1.16}$$

Proof. (i) *Approximation.* We define m, k, and so on as for Problem II.1.1$_m$, and approximate $(1.4)_1$ by implicit time-discretization.

Problem 1.1$_m$ *To find*

$$u_m^n \in V, \quad w_m^n \in L^2(\Omega), \quad \vec{z}_m^n \in L^2\left(\Omega;\mathbf{R}^N\right) \qquad \text{for } n = 1,\ldots,m, \tag{1.17}$$

such that, setting $w_m^0 := w^0$ *a.e. in* Ω, *for* $n = 1,\ldots,m$,

$$\frac{w_m^n - w_m^{n-1}}{k} - \nabla\cdot\vec{z}_m^n = f_m^n \qquad \text{in } V', \tag{1.18}$$

III.1 Doubly Nonlinear Parabolic Equations of First Type   71

$$\int_\Omega (w_m^n - \xi)(u_m^n - v)dx \geq 0 \qquad \forall v \in L^2(\Omega), \forall \xi \in \alpha(v), \tag{1.19}$$

$$\int_\Omega (\vec{z}_m^n - \vec{\eta}) \cdot (\nabla u_m^n - \vec{v}) \geq 0 \qquad \forall \vec{v} \in L^2(\Omega; \mathbf{R}^N), \forall \vec{\eta} \in \vec{\gamma}(\vec{v}). \tag{1.20}$$

The last two variational inequalities are respectively equivalent to $w_m^n \in \alpha(u_m^n)$ and $\vec{z}_m^n \in \vec{\gamma}(\nabla u_m^n)$ a.e. in Ω.

For this problem, existence of a solution can be proved step by step. Defining the (possibly multi-valued) operator $\mathcal{B}: V \to V'$

$$\mathcal{B}(v) := \alpha(v) - k\nabla \cdot \vec{\gamma}(\nabla v) \qquad \forall v \in V, \tag{1.21}$$

the system (1.18) through (1.20) can be written in the form $\mathcal{B}(u_m^n) \ni kf_m^n + w_m^{n-1}$ in V' for $n = 1, \ldots, m$.

By (1.13) and the *generalized Poincaré inequality* Theorem XI.9.3, it is easy to see that $\mathcal{B}$ is monotone and coercive in the sense of (XI.5.9). By Theorem XI.5.3, $\mathcal{B}$ is also maximal monotone, since α is defined on the whole V. Therefore, by Theorem XI.5.2, Problem 1.1_m has a solution.

(ii) *A Priori Estimates*. Let us multiply (1.18) by ku_m^n, integrate over Ω and sum for $n = 1, \ldots, \ell$, for any $\ell \in \{1, \ldots, m\}$. Defining b as in (II.1.3), we have

$$\sum_{n=1}^{\ell} \int_\Omega (w_m^n - w_m^{n-1}) u_m^n \, dx \geq \sum_{n=1}^{\ell} \int_\Omega \left[b(w_m^n) - b(w_m^{n-1}) \right] dx$$

$$= \int_\Omega \left[b(w_m^\ell) - b(w^0) \right] dx.$$

By (II.1.27) and (1.13), then one easily gets the following estimates:

$$\max_{n=1,\ldots,m} \|w_m^n\|_{L^2(\Omega)}, k \sum_{n=1}^{m} \|u_m^n\|_V^2, k \sum_{n=1}^{m} \|\vec{z}_m^n\|_{L^2(\Omega; \mathbf{R}^N)}^2 \tag{1.22}$$
$$\leq \text{Constant (independent of } m\text{)}.$$

By comparison in (1.18), one also has

$$k \sum_{n=1}^{m} \left\| \frac{w_m^n - w_m^{n-1}}{k} \right\|_{V'}^2 \leq \text{Constant}. \tag{1.23}$$

Let us define u_m and $\bar{u}_m$ as in (II.1.29) and (II.1.30), and $w_m, \bar{w}_m, \vec{z}_m$, and $\bar{f}_m$ similarly. The equation (1.18) then yields

$$\frac{\partial w_m}{\partial t} - \nabla \cdot \vec{z}_m = \bar{f}_m \qquad \text{in } V', \text{ a.e. in }]0, T[. \tag{1.24}$$

(iii) *Limit Procedure.* By the preceding estimates, there exist $u, w, \vec{z}$ such that, possibly taking $m \to \infty$ along a subsequence,

$$u_m \to u \quad \text{weakly in } L^2(0,T;V), \tag{1.25}$$

$$w_m \to w \quad \text{weakly star in } L^\infty\left(0,T;L^2(\Omega)\right) \cap H^1(0,T;V'), \tag{1.26}$$

$$\vec{z}_m \to \vec{z} \quad \text{weakly in } L^2\left(Q;\mathbf{R}^N\right). \tag{1.27}$$

By passing to the limit in (1.24), we get (1.8). By Lions-Aubin's Theorem XI.3.5 (ii), (1.25) and (1.26) yield $\iint_Q \bar{w}_m \bar{u}_m dxdt \to \iint_Q wu\,dxdt$. Hence, by writing (1.19) in terms of $\bar{w}_m$ and $\bar{u}_m$ and passing to the limit, we get (1.9).

In order to identify $\vec{z}$, we use the *equation + lower semicontinuity* technique, which we outlined in Sect. II.2, cf. (II.2.4). By applying Lemma II.3.2 with $H := L^2(\Omega)$ and $W := V'$, by (1.26) we have

$$\bar{w}_m(\cdot,T) = w_m(\cdot,T) \to w(\cdot,T) \quad \text{weakly in } L^2(\Omega).$$

By (1.8), (1.18), the lower semicontinuity of the convex functional $v \mapsto \int_\Omega b(v)dx$, the latter convergence, and Proposition XI.4.11, we have

$$\begin{aligned}
\limsup_{m\to\infty} \iint_Q \vec{z}_m \cdot \nabla \bar{u}_m dxdt &= \limsup_{m\to\infty} \int_0^T {}_{V'}\!\left\langle \bar{f}_m - \frac{\partial w_m}{\partial t}, \bar{u}_m \right\rangle_V dt \\
&\leq \lim_{m\to\infty} \int_0^T {}_{V'}\!\langle \bar{f}_m, \bar{u}_m \rangle_V dt - \liminf_{m\to\infty} \int_\Omega \left[b(\bar{w}_m(\cdot,T)) - b(w^0)\right] dx \\
&\leq \int_0^T {}_{V'}\!\langle f, u \rangle_V dt - \int_\Omega \left[b(w(\cdot,T)) - b(w^0)\right] dx \\
&= \int_0^T {}_{V'}\!\left\langle f - \frac{\partial w}{\partial t}, u \right\rangle_V dt = \iint_Q \vec{z} \cdot \nabla u\, dxdt.
\end{aligned} \tag{1.28}$$

Therefore, by writing (1.19) in terms of $\bar{z}_m$ and $\bar{u}_m$ and passing to the superior limit in (1.20), we get (1.10). □

Proposition 1.2 *(Regularity) Assume that (1.3), (1.12), (1.14), (1.15) are satisfied, and*

$$\begin{aligned}
&\exists c > 0 : \forall (u_i, w_i) \in \text{graph}(\alpha)\,(i = 1,2), \\
&(w_1 - w_2)(u_1 - u_2) \geq c(u_1 - u_2)^2,
\end{aligned} \tag{1.29}$$

$$\exists j : \mathbf{R}^N \to \mathbf{R} \text{ convex}: \vec{\gamma} = \partial j, \tag{1.30}$$

$$\exists \tilde{L}, \tilde{M} \in \mathbf{R}, (\tilde{L} > 0) : \forall \vec{v} \in \mathbf{R}^N, j(\vec{v}) \geq \tilde{L}|\vec{v}|^2 - \tilde{M},$$

$$f \in L^2(Q), \quad u^0 \in V. \tag{1.31}$$

Then Problem 1.1 has a solution such that

$$u \in H^1\left(0,T;L^2(\Omega)\right) \cap L^\infty(0,T;V). \tag{1.32}$$

Outline of the Proof. A priori estimates corresponding to the regularity (1.32) are easily derived multiplying (1.18) by $u_m^n - u_m^{n-1}$.

(Among other things, note that $j(\vec{\nabla}u^0) \in L^1(\Omega)$, for by (1.14) the growth of j is at most quadratic.) □

The solution of Problem 1.1 can be expected to be unique only under strong quantitative assumptions. See the counterexample of DiBenedetto and Showalter [191; Sect. 5] for a strictly related setting.

Theorem 1.3 *(Uniqueness) Assume that (1.3) is satisfied, and that either α is linear and strictly increasing, or $\vec{\gamma}$ is linear, symmetric, and positive definite. Then Problem 1.1 has at most one solution.*

Outline of the Proof. Let $(u_i, w_i, \vec{z}_i)$ ($i = 1, 2$) be two solutions, and take the difference of the corresponding equalities $(1.4)_1$. If α is linear, multiply this difference by $u_1 - u_2$; if instead $\vec{\gamma}$ is linear and invertible, integrate the difference of the equalities in time and multiply it by $u_1 - u_2$. In either case one easily gets $u_1 = u_2$ a.e. in Q, whence $w_1 = w_2$ and $\vec{z}_1 = \vec{z}_2$ a.e. in Q. □

Remarks. (i) Several results of this section can be extended to third order *pseudo-parabolic* inclusions of the form

$$A\frac{\partial u}{\partial t} + \frac{\partial \alpha(u)}{\partial t} - \nabla \cdot \vec{\gamma}(\nabla u) \ni f \qquad \text{in } Q, \qquad (1.33)$$

where $A: V \to V'$ is a linear elliptic operator. For instance, if A is self-adjoint, estimates can be derived multiplying (1.33) by u, as in Theorem 1.1. This condition is not needed to multiply the equation by $\partial u/\partial t$ as in Proposition 1.2, if $\vec{\gamma}$ is *cyclically monotone*. [4]

(ii) The previous discussion can be extended to an abstract setting, with α and $\nabla \cdot \vec{\gamma}(\nabla \cdot)$ replaced by maximal monotone operators in Banach spaces; cf. Di Benedetto and Showalter [191]. Theorem 1.1 can be extended assuming that α is cyclically monotone, whereas the other results hold essentially unchanged.

An abstract formulation of this sort also includes systems of P.D.E.s. □

Exercises.

1.1 If V is defined as in (II.1.6), what is changed in the interpretation of Problem 1.1 and in Theorem 1.1?

1.2 Justify the first inequality in (1.28).

1.3 Under which conditions is (1.21), that is, Problem 1.1_m, equivalent to the minimization of a convex functional? of which functional?

---

[4] This means that $\vec{\gamma}$ is the subdifferential of a (lower semicontinuous) convex functional; see Sect. XI.5.

1.4 Prove Proposition 1.2 by the procedure outlined in the text.

1.5 Formulate Problem 1.1 and related results for systems of P.D.E.s.

1.6 Detail the proof of Theorem 1.3.

1.7 Formulate an initial and boundary value problem associated with the pseudo-parabolic inclusion (1.33). Then prove an existence result. Assuming that either α or $\vec{\gamma}$ are linear, is the solution unique?

1.8 Let $\vec{\varphi}$ and α be maximal monotone graphs. Formulate an initial and boundary value problem for the inclusion $-(\partial/\partial t)\nabla \cdot \vec{\varphi}(\nabla u) + \alpha(u) \ni f$ in Q, and prove an existence result.

III.2 Doubly Nonlinear Parabolic Equations of Second Type

In this section we deal with the equation (1.2), which exhibits several analogies with (1.1), which we studied in the previous section. We assume that

$$u^0 \in L^2(\Omega), \qquad f \in L^2(0,T;V'), \qquad (2.1)$$

and introduce a precise formulation, assuming the homogeneous Dirichlet condition for u.

Problem 2.1 *To find*

$$u \in H^1\left(0,T;L^2(\Omega)\right) \cap L^2(0,T;V), \quad w \in L^2(Q), \quad \vec{z} \in L^2\left(Q;\mathbf{R}^N\right) \quad (2.2)$$

such that

$$\iint_Q (wv + \vec{z} \cdot \nabla v)dxdt = \int_0^T {}_{V'}\langle f, v\rangle_V \, dt \qquad \forall v \in L^2(0,T;V), \qquad (2.3)$$

$$\iint_Q (w - \xi)\left(\frac{\partial u}{\partial t} - v\right) dxdt \geq 0 \qquad \forall v \in L^2(Q), \forall \xi \in \alpha(v), \qquad (2.4)$$

$$\iint_Q (\vec{z} - \vec{\eta}) \cdot (\nabla u - \vec{v})dxdt \geq 0 \qquad \forall \vec{v} \in L^2\left(\Omega;\mathbf{R}^N\right), \forall \vec{\eta} \in \vec{\gamma}(\vec{v}), \qquad (2.5)$$

$$u(\cdot, 0) = u^0 \qquad \text{a.e. in } \Omega. \qquad (2.6)$$

The equation (2.3) yields the differential equation $(1.5)_1$; (2.4) and (2.5) are, respectively, equivalent to $(1.5)_2$ and $(1.5)_3$.

Theorem 2.1 *(Existence) Assume that (1.3), (1.14), (1.30), (1.31) are satisfied, and*

$$\exists L, M, \tilde{L}, \tilde{M} \in \mathbf{R}, (L, \tilde{L} > 0) : \forall (u,w) \in \text{graph}(\alpha),$$
$$wu \geq \tilde{L}|u|^2 - \tilde{M}|u|, \quad |w| \leq L|u| + M. \qquad (2.7)$$

Then Problem 2.1 has a solution such that $\vec{z} \in L^\infty\left(0, T; L^2\left(\Omega; \mathbf{R}^N\right)\right)$.

Proof. (i) *Approximation.* Once more we define m, k, and so on as for Problem II.1.1$_m$, and approximate the P.D.E. by implicit time-discretization.

Problem 2.1$_m$ *To find*

$$u_m^n \in V, \quad w_m^n \in L^2(\Omega), \quad \vec{z}_m^n \in L^2\left(\Omega; \mathbf{R}^N\right) \quad \text{for } n = 1, \ldots, m, \quad (2.8)$$

such that, setting $u_m^0 := u^0$ *a.e. in* Ω, *for* $n = 1, \ldots, m$,

$$w_m^n - \nabla \cdot \vec{z}_m^n = f_m^n \quad \text{in } V', \quad (2.9)$$

$$\int_\Omega (w_m^n - \xi) \left(\frac{u_m^n - u_m^{n-1}}{k} - v \right) dx \geq 0 \quad \forall v \in L^2(\Omega), \forall \xi \in \alpha(v), \quad (2.10)$$

$$\int_\Omega (\vec{z}_m^n - \vec{\eta}) \cdot (\nabla u_m^n - \vec{v}) \geq 0 \quad \forall \vec{v} \in L^2\left(\Omega; \mathbf{R}^N\right), \forall \vec{\eta} \in \vec{\gamma}(\vec{v}). \quad (2.11)$$

The last two inclusions are, respectively, equivalent to $w_m^n \in \alpha((u_m^n - u_m^{n-1})/k)$ and $\vec{z}_m^n \in \vec{\gamma}(\nabla u_m^n)$ a.e. in Ω.

By setting $\tilde{B}(v) := \alpha((u_m^n - u_m^{n-1})/k) - \nabla \cdot \vec{\gamma}(\nabla v)$ for any $v \in V$, the system (2.9) through (2.11) can be written in the form $\tilde{B}(u_m^n) \ni f_m^n$ in V' for $n = 1, \ldots, m$.

Theorem XI.5.2 can then be applied to get existence of a solution of Problem 2.1$_m$, as we did for Problem 1.1$_m$. An alternative argument is based on the fact that this problem is equivalent to the minimization of a convex lower semicontinuous functional. [5]

(ii) *A Priori Estimates.* Let us multiply (2.9) by $u_m^n - u_m^{n-1}$, integrate over Ω, and sum for $n = 1, \ldots, \ell$, for any $\ell \in \{1, \ldots, m\}$. By (1.30) and (2.7), one easily gets the following estimates:

$$k \sum_{n=1}^m \left\| \frac{u_m^n - u_m^{n-1}}{k} \right\|_{L^2(\Omega)}^2, \max_{n=1,\ldots,m} \|u_m^n\|_V \quad (2.12)$$
$$< \text{Constant (independent of } m\text{)},$$

whence, by (1.14),

$$\max_{n=1,\ldots,m} \|\vec{z}_m^n\|_{L^2(\Omega; \mathbf{R}^N)} \leq \text{Constant}. \quad (2.13)$$

---

[5] Cf. the functional J_m^n in Sect. II.1.

Again by (2.7) and (2.12), one gets

$$k \sum_{n=1}^{m} \|w_m^n\|_{L^2(\Omega)}^2 \leq \text{Constant}, \qquad (2.14)$$

whence, by comparison in (2.9),

$$k \sum_{n=1}^{m} \|\nabla \cdot \vec{z}_m^n\|_{L^2(\Omega)}^2 \leq \text{Constant}. \qquad (2.15)$$

Defining u_m, $\bar{u}_m$ and other interpolate functions as in (II.1.29) and (II.1.30), (2.9) yields

$$\bar{w}_m - \nabla \cdot \vec{z}_m = \bar{f}_m \qquad \text{in } V', \qquad (2.16)$$

and (2.10) reads $\bar{w}_m \in \alpha \left(\partial u_m / \partial t \right)$ a.e. in Q.

(iii) *Limit Procedure.* By the previous estimates there exist $u, w, \vec{z}$ such that, possibly taking $m \to \infty$ along a subsequence,

$$u_m \to u \qquad \text{weakly star in } H^1\left(0, T; L^2(\Omega)\right) \cap L^\infty(0, T; V), \qquad (2.17)$$

$$w_m \to w \qquad \text{weakly in } L^2(Q), \qquad (2.18)$$

$$\vec{z}_m \to \vec{z} \qquad \text{weakly star in } L^\infty\left(0, T; L^2\left(\Omega; \mathbf{R}^N\right)\right). \qquad (2.19)$$

Moreover, $\nabla \cdot \vec{z}_m \to \nabla \cdot \vec{z}$ weakly in $L^2(Q)$.

By taking the limit in (2.9), we get (2.3). By (2.17), $u_m \to u$ strongly in $L^2(Q)$; hence we have

$$\iint_Q \vec{z}_m \cdot \nabla u_m \, dx dt = -\iint_Q (\nabla \cdot \vec{z}_m) u_m \, dx dt$$

$$\to -\iint_Q (\nabla \cdot \vec{z}) u \, dx dt = \iint_Q \vec{z} \cdot \nabla u \, dx dt,$$

and by taking the limit in (2.11) we get (2.5).

In order to derive (2.4), we use the *equation + lower semicontinuity* technique (see Sect. II.2).

By applying Lemma 3.2 with $H := V$ and $W := L^2(\Omega)$, by (2.17) we have

$$\bar{u}_m(\cdot, T) \to \bar{u}(\cdot, T) \qquad \text{weakly in } V.$$

By (2.3) and (2.9), the latter convergence and the lower semicontinuity of j, we get

$$\limsup_{m \to \infty} \iint_Q \bar{w}_m \frac{\partial u_m}{\partial t} dx dt = \limsup_{m \to \infty} \iint_Q (\bar{f}_m + \nabla \cdot \vec{z}_m) \frac{\partial u_m}{\partial t} dx dt$$

$$\leq \lim_{m \to \infty} \iint_Q \bar{f}_m \frac{\partial u_m}{\partial t} dx dt - \liminf_{m \to \infty} \int_\Omega \left[j(\nabla \bar{u}_m(\cdot, T)) - j(\nabla u^0) \right] dx$$

$$\leq \iint_Q f \frac{\partial u}{\partial t} dx dt - \int_\Omega \left[j(\nabla u(\cdot, T)) - j(\nabla u^0) \right] dx$$

$$= \iint_Q (f + \nabla \cdot \vec{z}) \frac{\partial u}{\partial t} dx dt = \iint_Q w \frac{\partial u}{\partial t} dx dt.$$

Hence, by taking the superior limit in (2.10), we get (2.4). □

Now we prove existence of a solution, and also some further regularity, without assuming $\vec{\gamma}$ to be cyclically monotone.

Theorem 2.2 *(Existence and Regularity) Assume that (1.3), (2.7) are satisfied, and*

$$\exists c > 0 : \forall \vec{v}_1, \vec{v}_2 \in \mathbf{R}^N, \left[\vec{\gamma}(\vec{v}_1) - \vec{\gamma}(\vec{v}_2)\right] \cdot (\vec{v}_1 - \vec{v}_2) \geq c|\vec{v}_1 - \vec{v}_2|^2, \qquad (2.20)$$

$$u^0 \in V, \qquad \exists \vec{g}^0 \in \vec{\gamma}(\nabla u^0) \text{ such that } \nabla \cdot \vec{g}^0 \in L^2(\Omega), \qquad (2.21)$$

$$f \in H^1\left(0, T; L^2(\Omega)\right). \qquad (2.22)$$

Then Problem 2.1 has a solution such that

$$u \in W^{1,\infty}\left(0, T; L^2(\Omega)\right) \cap H^1(0, T; V), \quad w \in L^{\infty}\left(0, T; L^2(\Omega)\right). \qquad (2.23)$$

Outline of the Proof. Let us take the difference between $(2.9)_n$ and $(2.9)_{n-1}$, and multiply it by $u_m^n - u_m^{n-1}$. Notice that $w^0 = f(0) - \nabla \cdot \vec{g}^0 \in L^2(\Omega)$; hence, defining b as in (II.1.3), $\int_\Omega b(w^0) dx < +\infty$. One then gets a priori estimates corresponding to the regularity (2.23). □

Remarks. (i) Under the assumptions of Theorem 1.3 the solution of Problem 2.1 is unique.
(ii) The results of this section can easily be extended to third order *pseudo-parabolic* inclusions of the form

$$A \frac{\partial u}{\partial t} + \alpha \left(\frac{\partial u}{\partial t}\right) - \nabla \cdot \vec{\gamma}(\nabla u) \ni f \qquad \text{in } Q, \qquad (2.24)$$

where $A : V \to V'$ is a (nonnecessarily self-adjoint) linear elliptic operator.
(iii) The previous discussion can be extended to an abstract setting, with α and $\nabla \cdot \vec{\gamma}(\nabla \cdot)$ replaced by maximal monotone operators in Banach spaces. Theorem 2.1 can be extended, provided that the elliptic operator is cyclically monotone; to get regularity properties like (2.23), α must be cyclically monotone.
This also includes systems of P.D.E.s. □

Exercises.

2.1 Prove that Problem 2.1_m has a solution, detailing the two arguments outlined in the text.

2.2 Detail the argument of Theorem 2.2.

2.3 Prove that the solution of Problem 2.1 is unique, under the assumptions of Theorem 1.3.

2.4 Formulate an initial and boundary value problem associated with the pseudo-parabolic inclusion (2.24), and prove an existence result. Is the solution unique, if either α or $\vec{\gamma}$ are linear?

2.5 Let $\vec{\varphi}$ and α be maximal monotone graphs. Formulate an initial and boundary value problem for the inclusion $-\nabla \cdot \vec{\varphi}\left(\partial \nabla u/\partial t\right) + \alpha(u) \ni f$ in Q, and prove an existence result.

III.3 Other Nonlinear Parabolic Equations

The preceding techniques can be applied to several other equations; in this section we provide some examples. The arguments are only outlined.

We assume the homogeneous Dirichlet condition for u, and use the same notation as in Sect. III.1.

An Equation with Nonmonotone Elliptic Part. Let $\alpha, \tilde{\gamma} : \mathbf{R} \to 2^{\mathbf{R}}$ be maximal monotone graphs, and consider the inclusion

$$\frac{\partial}{\partial t}\alpha(u) - \Delta\tilde{\gamma}(u) \ni f \qquad \text{in } Q, \tag{3.1}$$

which is equivalent to a P.D.E. coupled with two variational inequalities.

The analysis of this inclusion is more delicate than that of (1.1), since the operator $-\Delta\tilde{\gamma}$ is nonmonotone in $L^2(\Omega)$ whenever $\tilde{\gamma}$ is nonlinear.

By setting $z := \tilde{\gamma}(u)$ and $\hat{\alpha} := \alpha \circ \tilde{\gamma}^{-1} : \text{Dom}(\hat{\alpha})(\subset \mathbf{R}) \to 2^{\mathbf{R}}$, (3.1) can be written in the form $\partial\hat{\alpha}(z)/\partial t - \Delta z \ni f$ in Q. If $\hat{\alpha}$ is maximal monotone, we can apply the results of Chap. II. However, this property may fail: take, for instance, $\alpha(\xi) := \xi + \text{sign}(\xi)$ and $\tilde{\gamma}(\xi) := 2\xi + \text{sign}(\xi)$ for any $\xi \in \mathbf{R}$.

In the alternative, one can directly use techniques similar to those of the previous sections. A priori estimates can be derived by multiplying the approximate equation by u_m. This can be accomplished if $\tilde{\gamma}$ is strongly monotone, cf. (II.2.13), and Lipschitz continuous; however, under this condition $\alpha \circ \tilde{\gamma}^{-1}$ is also maximal monotone, and this approach is equivalent to replacing u by $z := \tilde{\gamma}(u)$. Further regularity can be obtained by multiplying the approximate equation by $\partial\tilde{\gamma}(u_m)/\partial t$.

In the next section we prove existence of a solution for this equation by a technique that can also be extended to the vectorial case.

Another Equation with Nonmonotone Elliptic Part. Assume that $\alpha, \tilde{\gamma} : \mathbf{R} \to 2^{\mathbf{R}}$ are maximal monotone graphs, and consider the inclusion

$$\alpha\left(\frac{\partial u}{\partial t}\right) - \Delta\tilde{\gamma}(u) \ni f \qquad \text{in } Q, \tag{3.2}$$

which is also equivalent to a P.D.E. coupled with two variational inequalities.

We claim that the corresponding initial and boundary value problem [6] has a solution, whenever

$$u^0 \in V, \quad f \in L^2(Q), \tag{3.3}$$

α has affine growth at infinity, in the sense of (1.12), and

$$\exists C_1, C_2 \in \mathbf{R}(C_1 > 0) : \forall (u,w) \in \mathrm{graph}(\alpha), wu \geq C_1 u^2 - C_2 u, \tag{3.4}$$

$$\tilde{\gamma} \in C^1(\mathbf{R}); \quad \exists C_3, C_4 \in \mathbf{R}(C_3 > 0) : \forall v \in \mathbf{R}, C_3 \leq \tilde{\gamma}'(v) \leq C_4. \tag{3.5}$$

We just outline the argument. Implicit time-discretization can be used, and a priori estimates can be derived, multiplying the approximate equation by $\partial \tilde{\gamma}(u_m)/\partial t$. By (3.4) and (3.5), formally we have

$$\alpha\left(\frac{\partial u_m}{\partial t}\right) \frac{\partial}{\partial t} \tilde{\gamma}(u_m) \geq C_1 C_3 \left(\frac{\partial u_m}{\partial t}\right)^2 - C_2 C_4 \left|\frac{\partial u_m}{\partial t}\right|. \tag{3.6}$$

This yields a uniform bound for u_m and $\tilde{\gamma}(u_m)$ in $H^1\left(0, T; L^2(\Omega)\right) \cap L^\infty(0, T; V)$. Hence, possibly taking $m \to \infty$ along a subsequence,

$$\begin{aligned} & u_m \to u, \; \tilde{\gamma}(u_m) \to \tilde{\gamma}(u) \\ & \text{weakly star in } H^1\left(0, T; L^2(\Omega)\right) \cap L^\infty(0, T; V), \end{aligned} \tag{3.7}$$

$$u_m \to u, \tilde{\gamma}(u_m) \to \tilde{\gamma}(u), \tilde{\gamma}'(u_m) \to \tilde{\gamma}'(u) \quad \text{strongly in } L^2(Q),$$

$$\alpha\left(\frac{\partial u_m}{\partial t}\right) \to w \quad \text{weakly in } L^2(Q).$$

By comparison in the approximate equation, we get that $\Delta \tilde{\gamma}(u_m)$ is uniformly bounded in $L^2(Q)$. Hence

$$\Delta \tilde{\gamma}(u_m) \to \Delta \tilde{\gamma}(u) \quad \text{weakly in } L^2(Q).$$

Let us define $u_m, w_m, f_m, \bar{u}_m, \bar{w}_m, \bar{f}_m$ as in (II.1.29), (II.1.30). As $\tilde{\gamma}' > 0$, to prove that $w \in \alpha(\partial u/\partial t)$ it suffices to show that

$$\iint_Q (w - \eta)\left(\frac{\partial u}{\partial t} - v\right) \tilde{\gamma}'(u) dx dt \geq 0 \quad \forall v \in L^2(Q), \forall \eta \in \alpha(v). \tag{3.8}$$

Now we have

$$\iint_Q (\bar{w}_m - \eta)\left(\frac{\partial u_m}{\partial t} - v\right) \tilde{\gamma}'(u_m) dx dt \geq 0 \quad \forall v \in L^2(Q), \forall \eta \in \alpha(v), \tag{3.9}$$

---

[6] Unless not otherwise specified, we refer to possibly nonhomogeneous Dirichlet or Neumann or *mixed-type* boundary conditions.

and the by now usual *equation + lower semicontinuity* procedure yields

$$\limsup_{m \to \infty} \iint_Q \bar{w}_m \frac{\partial u_m}{\partial t} \tilde{\gamma}'(u_m) dx dt$$

$$= \limsup_{m \to \infty} \iint_Q [\bar{f}_m + \Delta \tilde{\gamma}(\bar{u}_m)] \frac{\partial}{\partial t} \tilde{\gamma}(u_m) dx dt$$

$$\leq \lim_{m \to \infty} \iint_Q \bar{f}_m \frac{\partial}{\partial t} \tilde{\gamma}(u_m) dx dt - \frac{1}{2} \liminf_{m \to \infty} \int_\Omega |\nabla \tilde{\gamma}(\bar{u}_m)|^2 dx \Big|_{t=0}^{t=T} \quad (3.10)$$

$$\leq \iint_Q f \frac{\partial}{\partial t} \tilde{\gamma}(u) dx dt - \frac{1}{2} \int_\Omega |\nabla \tilde{\gamma}(u)|^2 dx \Big|_{t=0}^{t=T}$$

$$= \iint_Q [f + \Delta \tilde{\gamma}(u)] \frac{\partial}{\partial t} \tilde{\gamma}(u) dx dt = \iint_Q w \frac{\partial u}{\partial t} \tilde{\gamma}'(u) dx dt.$$

Hence, by taking the superior limit in (3.9), we get (3.8). □

A Triply Nonlinear Equation. Let $\alpha, \beta, \tilde{\gamma} : \mathbf{R} \to 2^{\mathbf{R}}$ be maximal monotone graphs, and consider the inclusion

$$\alpha \left(\frac{\partial}{\partial t} \beta(u) \right) - \Delta \tilde{\gamma}(u) \ni f \quad \text{in } Q, \quad (3.11)$$

which is also equivalent to a P.D.E. coupled with three variational inequalities, one for each maximal monotone graph.

As for (3.1), here one can choose between two equivalent approaches. One can set $w := \beta(u)$, $\hat{\gamma} := \tilde{\gamma} \circ \beta^{-1}$, and write (3.11) as

$$\alpha \left(\frac{\partial w}{\partial t} \right) - \Delta \hat{\gamma}(w) \ni f \quad \text{in } Q, \quad (3.12)$$

which is of the form (3.1), provided that $\tilde{\gamma} \circ \beta^{-1}$ is a maximal monotone graph.

Equivalently, one can deal with (3.11) directly. Existence of a solution can be proved, if α and $\tilde{\gamma}$ have affine growth at infinity,

$$\beta \in C^1(\mathbf{R}); \quad \exists C_1, C_2 \in \mathbf{R}(C_1 > 0) : \forall v \in \mathbf{R}, C_1 \leq \beta'(v) \leq C_2, \quad (3.13)$$

and a similar condition holds for $\tilde{\gamma}$. Implicit time-discretization can be used, and a priori estimates can be derived multiplying the approximate equation by $\partial \tilde{\gamma}(u_m)/\partial t$. We omit details of this argument.

Another Triply Nonlinear Equation. Let $\alpha, \beta : \mathbf{R} \to 2^{\mathbf{R}}$ and $\vec{\tilde{\gamma}} : \mathbf{R}^N \to 2^{(\mathbf{R}^N)}$ be maximal monotone graphs, and consider the inclusion

$$\alpha \left(\frac{\partial}{\partial t} \beta(u) \right) - \nabla \cdot \vec{\tilde{\gamma}}(\nabla u) \ni f \quad \text{in } Q. \quad (3.14)$$

III.3 Other Nonlinear Parabolic Equations

This is equivalent to a P.D.E. coupled with three variational inequalities.

Existence of a solution can be proved if: (i) α and $\vec{\gamma}$ have affine growth at infinity, (ii) β fulfills (3.13), (iii) $\vec{\gamma}$ is the subdifferential of a convex function $j : \mathbf{R}^N \to \mathbf{R}$, (iv) $u^0 \in V$ and $f \in L^2(Q)$.

We just outline the argument. Implicit time-discretization can be used, and a priori estimates can be derived, multiplying the approximate equation by $\partial u_m/\partial t$. This yields a uniform bound for u_m and $\beta(u_m)$ in $H^1\left(0,T;L^2(\Omega)\right) \cap L^\infty(0,T;V)$. Hence, possibly taking $m \to \infty$ along a subsequence, we have

$$u_m \to u, \ \beta(u_m) \to \beta(u)$$

weakly star in $H^1\left(0,T;L^2(\Omega)\right) \cap L^\infty(0,T;V)$,

whence

$$u_m \to u, \ \frac{1}{\beta'(u_m)} \to \frac{1}{\beta'(u)} \quad \text{strongly in } L^2(Q).$$

Moreover, possibly extracting a further subsequence,

$$\alpha\left(\frac{\partial}{\partial t}\beta(u_m)\right) \ni \bar{w}_m \to w \quad \text{weakly in } L^2(Q).$$

As $\beta' > C_1 > 0$, to prove that $w \in \alpha\left(\partial\beta(u)/\partial t\right)$ it suffices to show that

$$\iint_Q (w-\hat{\eta})\left(\frac{\partial}{\partial t}\beta(u) - \eta\right)\frac{1}{\beta'(u)}\,dx\,dt \geq 0 \quad \forall \eta \in L^2(Q), \forall \hat{\eta} \in \alpha(\eta). \quad (3.15)$$

Now we have

$$\iint_Q (\bar{w}_m - \eta)\left(\frac{\partial}{\partial t}\beta(u_m) - v\right)\frac{1}{\beta'(u_m)}\,dx\,dt \geq 0 \quad (3.16)$$
$$\forall v \in L^2(Q), \forall \eta \in \alpha(v),$$

and the usual *equation + lower semicontinuity* procedure yields

$$\limsup_{m\to\infty} \iint_Q \bar{w}_m \frac{\partial \beta(u_m)}{\partial t}\frac{1}{\beta'(u_m)}\,dx\,dt$$
$$= \limsup_{m\to\infty} \iint_Q [\bar{f}_m + \nabla \cdot \vec{\gamma}(\nabla \bar{u}_m)]\frac{\partial u_m}{\partial t}\,dx\,dt$$
$$\leq \lim_{m\to\infty} \iint_Q \bar{f}_m \frac{\partial u_m}{\partial t}\,dx\,dt - \frac{1}{2}\liminf_{m\to\infty}\int_\Omega j(\nabla u_m)dx\bigg|_{t=0}^{t=T} \quad (3.17)$$
$$\leq \iint_Q f\frac{\partial u}{\partial t}\,dx\,dt - \frac{1}{2}\int_\Omega j(\nabla u)dx\bigg|_{t=0}^{t=T}$$
$$= \iint_Q [f + \nabla\cdot\vec{\gamma}(\nabla u)]\frac{\partial u}{\partial t}\,dx\,dt = \iint_Q w\frac{\partial\beta(u)}{\partial t}\frac{1}{\beta'(u)}\,dx\,dt.$$

Hence, by taking the superior limit in (3.16), we get (3.15). □

The Vectorial Case. Let $u : Q \to \mathbf{R}^M$ ($M > 1$), $\alpha, \beta, \tilde{\gamma}$ be maximal monotone graphs in $\mathbf{R}^M \times \mathbf{R}^M$, and $\vec{\tilde{\gamma}}$ be replaced by a graph in $\mathbf{R}^{(M \times N)} \times \mathbf{R}^{(M \times N)}$, namely, a possibly multi-valued function in the space of $M \times N$-matrices. Then one can consider vectorial equations, that is, systems of P.D.E.s, analogous to those of this section.

In this setting, the monotonicity of $\hat{\alpha} := \alpha \circ \tilde{\gamma}^{-1} : \text{Dom}(\hat{\alpha}) \, (\subset \mathbf{R}^M) \to 2^{(\mathbf{R}^M)}$ is a fairly restrictive hypothesis. A different approach to systems of the form (3.1) is described in the next section. For systems like (3.2) and (3.11), existence of a solution is an open question in the nonlinear case. In fact for instance (3.6) does not hold in general, and so it is not clear how a priori estimates might be derived.

Exercises.

3.1 Is it possible to derive a priori estimates for the equation (3.1), multiplying the time increment of the approximate equation by $\tilde{\gamma}(u_m^n)$?

3.2 Formulate an initial and boundary value problem for the pseudo-parabolic inclusion $A\partial u/\partial t + \partial \alpha(u)/\partial t - \Delta \tilde{\gamma}(u) \ni f$ in Q, where $A : V \to V'$ is a linear elliptic operator. Then prove existence of a solution. In particular, is it necessary to assume that A is self-adjoint?

3.3 Formulate an initial and boundary value problem for the inclusion (3.2). Then prove existence of a solution, under the assumptions (3.3) through (3.5), by means of the argument outlined in the text.

3.4 May existence of a solution be proved for the pseudo-parabolic inclusion, which is obtained by inserting the term $A\partial u/\partial t$ into the left hand side of (3.2)?

3.5 Formulate an initial and boundary value problem for the inclusion (3.11). Then prove existence of a solution, detailing the two procedures outlined in the text.

3.6 Discuss the problems obtained by adding a lower order term $g(u)$ to any of the equations (3.2), (3.11), or (3.14).

III.4 Use of Compactness by Strict Monotonicity

The method of *compactness by strict convexity* is introduced in Chap. X. [7] Here we illustrate some examples of how this technique can be used to study nonlinear P.D.E.s. This can also be extended to systems of equations.

Analysis of (3.1). Let $\alpha, \tilde{\gamma} : \mathbf{R} \to 2^\mathbf{R}$ be maximal monotone graphs, and consider the inclusion (3.1). We want to prove existence of a solution for the corresponding

---

[7] This chapter can be read independently, and the reader is referred to it.

initial and boundary value problem (with homogeneous Dirichlet condition for u, say), by an argument that can be extended to the vectorial setting.

Let
$$w^0 \in L^2(\Omega), \qquad f \in L^2(0,T;V'). \tag{4.1}$$

Problem 4.1 *To find $u, z \in L^2(0,T;V)$ and $w \in L^2(Q)$ such that*

$$\iint_Q \left(-w\frac{\partial v}{\partial t} + \nabla z \cdot \nabla v\right) dx dt$$
$$= \int_0^T {}_{V'}\langle f, v\rangle_V \, dt + \int_\Omega w^0 v(\cdot,0) dx \tag{4.2}$$
$$\forall v \in L^2(0,T;V) \cap H^1\left(0,T;L^2(\Omega)\right), \, v(\cdot,T) = 0,$$

$$w \in \alpha(u), \ z \in \tilde{\gamma}(u) \qquad \text{a.e. in } Q. \tag{4.3}$$

The interpretation of (4.2) is analogous to that of (II.1.9). In particular, (4.2) yields $w \in H^1(0,T;V')$, (3.1) in V' a.e. in $]0,T[$, and the initial condition $w(0) = w^0$ in the sense of the traces.

Theorem 4.1 *Assume that*

$$\alpha : \mathbf{R} \to 2^{\mathbf{R}} \text{ is a strictly monotone, maximal monotone graph,} \tag{4.4}$$

and that (1.12) and (3.5) are satisfied. Then there exists a solution of Problem 4.1 such that $w \in L^\infty\left(0,T;L^2(\Omega)\right)$.

Outline of the Proof. Approximation by implicit time-discretization can be used. Let us use notation like (II.1.29) and (II.1.30). By multiplying the approximate equation by the approximate solution $\bar{u}_m$, uniform estimates are easily derived for $w_m (\in \alpha(u_m))$ in $L^\infty\left(0,T;L^2(\Omega)\right)$ and for u_m (whence also for $\tilde{\gamma}(u_m)$, by (3.5)) in $L^2(0,T;V)$. By comparing the terms in the approximate equation, an estimate for w_m in $H^1(0,T;V')$ is then obtained. Hence, possibly taking $m \to \infty$ along a subsequence, we have

$$u_m \to u, \ \tilde{\gamma}(u_m) \to z \qquad \text{weakly in } L^2(0,T;V),$$

$$w_m \rightharpoonup w \qquad \text{weakly star in } L^\infty\left(0,T;L^2(\Omega)\right) \cap H^1(0,T;V'),$$

whence $w_m \to w$ strongly in $L^2(0,T;V')$, by Lions-Aubin's Theorem XI.3.5 (ii). Hence
$$\iint_Q \bar{u}_m \bar{w}_m \, dx dt \to \iint_Q uw \, dx dt. \tag{4.5}$$

By (4.4), α is the subdifferential of a lower semicontinuous, strictly convex function $\mathbf{R} \to \mathbf{R} \cup \{+\infty\}$. By Theorem X.3.1 we then get that $w \in \alpha(u)$ a.e. in Q and

$u_m \to u$ strongly in $L^1(Q)$. Hence, possibly extracting a further subsequence, $u_m \to u$ in measure in Q, and we get $z \in \tilde{\gamma}(u)$ a.e. in Q. □

Remark. The latter result can easily be extended to the vectorial case, assuming that α is the subdifferential of a *strictly convex* function $\mathbf{R}^M \to \mathbf{R} \cup \{+\infty\}$. □

Another P.D.E.. Let us assume that α is a maximal monotone graph, and

$$\vec{\gamma}: \mathbf{R} \times \mathbf{R}^N \to \mathbf{R}^n \text{ is a globally continuous function,} \tag{4.6}$$
$$\text{monotone with respect to the second argument,}$$

and consider the inclusion

$$\frac{\partial}{\partial t}\alpha(u) - \nabla \cdot \vec{\gamma}(u, \nabla u) \ni f \qquad \text{in } Q. \tag{4.7}$$

For a moment assume that α is single-valued, that $\beta := \alpha^{-1} \in C^1(\mathbf{R})$, and set $w := \alpha(u)$. Then $\vec{\gamma}(u, \nabla u) = \vec{\gamma}(\beta(w), \beta'(w)\nabla w) =: \hat{\vec{\gamma}}(w, \nabla w)$, and (4.7) is equivalent to

$$\frac{\partial w}{\partial t} - \nabla \cdot \hat{\vec{\gamma}}(w, \nabla w) = f \qquad \text{in } Q. \tag{4.8}$$

This equation can be approximated by time-discretization, and a priori estimates can be derived by multiplying the discretized equation by w_m. Existence of a solution for the corresponding initial and boundary value problem (with the homogeneous Dirichlet condition for w, say) can then be proved, assuming that $\hat{\vec{\gamma}}$ is globally continuous, has affine growth with respect to both arguments, and is coercive with respect to the second one. In particular the *equation + lower semicontinuity technique* can be used to pass to the limit in the $\hat{\vec{\gamma}}$-term.

Another Method. Now we prove existence of a solution for the latter problem, by means of a procedure based on *compactness by strict convexity*, without requiring β to be of class C^1.

Problem 4.2 *To find $u \in L^2(0, T; V)$ and $w \in L^2(Q)$ such that*

$$\iint_Q \left(-w\frac{\partial v}{\partial t} + \vec{\gamma}(u, \nabla u) \cdot \nabla v\right) dxdt$$
$$= \int_0^T {}_{V'}\langle f, v\rangle_V \, dt + \int_\Omega w^0 v(\cdot, 0) dx \tag{4.9}$$
$$\forall v \in L^2(0, T; V) \cap H^1\left(0, T; L^2(\Omega)\right), v(\cdot, T) = 0,$$

$$w \in \alpha(u) \qquad \text{a.e. in } Q. \tag{4.10}$$

The interpretation of (4.9) is analogous to that of (II.1.9).

III.4 Use of Compactness by Strict Monotonicity

Theorem 4.2 *Assume that (4.1), (4.4), and (4.6) are satisfied, and that*

$$\exists C > 0 : \forall (v, \vec{\eta}) \in \mathbf{R} \times \mathbf{R}^N, |\vec{\gamma}(v, \vec{\eta})| \leq C|\vec{\eta}|, \tag{4.11}$$

$$\exists L, M \in \mathbf{R}(L > 0) : \forall (v, \vec{\eta}) \in \mathbf{R} \times \mathbf{R}^N, \vec{\gamma}(v, \vec{\eta}) \cdot \vec{\eta} \geq L|\vec{\eta}|^2 - M|\vec{\eta}|. \tag{4.12}$$

Then there exists a solution of Problem 4.2 such that $w \in L^\infty\left(0, T; L^2(\Omega)\right)$.

Outline of the Proof. Approximation by time-discretization can be used. By multiplying the approximate equation by the (piecewise constant) approximate solution $\bar{u}_m$, one easily gets the estimates

$$\|w_m\|_{L^\infty(0,T;L^2(\Omega))}, \|\bar{u}_m\|_{L^2(0,T;V)} \leq \text{Constant (independent of } m\text{)}, \tag{4.13}$$

whence by (4.11)

$$\|\vec{\gamma}(\bar{u}_m, \nabla \bar{u}_m)\|_{L^2(Q;\mathbf{R}^N)} \leq \text{Constant}. \tag{4.14}$$

A comparison in the approximate equation then yields

$$\|w_m\|_{H^1(0,T;V')} \leq \text{Constant}. \tag{4.15}$$

Hence there exist $u, w, \vec{z}$ such that, possibly taking $m \to \infty$ along a subsequence,

$$\bar{u}_m \to u \quad \text{weakly in } L^2(0, T; V), \tag{4.16}$$

$$w_m \to w \quad \text{weakly star in } L^\infty\left(0, T; L^2(\Omega)\right) \cap H^1(0, T; V'), \tag{4.17}$$

$$\vec{\gamma}(\bar{u}_m, \nabla \bar{u}_m) \to \vec{z} \quad \text{weakly in } L^2\left(Q; \mathbf{R}^N\right). \tag{4.18}$$

We then have (4.5), which yields (4.10) by a standard monotonicity technique; cf. Lemma XI.5.1.

By taking the limit in the approximate equation, we get (4.9), with $\vec{z}$ in place of $\vec{\gamma}(\bar{u}, \nabla \bar{u})$. To pass to the limit in $\vec{\gamma}$, we use the monotonicity with respect to the second argument. By (4.5) and Theorem X.3.1, we have

$$\bar{u}_m \to u \quad \text{strongly in } L^2(Q). \tag{4.19}$$

By the *equation + lower semicontinuity* technique, cf. (II.2.4), one can also show that

$$\limsup_{m \to \infty} \iint_Q \vec{\gamma}(\bar{u}_m, \nabla \bar{u}_m) \cdot \nabla \bar{u}_m \, dx dt \leq \iint_Q \vec{z} \cdot \nabla u \, dx dt. \tag{4.20}$$

By monotonicity, we have

$$\iint_Q \left[\vec{\gamma}(\bar{u}_m, \nabla \bar{u}_m) - \vec{\gamma}(\bar{u}_m, \vec{\xi})\right] \cdot (\nabla \bar{u}_m - \vec{\xi}) dx dt \geq 0$$

$$\forall \vec{\xi} \in L^2\left(Q; \mathbf{R}^N\right). \tag{4.21}$$

By (4.19), $\vec{\gamma}(\bar{u}_m, \vec{\xi}) \to \vec{\gamma}(u, \vec{\xi})$ strongly in $L^2(Q; \mathbf{R}^N)$. Hence, passing to the superior limit as $m \to \infty$, we get

$$\iint_Q \left[\vec{z} - \vec{\gamma}(u, \vec{\xi})\right] \cdot (\nabla u - \vec{\xi}) dx dt \geq 0 \qquad \forall \vec{\xi} \in L^2(Q; \mathbf{R}^N), \qquad (4.22)$$

which is equivalent to $\vec{z} = \vec{\gamma}(u, \nabla u)$ a.e. in Q. □

Remarks. (i) This theorem can also be extended to the vectorial case, assuming that α is the subgradient of a *strictly convex* function $\mathbf{R}^M \to \mathbf{R} \cup \{+\infty\}$.
(ii) The equation

$$\alpha\left(\frac{\partial u}{\partial t}\right) - \nabla \cdot \vec{\gamma}(u, \nabla u) \ni f \qquad \text{in } Q \qquad (4.23)$$

looks more challenging. □

Other P.D.E.s with Lower Order Nonlinearities. Let $\alpha : \mathbf{R}^2 \to \mathbf{R}$ be monotone with respect to the second argument and globally continuous, $\vec{\gamma} : \mathbf{R}^N \to 2^{(\mathbf{R}^N)}$ be a maximal monotone graph, and consider the inclusion

$$\alpha\left(u, \frac{\partial u}{\partial t}\right) - \nabla \cdot \vec{\gamma}(\nabla u) \ni f \qquad \text{in } Q. \qquad (4.24)$$

Here a priori estimates can be derived multiplying the approximate equation by $\partial u_m/\partial t$, assuming that $\vec{\gamma}$ is the subdifferential of a convex function $j : \mathbf{R}^N \to \mathbf{R}$. Existence of a solution can then be proved. We omit the other hypotheses and details of the argument. This can easily be extended to the vectorial case.

Now let $\tilde{\gamma} : \mathbf{R} \to 2^{\mathbf{R}}$ be a maximal monotone graph, and consider the inclusion

$$\alpha\left(u, \frac{\partial u}{\partial t}\right) - \Delta \tilde{\gamma}(u) \ni f \qquad \text{in } Q. \qquad (4.25)$$

This inclusion looks less simple than the previous one, as the operator $-\Delta \tilde{\gamma}$ is not monotone in $L^2(\Omega)$. Nevertheless, in the scalar case, it is possible to prove existence of a solution, since a priori estimates can be derived multiplying the approximate equation by $\partial \tilde{\gamma}(u_m)/\partial t$. We omit details and just note that, assuming (3.5) and a condition similar to (3.4), an inequality analogous to (3.6) holds.

As we already pointed out, in the vectorial case a condition like (3.6) would require stronger assumptions to be fulfilled, and it does not seem straightforward to generalize the proof of existence of a solution.

Exercises.

4.1 Formulate an initial and boundary value problem for (4.24). Then prove existence, by means of the procedures outlined in the text.

Analogous questions for (4.25).

4.2 Discuss the problems obtained by adding a lower order term $g(u)$ to any of the equations (4.7), (4.24) or (4.25).

III.5 Comments

Quasilinear parabolic equations containing two nonlinear monotone operators occur in several models of interest for applications (see, e.g., Sects. V.3 and V.5). However, here we went beyond the analysis of the models of phase transitions which we introduce in Chaps. IV and V. We confined ourselves to some *model problems;* these developments might be extended in several ways, for instance including lower order terms.

In this chapter, we used compactness and monotonicity techniques similar to those of Chap. II. Lions [351] is a standard reference for such methods. We also applied *compactness by strict convexity* results, cf. Chap. X.

Several works have been devoted to doubly nonlinear parabolic equations like (1.1). For instance, see Raviart [462], Lions [351; Sect. 4.1], Grange and Mignot [272], Barbu [53], DiBenedetto and Showalter [191], Alt and Luckhaus [12], and Bernis [74]. Most of these papers deal with the abstract formulation in Hilbert and Banach spaces.

Doubly nonlinear parabolic equations like (1.2) seem to have been the object of less attention, although they have applications. For instance, they may represent a *gradient flow* [8] in presence of a *pseudo-potential* of dissipation $\Phi\left(\partial u/\partial t\right)$, with Φ such that $\partial\Phi = \alpha$. [9] These equations have been studied in Banach and Hilbert spaces by Arai [25], Barbu [52], Colli and V. [143], and Colli [139]. Further references can be found the latter papers. See Bénilan and Ha [68], and Ha [286] for a different approach based on semigroups of nonlinear contractions in L^∞.

Let I_{K_i} be the characteristic function of a nonempty, closed, convex set $K_i \subset \mathbf{R}^M$ ($i = 1, 2$), and $w, u : [0, T] \to \mathbf{R}^M$. Doubly nonlinear O.D.E.s of the form

$$\left(\partial I_{K_1}\right)^{-1}\left(\frac{du}{dt}\right) + \partial I_{K_2}(u) \ni w \qquad \text{in }]0, T[. \tag{5.1}$$

coupled with an initial condition for u, can be used to represent (either scalar or vectorial) discontinuous *hysteresis operators* $w \mapsto u$; see V. [564; Sects. VI.3, VI.4].

---

[8] This concept is illustrated in Sect. VII.6.

[9] See, e.g., Germain [262; Sect. VII.3].

	$\frac{\partial}{\partial t}\alpha(u) - \nabla \cdot \vec{\gamma}(\nabla u) \ni f$	$\alpha\left(\frac{\partial u}{\partial t}\right) - \nabla \cdot \vec{\gamma}(\nabla u) \ni f$
1st Estimate hypoth.	$\alpha^{-1}, \vec{\gamma}$ coercive	$\alpha, \vec{\gamma}$ coercive
1st Estimate proced.	eq. $\times u$	eq. $\times \frac{\partial u}{\partial t}$
2nd Estimate hypoth.	$\vec{\gamma}$ linearly bounded	α linearly bounded
2nd Estimate proced.	comparison in equation	comparison in equation
Limit in α procedure	compactness	equation + semicont.
Limit in $\vec{\gamma}$ procedure	equation + semicont.	compactness
Regularity procedure	eq. $\times \frac{\partial u}{\partial t}$	$\frac{\partial}{\partial t}$eq. $\times \frac{\partial u}{\partial t}$

Table 1. Comparison of the existence results of Theorems 1.1 and 2.1. The last line refers to a procedure yielding further regularity.

Some shortwriting is used. For instance, $\frac{\partial}{\partial t}$eq. $\times \frac{\partial u}{\partial t}$ stands for the following procedure: multiply the time incremental ratio of the discretized equation by the time incremental ratio of the approximate u, and then integrate in space and time. *Equation + semicont.* stays for the procedure outlined in Sect. II.2, cf. (II.2.4).

Equation	Main hypotheses	Estim. Proced.
$\frac{\partial}{\partial t}\alpha(u) - \Delta\tilde{\gamma}(u) \ni f$	$0 \leq C_1 \leq \tilde{\gamma}' \leq C_2$ α with affine growth at ∞	eq. $\times u$
$\alpha\left(\frac{\partial u}{\partial t}\right) - \Delta\tilde{\gamma}(u) \ni f$	$0 \leq C_1 \leq \tilde{\gamma}' \leq C_2$ $\alpha(v)v \geq Cv^2 - \hat{C}v \quad \forall v$ α with affine growth at ∞	eq. $\times \frac{\partial}{\partial t}\tilde{\gamma}(u)$
$\alpha\left(\frac{\partial}{\partial t}\beta(u)\right) - \Delta\tilde{\gamma}(u) \ni f$	$0 \leq C_1 \leq \beta', \gamma' \leq C_2$ $\alpha(v)v \geq Cv^2 - \hat{C}v \quad \forall v$ α with affine growth at ∞	eq. $\times \frac{\partial}{\partial t}\tilde{\gamma}(u)$
$\alpha\left(\frac{\partial}{\partial t}\beta(u)\right) - \nabla \cdot \vec{\gamma}(\nabla u) \ni f$	$0 \leq C_1 \leq \beta' \leq C_2$ $\alpha(v)v \geq Cv^2 - \hat{C}v \quad \forall v$ $\alpha, \vec{\gamma}$ with affine growth at ∞ $\vec{\gamma} = \partial j \ (j: \mathbf{R}^N \to \mathbf{R}$ convex$)$	eq. $\times \frac{\partial}{\partial t}\tilde{\gamma}(u)$

Table 2. Overview of some of the equations studied in Sect. III.3. $\alpha, \beta, \tilde{\gamma}, \vec{\gamma}$ are maximal monotone graphs.

Existence of a solution can also be proved for systems of the form

$$\begin{cases} \dfrac{\partial}{\partial t}\alpha(u) - \nabla \cdot \vec{\gamma}(\nabla u) \ni w & \text{in } Q, \\ \dfrac{\partial}{\partial t}(w+u) - \Delta w = f & \text{in } Q, \end{cases} \quad (5.2)$$

or for the analogous problem obtained replacing $\partial\alpha(u)/\partial t$ by $\alpha\left(\partial u/\partial t\right)$; of course either of these systems must be coupled with appropriate boundary and initial conditions. This is based on techniques similar to those we used in Sects. III.1, III.2. On the other hand, the system obtained coupling (5.1) and (5.2)$_2$ seems harder to be treated.

Both Theorem 1.1 and 2.1 also cover degenerate elliptic-parabolic equations, including the case in which both α and $\vec{\gamma}$ vanish in some open set. A *modified porous medium equation* of the form (3.2) was studied by Barenblatt and others; see, for example, Barenblatt [57], Hulshof and Vázquez [306] and references therein.

An Overview. The main results and techniques of this chapter are displayed in Tables 1 and 2.

IV. The Stefan Problem

Outline

The Stefan model of phase transition in solid–liquid systems is introduced. This accounts for heat diffusion in each phase and exchange of latent heat at the solid–liquid interface. Its strong formulation is a *free boundary problem,* since the interface evolution is a priori unknown. Formulations in one and in several space dimensions are derived.

The classical *Gibbs-Thomson law* accounting for *surface tension* is also introduced. This contains a parameter, the surface tension coefficient σ, which determines a *mesoscopic length scale.*

The onset of singularities in the strong formulation of the Stefan problem leads one to consider a weak formulation, which accounts for the formation of the so-called *mushy region.* The two formulations are compared.

Two classical free boundary problems arising in fluid dynamics are outlined: the *Muskat* and *Hele-Shaw* problems. A vectorial Stefan-type model representing ferromagnetic evolution is also introduced.

Finally, the development of the research on the Stefan problem is briefly reviewed.

Prerequisites. Calculus and basic notions of thermodynamics are applied. The definition of derivative in the sense of distributions is used in Sect. IV.1.

IV.1 Strong Formulation of the Stefan Problem

This and the next chapter are devoted to the derivation of some models of phase transitions. In this section we introduce the strong formulation of the *Stefan problem.*

The Heat Equation. Let us consider a system composed of a homogeneous and isotropic material, capable of attaining two phases, liquid and solid, say. [1]

---

[1] Here we only deal with solid–liquid and liquid-solid transitions. Our analysis might be applied to other *changes of aggregate state* as well, but excludes a multitude of other processes that physicists label as phase transitions.

Incidentally we note that some authors use the latter term dealing with *stationary* multi-phase systems. At variance with that, we speak of phase transitions only referring to *processes*.

IV.1 Strong Formulation of the Stefan Problem

Let us denote by Ω a bounded domain of $\mathbf{R}^3$ occupied by this material, fix any $T > 0$ and set $Q := \Omega \times]0, T[$. We label by 1 and 2 quantities relative to the liquid and solid phases, respectively, and use the following notation:

Q_i: (open) subset of Q corresponding to the phase i,

$S := \partial Q_1 \cap \partial Q_2$: (possibly disconnected) space-time manifold that separates the phases,

$S_t := S \cap (\Omega \times \{t\})$: configuration of S at an instant $t \in [0,T]$,

u: density of *internal energy* (namely, internal energy per unit volume),

θ: relative temperature (namely, the difference between the actual absolute temperature τ and the value τ_E at which a planar solid–liquid interface is at equilibrium),

$\vec{q}$: heat flux (per unit surface),

$C_{Vi}(\theta)$: heat capacity per unit volume (namely, the heat needed to increase the temperature of a unit volume by one degree),

$k_i(\theta)$: thermal conductivity,

$L(\theta)$: density of *latent heat* of phase transition (namely, the heat exchanged by phase transition of a unit volume). [2]

We assume that both phases are incompressible, so that processes occur at constant volume. [3] The definition of C_{Vi} and, neglecting convection, the classical *Fourier conduction law* respectively yield

$$\frac{\partial u}{\partial t} = C_{Vi}(\theta) \frac{\partial \theta}{\partial t} \quad \text{in } Q_i \ (i = 1, 2), \tag{1.1}$$

$$\vec{q} = -k_i(\theta) \nabla \theta \quad \text{in } Q_i \ (i = 1, 2). \tag{1.2}$$

If neither heat sources nor sinks are present, in each phase the energy balance reads

$$\frac{\partial u}{\partial t} = -\nabla \cdot \vec{q} \quad \text{in } Q_i \ (i = 1, 2). \tag{1.3}$$

From these relations we get the homogeneous *heat equation* in the interior of each phase:

$$C_{Vi}(\theta) \frac{\partial \theta}{\partial t} - \nabla \cdot [k_i(\theta) \nabla \theta] = 0 \quad \text{in } Q_i \ (i = 1, 2). \tag{1.4}$$

Interface Conditions. [4] We assume that the solid which is initially present and that which is formed in the process are all in the crystalline state, so that they contain no latent heat. We denote by $\vec{v}$ the normal velocity of the interface S_t (assumed to

---

[2] Here it is implicitly assumed that phase transitions occur at constant temperature; this is explicitly required in (1.8).

[3] Most of these developments hold also for systems at constant pressure, with just minor changes in the terminology.

[4] We often call *interface* the surface that separates the solid from the liquid.

be *smooth*), [5] by $\vec{\nu} \in \mathbf{R}^3$ a unit vector field normal to $\mathcal{S}_t$, and by $\partial/\partial\nu := \vec{\nu} \cdot \nabla$ the corresponding directional derivative.

Let us consider an element dS of interface that moves with velocity $\vec{v}$, and denote by $\vec{q}_1$ the heat flux (per unit surface) contributed by the liquid phase, and by $\vec{q}_2$ the heat flux (per unit surface) absorbed by the solid phase. Latent heat is either absorbed or released at a rate $L(\theta)\vec{v} \cdot \vec{\nu} dS$, and this equals the heat exchanged by the interface itself by conduction: $(\vec{q}_1 \cdot \vec{\nu} - \vec{q}_2 \cdot \vec{\nu})dS$ (signs can be checked distinguishing melting from solidification). Therefore

$$\vec{q}_1 \cdot \vec{\nu} - \vec{q}_2 \cdot \vec{\nu} = L(\theta)\vec{v} \cdot \vec{\nu} \qquad \text{on } \mathcal{S}. \tag{1.5}$$

Therefore the front velocity and the total exchanged heat flux have the same orientation. By (1.2), this yields the classical *Stefan condition*

$$k_1(\theta)\frac{\partial \theta_1}{\partial \nu} - k_2(\theta)\frac{\partial \theta_2}{\partial \nu} = -L(\theta)\vec{v} \cdot \vec{\nu} \qquad \text{on } \mathcal{S}, \tag{1.6}$$

where we denote by $\partial \theta_i / \partial \nu$ the normal derivative of θ relative to the phase labelled by i ($= 1, 2$). [6]

Local Thermodynamical Equilibrium. Although the heat equation (1.4) describes nonequilibrium, here we assume *local thermodynamical equilibrium*. By this we mean that in a neighbourhood of each point the system is *close* to equilibrium; that is, the state variables attain values that are *close* to those of equilibrium. [7] Under this hypothesis, the thermodynamical variables can be assumed to be pointwise related by the same laws as at equilibrium, in particular

$$\theta \text{ is continuous across } \mathcal{S}, \tag{1.7}$$

as any discontinuity would trigger a diverging heat flux. For the moment we neglect surface tension effects. For a homogeneous material, (1.7) yields

$$\theta = 0 \qquad \text{on } \mathcal{S}; \tag{1.8}$$

hence in (1.6) we can replace $L(\theta)$ by $L := L(0)$ and $k_i(\theta)$ by $k_i := k_i(0)$ ($i = 1, 2$).

The evolution of the solid–liquid interface is unknown. In principle, this lack of information is compensated by *two* quantitative conditions, namely, (1.6) and (1.8), at the *free boundary* $\mathcal{S}$. The preceding conditions must be coupled with conditions

---

[5] Phase interfaces are not material surfaces: their evolution does not represent motion of particles. Hence only the normal component of their speed has physical meaning.

[6] If $f \in C^1(Q)$ is such that $\mathcal{S} = \{(x, y, z, t) \in Q : f(x, y, z, t) = 0\}$, then $\nabla f \cdot \vec{v} + \partial f/\partial t = 0$ on $\mathcal{S}$. Hence (1.6) is equivalent to $\left[k_1(\theta)\nabla\theta_1 - k_2(\theta)\nabla\theta_2\right] \cdot \nabla f = L(\theta)\partial f/\partial t$ on $\mathcal{S}$.

[7] The specific criterion of closeness depends on the material, and its determination is left to the physicist.

on the initial value of θ and on the initial phase configuration. Boundary condition must also be imposed; for instance, one might choose a partition $\{\Gamma_1, \Gamma_2\}$ of the boundary Γ of Ω, and prescribe θ on $\Gamma_1 \times]0, T[$, $\partial\theta/\partial\vec{\nu}$ on $\Gamma_2 \times]0, T[$.

As an example we consider the following model problem.

Problem 1.1 (*Strong Formulation of the Three-Dimensional Two-Phase Stefan Problem*)[8] *To find $\theta \in C^0(\bar{Q})$ and a partition $\{Q_1, Q_2, S\}$ of Q such that:*
(i) Q_1 and Q_2 are open sets;
(ii) $S \subset Q$ is a smooth 3-dimensional manifold, and $S_t := S \cap (\Omega \times \{t\})$ is a (possibly disconnected) smooth surface, for any $t \in]0, T[$;
(iii) θ is smooth in Q_1 and in Q_2, and $\partial\theta/\partial\nu$ exists on both sides of S;
(iv) the equations (1.4), (1.6), (1.7) and (1.8) are fulfilled;
(v) $\partial\theta/\partial\vec{\nu}$ equals a given field on $\Gamma_2 \times]0, T[$;
(vi) θ equals a given field on $\Omega \times \{0\}$ and on $\Gamma_1 \times]0, T[$;
(vii) $S \cap (\Omega \times \{0\})$ is prescribed.

If the temperature vanishes identically in one of the phases, a problem like this is often called a *one-phase* Stefan problem.

Notice that, although in our derivation we assumed local thermodynamical equilibrium, Problem 1.1 can also be formulated in presence of either *undercooled* or *superheated* regions, which are respectively characterized by liquid at $\theta < 0$ or solid at $\theta > 0$. These states are *metastable*.[9] The occurrence of such states can be excluded, by assuming natural sign conditions on the initial and boundary data. We occasionally refer to the latter setting as the *classical formulation of the Stefan problem*.

Ill-Posedness. We do not detail the preceding formulation, which in general is not well-posed. In fact the solution of Problem 1.1 may not be unique. For instance, if a negative temperature is imposed on a part of the fixed boundary in contact with the liquid phase, this model allows for undercooling of the liquid, as well as for nucleation of a new solid phase.

This problem may have no global-in-time solution, for it does not account for possible discontinuities in the evolution of the solid–liquid interface.[10] Discontinuities may occur in several ways: two connected components may merge, or conversely a connected component may split into two components; a new phase may appear (*nucleation*), or conversely a phase may vanish (*annihilation*); the order of connection may change, and so on. We see in the following that the interface may exhibit discontinuities also in the one-dimensional setting.

---

[8] Often one speaks of a problem in *several* space dimensions, since all of this discussion holds essentially unchanged in any number of dimensions.

[9] We illustrate this concept in Chaps. VI and VII.

[10] However, Meirmanov [385, 386] has proved that, under natural assumptions, a solution exists in a small time interval.

One-Dimensional Stefan Problem. Let us now assume that one of the dimensions of our system prevails over the others, so that Ω can be represented as one-dimensional: $\Omega =]a, b[$, say. Let $a < s^0 < b$, and at $t = 0$ let the interval $]a, s^0[$ ($]s^0, b[$, resp.) correspond to the solid (liquid, resp.) phase.

If we exclude the formation of new phases, the phase interface S coincides with the graph of a function $s : [0, T] \to [a, b]$ such that $s(0) = s^0$. Hence $x > s(t)$ in Q_1 and $x < s(t)$ in Q_2. Assume that

$$C_{Vi} \in C^0(\mathbf{R}), \quad k_i \in C^1(\mathbf{R}), \quad C_{Vi}, k_i > 0 \quad (i = 1, 2),$$
$$\theta_a, \theta_b \in C^0([0, T]), \quad \theta_a < 0, \quad \theta_b > 0, \quad a < s^0 < b, \quad (1.9)$$
$$\theta^0 \in C^0([a, b]), \quad \theta^0 < 0 \text{ in }]a, s^0[, \quad \theta^0 > 0 \text{ in }]s^0, b[.$$

The equations (1.4), (1.6), (1.7), and (1.8) coupled with natural initial and boundary conditions yield the following problem; cf. Fig. 1.

Problem 1.2 *(Strong Formulation of the One-Dimensional Two-Phase Stefan Problem) To find $s \in C^0([0, T]) \cap C^1(]0, T[)$ and $\theta \in C^0(\bar{Q})$ such that, setting*

$$Q_1 := \{(x, t) \in Q : x > s(t)\}, \quad Q_2 := \{(x, t) \in Q : x < s(t)\},$$

$\partial\theta/\partial t, \partial^2\theta/\partial x^2 \in C^0(Q_i)$ $(i = 1, 2)$, *the limits $\partial\theta/\partial x(s(t) \pm 0, t)$ exist for any $t \in]0, T[$, and*

$$C_{Vi}(\theta)\frac{\partial\theta}{\partial t} - \frac{\partial}{\partial x}\left[k_i(\theta)\frac{\partial\theta}{\partial x}\right] = 0 \quad \text{in } Q_i \ (i = 1, 2), \quad (1.10)$$

$$\left(k_1(\theta)\frac{\partial\theta}{\partial x}\right)(s(t) + 0, t) - \left(k_2(\theta)\frac{\partial\theta}{\partial x}\right)(s(t) - 0, t) = -L\frac{ds}{dt}(t) \quad (1.11)$$
$$\text{for } 0 < t < T,$$

$$\theta(s(t), t) = 0 \quad \text{for } 0 < t < T, \quad (1.12)$$

$$\theta(a, t) = \theta_a(t), \quad \theta(b, t) = \theta_b(t) \quad \text{for } 0 < t < T, \quad (1.13)$$

$$s(0) = s^0, \quad \theta(x, 0) = \theta^0(x) \quad \text{for } a < x < b. \quad (1.14)$$

Radially Symmetric Systems. Let us consider a radially symmetric system in $\mathbf{R}^3$, which consists of a solid ball (or shell) surrounded by liquid. Denoting the radial coordinate by r, we have $\theta = \theta(r, t)$. Here (1.4) reads

$$C_{Vi}(\theta)\frac{\partial\theta}{\partial t} - \frac{1}{r^2}\frac{\partial}{\partial r}\left[k_i(\theta)r^2\frac{\partial\theta}{\partial r}\right] = 0 \quad \text{in } Q_i \ (i = 1, 2), \quad (1.15)$$

and must be coupled with the condition $\partial\theta/\partial r(0, \cdot) = 0$ in $]0, T[$, if $0 \in \Omega$.

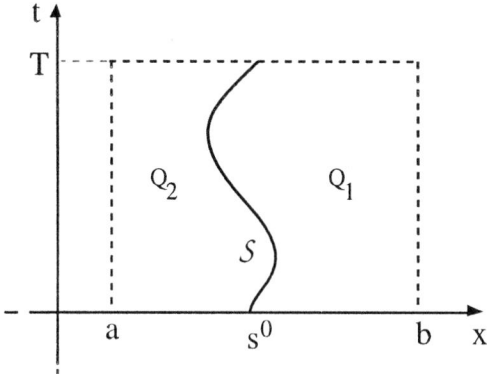

Figure 1. *One-Dimensional Two-Phase Stefan Problem.*

The free boundary is characterized by a scalar function, $r = s(t)$, and (1.11) is replaced by

$$\left(k_1(\theta)\frac{\partial \theta}{\partial r}\right)(s(t)+0,t) - \left(k_2(\theta)\frac{\partial \theta}{\partial r}\right)(s(t)-0,t) = -L\frac{ds}{dt}(t) \quad (1.16)$$

$$\text{for } 0 < t < T,$$

whereas (1.12) is unchanged. The initial and boundary conditions are obvious.

Phase Annihilation and Nucleation. Now we illustrate the onset of irregularity in the one-dimensional setting, in connection with phase *annihilation* or *nucleation*. For instance, let us consider annihilation of a solid phase. Let $s_1, s_2 : t \mapsto x$ and $\tilde{t} > 0$ be such that $s_1(t) < s_2(t)$ for any $t \in [0, \tilde{t}[$, $s_1(\tilde{t}) = s_2(\tilde{t})$, and let the space-time domain $\{(x,t) \in Q : s_1(t) < x < s_2(t)\}$ correspond to the solid phase.

If the graphs of s_1 and s_2 merge smoothly (i.e. parallely to the x-axis, cf. Fig. 2(a)), then the Stefan condition (1.6) yields

$$\left(k_1(\theta)\frac{\partial \theta}{\partial x}\right)(s_1(\tilde{t}-0), \tilde{t}-0) = -L\frac{ds_1}{dt}(\tilde{t}-0) = -\infty,$$

$$\left(k_2(\theta)\frac{\partial \theta}{\partial x}\right)(s_2(\tilde{t}-0), \tilde{t}-0) = -L\frac{ds_2}{dt}(\tilde{t}-0) = +\infty. \quad (1.17)$$

In order to get a smooth temperature gradient, by the Stefan condition both graphs of s_1 and s_2 must meet parallel to the t-axis, cf. Fig. 2(b), so that

$$\left(k_i(\theta)\frac{\partial \theta}{\partial x}\right)(s_i(\tilde{t}-0), \tilde{t}-0) = -L\frac{ds_i}{dt}(\tilde{t}-0) = 0 \quad (i=1,2). \quad (1.18)$$

The *generic* behaviour is illustrated in Fig. 2(c). In this setting the graphs of s_1 and s_2 merge forming a corner, and at that point the temperature gradient is discontinuous in space and in time.

Nucleation exhibits a similar behaviour, as is easy to see.

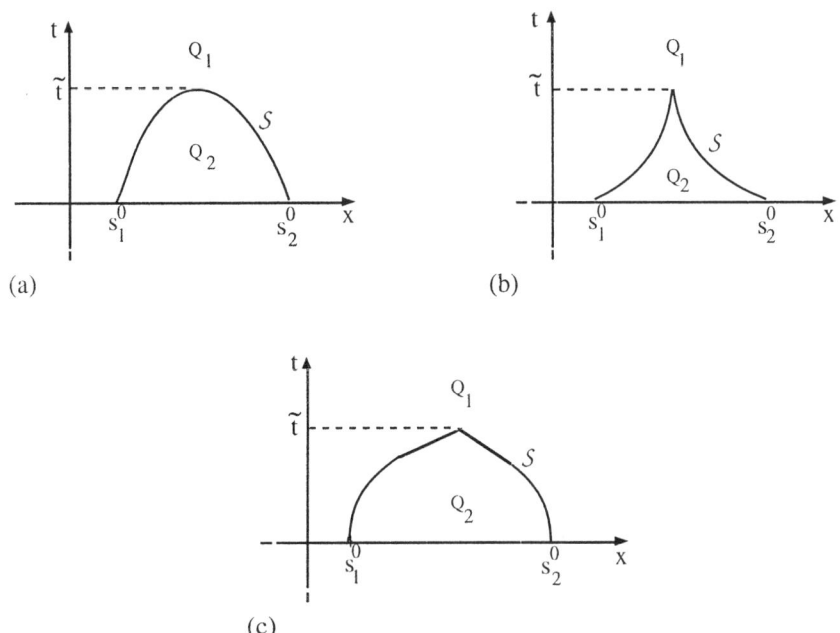

Figure 2. Possible behaviours of the interface at solid annihilation, in the one-dimensional setting. See text.

IV.2 Surface Tension

In this section we modify the Stefan problem, to account for surface tension. At first, we discuss the condition (1.8).

Undercooling and Superheating. So far we dealt with phase transition in a pure material, assumed local equilibrium, and neglected surface tension. If we drop these restrictions, then θ no longer vanishes at the solid–liquid interface, and (1.8) must be replaced by a law of the form

$$\theta = \theta_{\text{s.t.}} + \theta_{\text{n.e.}} + \theta_{\text{imp.}} \qquad \text{on } \mathcal{S}. \qquad (2.1)$$

The term $\theta_{\text{s.t.}}$ accounts for *surface tension*, and is proportional to the mean curvature of the interface; $\theta_{\text{n.e.}}$ is related to *nonequilibrium*, and depends on the rate of phase transition; $\theta_{\text{imp.}}$ is due to the presence of other components (so-called *impurities*). Here and in Chap. VIII we just deal with the term $\theta_{\text{s.t.}}$; $\theta_{\text{n.e.}}$ and $\theta_{\text{imp.}}$ are studied in Sects. V.1 and V.2.

By (2.1) and the continuity of the temperature at the interface, undercooling or superheating appears in the interior of the phases, *near* the interface. However, as we see in Chap. VII, undercooling and superheating may also occur in the bulk *far* from any interface, and be quantitatively very significant.

The Gibbs-Thomson Law. We want to include surface tension effects in our model. We assume that the phases are separated by a *smooth* interface S (the C^2-regularity with respect to the space variables suffices), and replace the condition (1.8) by the classical *Gibbs-Thomson law*

$$\theta = -\frac{2\sigma T_E}{L}\kappa \qquad \text{on } S. \tag{2.2}$$

σ is the surface tension coefficient, which is here supposed to be constant, like L; by κ we denote the *mean curvature* of S_t, which is assumed to be positive for a convex solid phase.

A similar law, named after Young and Laplace, applies to pressure. Denoting by p_i ($i = 1, 2$) the limit value of the pressure on S from the phase labelled by i, the negative of the pressure discontinuity $p_1 - p_2$ across the solid–liquid interface is proportional to the mean curvature:

$$p_1 - p_2 = -2\sigma\kappa \qquad \text{on } S. \tag{2.3}$$

In the following we do not use this equation, since we assumed both phases to be incompressible.

Contact Angle Condition. For any $(x,t) \in S \cap (\Gamma \times]0, T[)$, let us denote by $\omega(x,t)$ the angle formed by the normal to S, oriented towards the liquid phase $\Omega_1(t)$, and the outward normal to Ω at x. We then impose the *contact angle condition*

$$\cos\omega = h \qquad \text{on } S \cap (\Gamma \times]0, T[), \tag{2.4}$$

where $h \in [-1, 1]$ depends on the external material with which the system is in contact. In particular, $h = 0$ if Ω is surrounded by the vacuum.

In Sect. VI.4 we derive (2.2) and (2.4) by minimizing the free energy potential.

The surface tension has important consequences far from solid–liquid interfaces. In particular, it is responsible for the high undercooling which is required for solid *nucleation*, [11] as we see in Chap. VII.

---

[11] See e.g. Chalmers [129; Chap. 3], Flemings [240; Chap. 9], and Woodruff [583; Chap. 2].

Problem 2.1 *(Stefan-Gibbs-Thomson Problem)* [12] *To find $\theta \in C^0(\bar{Q})$ and a partition $\{Q_1, Q_2, S\}$ of Q such that*
  (i) Q_1 *and* Q_2 *are open sets;*
  (ii) $S \subset Q$ *is a smooth 3-dimensional manifold, and $S_t := S \cap (\Omega \times \{t\})$ is a (possibly disconnected) smooth surface, for any $t \in]0, T[$;*
  (iii) θ *is smooth in Q_1 and in Q_2, and $\partial \theta / \partial \nu$ exists on both sides of S;*
  (iv) *the equations (1.4), (1.6), (1.7), (2.2) and (2.4) are fulfilled;*
  (v) $\partial \theta / \partial \vec{\nu}$ *equals a given field on $\Gamma_2 \times]0, T[$;*
  (vi) θ *equals a given field on $\Omega \times \{0\}$ and on $\Gamma_1 \times]0, T[$;*
  (vii) $\overline{S} \cap (\Omega \times \{0\})$ *is prescribed.*

In Chap. VIII we amend this problem. In a radially symmetric system, the Gibbs-Thomson law (2.2) can easily be expressed in terms of the function that represents the evolution of the phase interface.

For a solid ball $\Omega_2(t) := \{x \in \mathbf{R}^3 : |x| < s(t)\}$ surrounded by a liquid phase $\Omega_1(t) := \{x \in \mathbf{R}^3 : s(t) < |x| < b\}$ ($b > 0$), (2.2) reads

$$\theta(s(t), t) = -\frac{2\sigma \tau_E}{L s(t)} \qquad \text{for } 0 < t < T. \tag{2.5}$$

This law must be coupled with (1.15), (1.16), the condition of continuity on S, and appropriate initial and boundary conditions.

IV.3 Length Scales and Mushy Region

Scaling. This is an important issue in modelling.

"In the process of scaling one attempts to select intrinsic reference quantities, so that each term in the dimensional equation transforms into the product of a constant dimensional factor, the *scale*, which closely estimates the term's order of magnitude, and a dimensionless factor of unit order of magnitude." "A function f is said to have *order of magnitude* 10^n, n being an integer, if $3 \cdot 10^{n-1} < \sup |f| < 3 \cdot 10^n$."
[13] To avoid proliferation of scales, we interpret this concept in a broader sense: quantities that differ by a multiplicative factor up to (say) 10^2 are regarded as having the same order of magnitude.

We are mainly concerned with *length* scales. However, in diffusion processes governed by parabolic equations, any length scale ξ determines a time scale τ, because of the scaling relation $\xi^2/\tau = $ constant.

---

[12] This is usually named the *Stefan problem with surface tension,* or *Stefan problem with Gibbs-Thomson law.* We suggest the preceding denomination, without implying that either Stefan or Gibbs or Thomson ever considered this problem!

[13] Adapted from Lin and Segel [347; p. 211-214].

Mesoscopic Length Scale. The characteristic length scale of surface tension phenomena is essentially determined by the surface tension coefficient σ, which depends on the material. Using the calorie [cal], the centimeter [cm], and the Kelvin degree [K] as measure units, for water at room pressure at about $0^0 C$ we have

$$\sigma \sim 1.8 \times 10^{-6} \text{ cal cm}^{-2}, \quad L \sim 80 \text{ cal cm}^{-3}, \quad \tau_E \sim 273 \text{ K}, \quad (3.1)$$

whence $2\sigma\tau_E/L \sim 1.2 \times 10^{-5}$ cm K.

We use Kelvin degrees for temperature, so that σ determines the *mesoscopic length scale* 10^{-5} cm. By this term we mean that this scale is intermediate between the macroscopic scale of laboratory experiments and that of molecular phenomena. The passage from the finer length scale to the macroscopic scale must be provided by an averaging procedure; see Sect. IX.1.

On the macroscopic length scale, the condition (1.8) is a good approximation of the Gibbs-Thomson law (2.2), because of the smallness of the coefficient $2\sigma\tau_E/L$. On the other hand, for *mesoscopic interfaces* it is important to deal with (2.2): whenever κ is of the order of 10^{-5} cm, the temperature becomes of order 1. As we see in Chaps. VII and VIII, this is especially evident in connection with phase nucleation.

At the length scale of 10^{-7} cm (i.e., almost at the atomic scale), a thin transition layer between solid and liquid phases can be distinguished. For instance, the so-called *phase-field model* deals with such a scale. [14] However, here we use the intermediate mesoscopic length scale at which the Gibbs-Thomson law can be assumed. This avoids any conceptual difficulty that might arise by using a continuous model at too tiny a length scale.

Mushy Region and Two-Scale Model. In view of the weak formulation of the Stefan problem which we introduce in the next section, let us define the *phase function*

$$\chi := 1 \quad \text{in } Q_1, \qquad \chi := -1 \quad \text{in } Q_2. \quad (3.2)$$

Setting $\varphi := (\chi + 1)/2$, this is obviously equivalent to $\varphi = 1$ in the liquid and $\varphi = 0$ in the solid. [15] Now we drop the requirement that the phases be separated by a sharp interface S, and allow χ to attain intermediate values between -1 and 1.

This can be interpreted according to the following *two-scale model*. It is assumed that at the mesoscopic length scale just pure phases (either liquid or solid) can be observed; that is, $\chi = \pm 1$. However, very fine solid–liquid mixtures are allowed to appear at the macroscopic scale. A set where this occurs is characterized by $-1 < \chi < 1$, corresponding to a liquid concentration $0 < \varphi < 1$, and is called a

---

[14] See e.g. the works quoted in Sect. VIII.6.
[15] We use χ rather than φ, since this simplifies some calculations in Chap. VIII. $-\varphi$ plays the role of an *order parameter*, in the sense in which this term is used in the Landau theory of phase transitions; see, e.g., Landau and Lifshitz [340].

mushy (or *slushy*) *region*. Zones of this sort may appear if there is a distributed heat source, or if the latent heat depends on space; see Sect. IV.9 for some references.

IV.4 Weak Formulation of the Stefan Problem

As we saw, either global-in-time existence and uniqueness of the strong solution of the Stefan problem may fail. This leads us to consider a *weak formulation* of the problem.

Here we allow for occurrence of a mushy region, which can be interpreted according to the two-scale model that we mentioned previously.

An Equation in the Sense of Distributions. Let us define $C_V(\theta, \chi)$ and $k(\theta, \chi)$ as follows:

$$\begin{cases} C_V(\theta, \chi) := C_{V1}(\theta)\frac{1+\chi}{2} + C_{V2}(\theta)\frac{1-\chi}{2}, \\ k(\theta, \chi) := k_1(\theta)\frac{1+\chi}{2} + k_2(\theta)\frac{1-\chi}{2} \qquad \forall \theta \in \mathbf{R}, \forall \chi \in [-1, 1]. \end{cases} \qquad (4.1)$$

It is immaterial how these functions are defined for $-1 < \chi < 1$; in fact, the mushy region is isothermal, and so no heat diffusion can occur there.

Proposition 4.1 *(Equation in the Sense of Distributions) Assume that:*
(i) $S \subset Q$ is a smooth 3-dimensional manifold, and $S_t := S \cap (\Omega \times \{t\})$ is a (possibly disconnected) smooth surface, for any $t \in]0, T[$;
(ii) $\theta \in C^0(\bar{Q})$, $\partial\theta/\partial t$, $\partial^2\theta/\partial x_i \partial x_j \in L^1(Q \setminus S)$, for $i, j = 1, 2, 3$, the function $L : \mathbf{R} \to \mathbf{R}$ is Lipschitz continuous; [16]
(iii) the trace of $\partial\theta/\partial\nu$ exists on both sides of S.
Let χ be defined as in (3.2). Then the system (1.4), (1.6), (1.7) is (formally) equivalent to (1.7) coupled with the equation [17]

$$C_V(\theta, \chi)\frac{\partial\theta}{\partial t} + \frac{L(\theta)}{2}\frac{\partial\chi}{\partial t} - \nabla \cdot [k(\theta, \chi)\nabla\theta] = 0 \qquad \text{in } \mathcal{D}'(Q). \qquad (4.2)$$

Proof. Let us denote by $\vec{n} := (\vec{n}_x, n_t) \in \mathbf{R}^4$ the unit vector field normal to S, oriented towards Q_1. Note that $n_t = -\vec{v} \cdot \vec{n}_x$, see the footnote after (1.6), and $\vec{n}_x$ is parallel to $\vec{\nu}$. The Stefan condition (1.6) can then be written in the form

$$[k_1(\theta)\nabla\theta_1 - k_2(\theta)\nabla\theta_2] \cdot \vec{n}_x = L(\theta)n_t \qquad \text{on } S. \qquad (4.3)$$

---
[16] Note that $L^1(Q \setminus S) \neq L^1(Q)$, since $\mathcal{D}'(Q \setminus S) \neq \mathcal{D}'(Q)$.
[17] See Sect. XI.1 for the definition of differentiation in the sense of distributions.

Let us denote the duality pairing between $\mathcal{D}'(Q)$ and $\mathcal{D}(Q)$ by $\langle \cdot, \cdot \rangle$. A simple calculation yields

$$\begin{aligned}
&\left\langle C_V(\theta, \chi) \frac{\partial \theta}{\partial t} + \frac{L(\theta)}{2} \frac{\partial \chi}{\partial t} - \nabla \cdot [k(\theta, \chi) \nabla \theta], \varphi \right\rangle \\
&= \iint_Q \left\{ C_V(\theta, \chi) \frac{\partial \theta}{\partial t} \varphi - \chi \frac{\partial}{\partial t} \frac{L(\theta)\varphi}{2} + k(\theta, \chi) \nabla \theta \cdot \nabla \varphi \right\} dx dt \\
&= \iint_{Q \setminus S} \left\{ C_V(\theta, \chi) \frac{\partial \theta}{\partial t} - \nabla \cdot [k(\theta, \chi) \nabla \theta] \right\} \varphi dx dt \\
&\quad + \int_S \left\{ L(\theta) n_t - \vec{n}_x \cdot [k_1(\theta) \nabla \theta_1 - k_2(\theta) \nabla \theta_2] \right\} \varphi dS \qquad \forall \varphi \in \mathcal{D}(Q).
\end{aligned} \qquad (4.4)$$

Notice that $\partial \theta / \partial t$ and $\nabla \theta$ are locally integrable; in fact, unlike $\partial \chi / \partial t$, these derivatives cannot exhibit any *Dirac-type* measure on S, as θ has been assumed continuous in Q.

The last two integrals of (4.4) vanish for any test function φ iff the heat equation (1.4) and (4.3) are fulfilled. □

More generally, one can consider a space dependent heat source or sink of intensity (i.e., heat produced per unit volume) $f(x,t)$, and replace (4.2) by

$$C_V(\theta, \chi) \frac{\partial \theta}{\partial t} + \frac{L(\theta)}{2} \frac{\partial \chi}{\partial t} - \nabla \cdot [k(\theta, \chi) \nabla \theta] = f \qquad \text{in } \mathcal{D}'(Q). \qquad (4.5)$$

Energy Balance. The weak equation (4.5) can be derived directly from physical principles, independently of the strong formulation.

We assume that the density of internal energy u is a known function of the state variables θ and χ, which is characteristic of the material. By this we mean that there exists a function $\hat{u} : \mathbf{R} \times [-1, 1] \to \mathbf{R}$ such that, for any process evolving through local equilibrium states,

$$u(x,t) = \hat{u}(\theta(x,t), \chi(x,t)) \qquad \forall (x,t) \in Q.$$

We assume that $\hat{u}$ is differentiable, and set

$$C_V(\theta, \chi) := \frac{\partial \hat{u}}{\partial \theta}(\theta, \chi), \quad L(\theta, \chi) := 2 \frac{\partial \hat{u}}{\partial \chi}(\theta, \chi) \qquad \forall (\theta, \chi) \in \mathbf{R} \times [-1, 1]. \quad (4.6)$$

The dependence of L on χ has a physical meaning only in presence of a mushy region. However, even in that case, it is usually neglected.

By (4.6), for any process at constant volume we have

$$du = C_V(\theta, \chi) d\theta + \frac{L(\theta)}{2} d\chi \qquad \text{in } Q. \qquad (4.7)$$

In presence of a distributed heat source f, the *energy balance* reads

$$\frac{\partial u}{\partial t} + \nabla \cdot \vec{q} = f \qquad \text{in } \mathcal{D}'(Q). \tag{4.8}$$

The equations (4.7), (4.8) and the Fourier law (1.2) yield (4.5) (here with $L(\theta, \chi)$ in place of $L(\theta)$).

Note that the preceding argument does not require the existence of any regular solid–liquid interface, and can be applied even if a *mushy region* is present. This encourages us to regard the weak formulation as more fundamental than the strong one, at variance with a customary viewpoint. [18]

The Temperature-Phase Rule. In the absence of internal sources, assuming obvious sign conditions on the initial and boundary data, by the maximum principle (1.4) and (1.8) yield the *temperature-phase rule*

$$\theta \geq 0 \quad \text{in } Q_1, \qquad \theta \leq 0 \quad \text{in } Q_2. \tag{4.9}$$

So *undercooling* and *superheating* effects are here excluded, because of the assumptions on the sign of the data, [19] and liquid and solid can coexist just at vanishing temperature:

$$\theta = 0 \qquad \text{where } -1 < \chi < 1. \tag{4.10}$$

Defining the *sign graph* as in (XI.5.3), the conditions (4.9) and (4.10) can be written in the equivalent form

$$\chi \in \text{sign}(\theta) \qquad \text{in } Q. \tag{4.11}$$

The inclusion (4.11) is equivalent to a constitutive law relating internal energy and temperature.

The system (4.5), (4.11) must be coupled with an initial condition for u and with boundary conditions either for θ or for its normal derivative. This constitutes the *weak formulation* of the *two-phase Stefan problem* in several space dimensions.

The presence of a source term in the energy equation (4.5) excludes the possibility of applying the maximum principle, which is at the basis of the temperature-phase rule (4.9). However the latter condition can still be assumed, if one allows new phases to be nucleated at the interior of those initially present. This may cause some (mainly formal) difficulties in the strong formulation, whereas it is immaterial in the weak formulation. As we discuss in Sect. IV.6, the strong and weak formulations may describe completely different processes.

---

[18] Here we refer to the case in which the strong solution exhibits neither undercooling nor superheating; otherwise the strong and the weak formulation are not equivalent; see Sect. IV.6.

[19] Note that, in the strong formulation, undercooling and superheating may appear in the interior of the phases (not at the interface).

Reformulation of the Problem. By (4.11), phase transition only occurs at $\theta = 0$; hence we can replace $L(\theta)$ by $L(0)$ in (4.5). Setting

$$\varphi(\xi) := \int_0^\xi C_V(\eta, \operatorname{sign}(\eta)) d\eta, \quad \psi(\xi) := \int_0^\xi k(\eta, \operatorname{sign}(\eta)) d\eta \quad \forall \xi \in \mathbf{R}, \quad (4.12)$$

(4.5) reads

$$\frac{\partial}{\partial t}\varphi(\theta) + \frac{L(0)}{2}\frac{\partial \chi}{\partial t} - \Delta\psi(\theta) = f \quad \text{in } \mathcal{D}'(Q). \quad (4.13)$$

Note that φ and ψ can be inverted, as $C_V > 0$ and $k > 0$, and $\operatorname{sign}(\theta) = \operatorname{sign}(\psi(\theta))$. Setting $\tilde{\theta} := \psi(\theta)$, the system (4.11), (4.13) is then equivalent to

$$\begin{cases} \dfrac{\partial \tilde{u}}{\partial t} - \Delta\tilde{\theta} = f & \text{in } \mathcal{D}'(Q), \\ \tilde{u} \in \alpha(\tilde{\theta}) := \varphi\left(\psi^{-1}(\tilde{\theta})\right) + \dfrac{L(0)}{2}\operatorname{sign}(\tilde{\theta}) & \text{in } Q, \end{cases} \quad (4.14)$$

α is a *maximal monotone graph* [20] of the form (II.1.2). Hence the weak Stefan problem can be formulated as Problem II.1.1, and the results of Chap. II can be applied.

Remark. In the one-dimensional setting that we introduced in the previous section, the phase function χ can be related to the interface function s, instead of the temperature. Thus (4.11) can be replaced by

$$\chi(x,t) \in \operatorname{sign}(x - s(t)) \quad \forall (x,t) \in Q. \quad (4.15)$$

This is still in the framework of a formulation of strong type, because of the occurrence of the interface function s. □

Exercises.

4.1 Provide a weak formulation of the one-dimensional Stefan problem, with (4.15) in place of (4.11). Then prove existence of a solution, by the usual procedure of approximation by time-discretization, derivation of a priori estimates, and passage to the limit.

4.2 Discuss the extension of the previous result to cases in which C_V or k or L are not assumed to be constant.

---

[20] Maximal monotone graphs are introduced in Sect. XI.5.

IV.5 On the Analysis of the Stefan Problem

As we said, the results of Chap. II can be applied to the analysis of the weak formulation of the Stefan problem. In this section we briefly deal with an aspect of the latter. As for the analysis of the strong formulation, we refer the reader to the monographs indicated in the *Book Selection,* in particular Meirmanov [388], as well as to the many papers that have been devoted to that subject.

Two-Phase Stefan Problem. The time-integral transformation can be applied to the weak formulation (4.5), (4.11) of the Stefan problem, under certain restrictions. Let us set

$$z(\cdot, t) := \int_0^t u(\cdot, \tau)d\tau, \quad F(\cdot, t) := \int_0^t f(\cdot, \tau)d\tau + w^0 \quad \text{in } Q, \quad (5.1)$$

and assume that C_V and k are (positive) constant. (If their dependence on the temperature and on the phase were taken into account, difficulties would arise in integrating the equation in time.) Let us define the sign graph as in (XI.5.3). The system (4.5), (4.11) is equivalent to

$$\begin{cases} C_V \dfrac{\partial z}{\partial t} + \dfrac{L}{2}\chi - k\Delta z = F \\ \chi \in \text{sign}\left(\dfrac{\partial z}{\partial t}\right) \end{cases} \quad \text{in } Q. \quad (5.2)$$

This equation can be assumed to hold pointwise, since the integration has removed the time derivative of the phase function. As the sign graph is the *subdifferential* of the absolute value function, (5.2) is also equivalent to the following *variational inequality:*

$$\iint_Q \left[\left(C_V \dfrac{\partial z}{\partial t} - k\Delta z - F\right)\left(\dfrac{\partial z}{\partial t} - v\right) + \dfrac{L}{2}\left(\left|\dfrac{\partial z}{\partial t}\right| - |v|\right)\right]dxdt \leq 0, \quad (5.3)$$
$$\forall v : Q \to \mathbf{R}.$$

Theorem II.5.1 can then be applied.

One-Phase Stefan Problem. As we saw, this problem is characterized by the fact that the temperature vanishes in one of the two phases (the solid, say). Solidification can be excluded, if $f \geq 0$ in Q and the initial and boundary data fulfill obvious sign conditions. If we exclude the initial presence of a liquid connected component surrounded by solid, we have $\theta > 0$ in the liquid phase Q_1; that is,

$$Q_1 = \left\{(x, t) \in Q : \dfrac{\partial z}{\partial t} > 0\right\}, \quad Q_2 = \left\{(x, t) \in Q : \dfrac{\partial z}{\partial t} = 0\right\}.$$

This entails that the interface is monotone. It is then easy to check that

$$Q_1 = \{(x,t) \in Q : z > 0\}, \qquad Q_2 = \{(x,t) \in Q : z = 0\},$$

that is, $\chi \in \text{sign}(z)$. Hence (5.2) is formally equivalent to [21]

$$\begin{cases} C_V \dfrac{\partial z}{\partial t} + \dfrac{L}{2}\chi - k\Delta z = F \\ \chi \in \text{sign}(z) \end{cases} \quad \text{in } Q, \tag{5.4}$$

$$\iint_Q \left[\left(C_V \dfrac{\partial z}{\partial t} - k\Delta z - F\right)(z-v) + \dfrac{L}{2}(|z|-|v|)\right]dxdt \leq 0 \tag{5.5}$$
$$\forall v : Q \to \mathbf{R}.$$

This variational inequality can be coupled with the initial condition "$z = 0$ in Ω," and either Dirichlet, Neumann, or mixed boundary conditions. By using the techniques of Chap. II, it is easy to prove that this problem is well-posed in the space $L^2(0,T;H^1(\Omega)) \cap H^1(0,T;H^{-1}(\Omega))$ if $f \in L^2(0,T;H^{-1}(\Omega))$, and in $H^1(0,T;L^2(\Omega)) \cap L^2(0,T;H^2(\Omega))$ if $f \in L^2(Q)$. These results, respectively, correspond to $\theta = \partial z/\partial t \in L^2(0,T;H^{-1}(\Omega))$ and $\theta \in L^2(Q)$.

An Obstacle Problem. Let us define the *maximal monotone multi-valued graph*

$$\gamma(\xi) := \begin{cases}]-\infty, 0] & \text{if } \xi = 0, \\ \{0\} & \text{if } \xi \geq 0, \end{cases}$$

and replace $\text{sign}(z)$ by $1 + \gamma(z)$ in $(5.4)_2$:

$$\begin{cases} C_V \dfrac{\partial z}{\partial t} + \dfrac{L}{2}\chi - k\Delta z = F \\ \chi \in 1 + \gamma(z) \end{cases} \quad \text{in } Q. \tag{5.6}$$

As γ is the subdifferential of $I_{\mathbf{R}^+}$ (the *indicator function* of $\mathbf{R}^1$: $I_{\mathbf{R}^+}(v) := 0$ if $v \geq 0$, $I_{\mathbf{R}^+}(v) := +\infty$ if $v < 0$), the latter inclusion is equivalent to the following variational inequality

$$\begin{cases} z \geq 0; \forall v : Q \to \mathbf{R}^+, \\ \iint_Q \left(C_V \dfrac{\partial z}{\partial t} + \dfrac{L}{2} - k\Delta z - F\right)(z-v)dxdt \leq 0 \end{cases} \tag{5.7}$$

---

[21] In the terminology of Duvaut and Lions [205], (5.3) and (5.5) are variational inequalities of second and first type, respectively.

and to the following *obstacle problem* as well

$$\begin{cases} z \geq 0 & \text{in } Q, \\ C_V \dfrac{\partial z}{\partial t} + \dfrac{L}{2} - k\Delta z + \dfrac{L}{2} - F \geq 0 & \text{in } Q, \\ z \left(C_V \dfrac{\partial z}{\partial t} + \dfrac{L}{2} - k\Delta z + \dfrac{L}{2} - F \right) = 0 & \text{in } Q. \end{cases} \quad (5.8)$$

We claim that, under suitable initial and boundary conditions, (5.4) is equivalent to (5.6). In fact, under obvious conditions on the data, (5.4) has a (unique) non-negative solution z. Let us denote by sign_+ (γ_+, respect.) the restriction of sign (γ, respect.) to $\mathbf{R}^+$. Since $\text{sign}_+ \subset 1 + \gamma_+$, the solution of (5.4) solves (5.6). The equivalence then follows from the uniqueness of the solution of the latter inclusion, which can easily be checked.

IV.6 Comparison between Strong and Weak Formulations

The basic difference between the *strong* formulation of the *Stefan problem* (S.S.P.), cf. Problem 1.1, and the *weak* formulation (W.S.P.), cf. (4.14), is that in the former it is assumed that the phases are separated by a *(smooth)* interface, whereas in the latter no interface is supposed to exist.

This terminology is customary but misleading, since the two problems may have different solutions. In fact, the solution of the S.S.P. can exhibit undercooling and superheating, but no mushy region; on the other hand, the W.S.P. can represent the occurrence of a mushy region, but neither undercooling nor superheating. As we saw, the W.S.P. can be derived from the S.S.P. whenever metastability is excluded; the converse holds under regularity properties and in the absence of mushy region. Schematically,

$$\text{S.S.P. and no metastability} \implies \text{W.S.P.};$$

under regularity conditions,

$$\text{W.S.P. and no mushy region} \implies \text{S.S.P.}.$$

In other settings, for example, quasilinear hyperbolic equations, the strong solution is always a weak solution, and the converse holds under regularity conditions. This is also the relation between the *strong equations* (1.4), (1.6), (1.7) and the *weak equations* (1.7), (4.2); either pair of equations accounts for the energy balance and the Fourier law in the whole system. But, despite the terminology,

in general, the S.S.P. and the W.S.P. are not

different formulations of the same problem.

This fact is related to the characterization of the phases:

in the S.S.P. the phases are determined globally by the interface,

in the W.S.P. the phases are characterized pointwise by the sign of θ.

This raises the questions:
(i) Is it possible to account for the mushy region in the S.S.P.?
(ii) Is it possible to include undercooling and superheating in the W.S.P.?

The first question leads to the formulation of a *three-phase problem*, which we now outline in the one-dimensional setting. The second question is crucial for most of the physically justified extensions of the Stefan model, as it appears by (2.1). This issue is studied in Chap. VIII, where surface tension effects are also inserted into the model.

One-Dimensional Three-Phase Problem. Let us consider a three-phase system occupying an interval $]a,b[$: a solid phase $]a, s_1(t)[$, a mushy region $]s_1(t), s_2(t)[$, and a liquid phase $]s_2(t), b[$, with $a \leq s_1(t) \leq s_2(t) \leq b$, for any $t \in]0, T[$.

As we said, the mushy region is partially crystallized. At local equilibrium, there the temperature gradient vanishes; hence the mush cannot advance through a pure phase, if no distributed heat sources or sinks are present and if the latent heat L is independent of x. Thus $(ds_1/dt)(t) \geq 0$ and $(ds_2/dt)(t) \leq 0$, as long as $s_1(t) < s_2(t)$.

At the solid–mush interface $\mathcal{S}_1$ of equation $x = s_1(t)$, the density of latent heat of crystallization equals the jump of internal energy $[1 + \chi(s_1(t) + 0, t)]L(0)/2$ ($\in [0, L(0)]$). As $(\partial \theta/\partial x)(s_1(t) + 0, t) = 0$, at this interface the Stefan condition (1.11) must be replaced by

$$k_2(0)\frac{\partial \theta}{\partial x}(s_1(t) - 0, t) = \frac{1 + \chi(s_1(t) + 0, t)}{2} L(0) \frac{ds_1}{dt}(t) \qquad \text{for } 0 < t < T. \quad (6.1)$$

Analogously, at the liquid–mush interface $\mathcal{S}_2$ of equation $x = s_2(t)$, the density of latent heat of fusion equals $[1 - \chi(s_2(t) - 0, t)]L(0)/2$ ($\in [0, L(0)]$), and the Stefan condition takes the form

$$k_1(0)\frac{\partial \theta}{\partial x}(s_2(t) + 0, t) = -\frac{1 - \chi(s_2(t) - 0, t)}{2} L(0) \frac{ds_2}{dt}(t) \quad (6.2)$$
$$\text{for } 0 < t < T.$$

One should also account for the possible merging of the free boundaries $\mathcal{S}_1$ and $\mathcal{S}_2$; see Fig. 3. This problem may be labelled as a formulation of *strong* type, since the two free boundaries appear explicitly. It can be extended to several dimensions of space. One can allow for the presence of a distributed heat source, and in this case the mushy region can also expand.

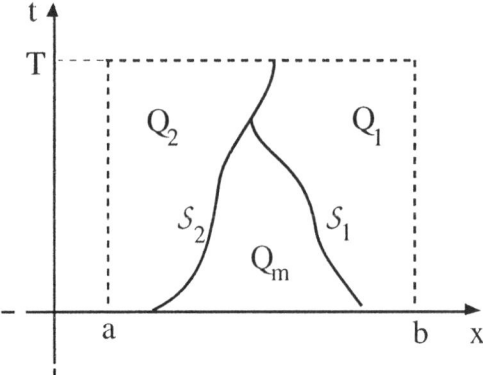

Figure 3. One-dimensional *three-phase Stefan problem.* Q_1, Q_2, and Q_m, respectively, represent the liquid, solid, and mushy phases.

A Highly Irregular Phase Interface. In presence of heat-distributed sources, in the W.S.P. the interface can degenerate into a three-dimensional region, as it appears in the following example, which is as simple as it is pathological.

Let us consider a solid system initially at a uniform temperature $\theta(\cdot, 0) = \theta^0 < 0$, that evolves under the action of a uniform heat source of intensity $f = 1$ (this can be accomplished by infrared radiation, for instance), with no heat flux across the fixed boundary Γ. Then the temperature remains uniform in Ω, and the equation (4.5) is reduced to the following O.D.E.:

$$C_V(\theta, \chi)\frac{d\theta}{dt} + \frac{L(\theta)}{2}\frac{d\chi}{dt} = 1 \qquad \text{in } [0, T], \tag{6.3}$$

which must still be coupled with (4.11).

As θ vanishes, melting starts and χ increases smoothly from -1 to 1, uniformly in Ω. Thus the whole Ω becomes a mushy region. As χ reaches the value 1, this mushy region becomes liquid; then the temperature increases again.

Thus, according to the W.S.P., for some time there is no phase interface; at some instant the interface appears in a highly degenerate form, invading the whole system; after some more time, the interface disappears instantaneously. The S.S.P. provides a different picture: here no mushy region can appear, θ becomes positive and increases indefinitely, yielding a superheated solid.

Comparison of Analytical Properties. The S.S.P. consists of nondegenerate equations set in an unknown domain; hence it is a genuine free boundary problem. On the other hand, in the W.S.P. the domain is fixed but the equation is degenerate.

The one-dimensional S.S.P. is well-posed. [22] Under natural regularity conditions, it has been proved by Meirmanov [385, 386] that in several space dimensions

---

[22] See, e.g., Meirmanov [388].

the S.S.P. has a solution in a *small* time interval, which depends on the data. However, if one excludes special settings, the solution of the S.S.P. fails after some time, even if the heat source term f vanishes identically. In fact discontinuities occur as the *topological properties* of the interface change. For instance, a connected component may split into two components, or conversely the latter may merge into a single one; a simply connected component may form a ring, or conversely; and so on. On the other hand the weak solution exists globally in time. As we saw, in generic cases discontinuities then occur for the temperature gradient, cf. Fig. 2(c).

The W.S.P. is of the form (II.1.2) (with α as in (II.1.1)), for which several results are known; see Chap. II. In particular, this problem is well-posed in any number of space dimensions, and regularity results hold. Some approximation methods used for the W.S.P. are not only natural from the mathematical viewpoint, but also reflect the physics of the phenomenon. For instance, in several materials phase transitions occur within a temperature range, [23] in which the heat capacity is very large but not infinite; this suggests smoothing the graph which represents the enthalpy versus temperature relation. The W.S.P. can also be solved numerically by standard techniques; however, the numerical determination of the moving interface requires special caution, since the equation is degenerate.

The Choice. Of course the main question should be: which one of the two models makes more sense?

This depends on the nucleation behaviour of the specific material. If nucleation occurs with negligible undercooling, a mushy region is formed consistently with the W.S.P.. On the other hand, if nucleation requires some undercooling, the temperature behaves as predicted by the strong model, until the nucleation threshold is attained. But for later times the physical evolution diverges from that prescribed by this model. Thus the two models represent extreme nucleation behaviours, and in Chap. IX we propose a different approach, based on the analysis of nucleation.

In conclusion, the S.S.P. and the W.S.P. are nonequivalent models of phase transitions. In the one-dimensional case the S.S.P. has good analytical properties, and can be extended to account for several physical effects. On the other hand, the W.S.P. is satisfactory in any number of dimensions of space. It would be suitable to dispose of a more general model, capable of representing intermediate nucleation behaviours.

---

[23] This behaviour is typical of organic substances. Actually, phase transitions have important applications to food preservation.

IV.7 The Muskat and Hele-Shaw Problems

In this section we outline two free boundary problems arising in fluid-dynamics, which exhibit some analogies with the Stefan model.

The Muskat Problem. Let a porous region $\Omega \subset \mathbf{R}^3$ be occupied by two immiscible viscous fluids, which are displaced by either injecting or extracting some fluid through the boundary. We label quantities relative to the more viscous (less viscous, respect.) fluid by $i = 1$ ($i = 2$, respect.), and use the following notation:

Q_i: (open) subset of Q corresponding to the fluid i,
$\Omega_i(t)$: region occupied by the fluid i at an instant $t \in [0, T]$,
$\mathcal{S} := \partial Q_1 \cap \partial Q_2$: (possibly disconnected) space-time manifold that separates the fluids,
$\mathcal{S}_t := \mathcal{S} \cap (\Omega \times \{t\})$: configuration of $\mathcal{S}$ at an instant $t \in [0, T]$,
$\vec{v}$: velocity of the fluid,
f: intensity of a distributed source or sink,
k_i: *mobility coefficient* (inversely proportional to the viscosity),
p_i: limit of p on $\mathcal{S}$ from the fluid labelled by i,
$\vec{\nu} \in \mathbf{R}^2$: unit vector field normal to $\mathcal{S}_t$.

We neglect the gravity force, and assume that the medium is saturated and that the fluids are incompressible. The *Darcy law*, the mass balance, and the law of conservation of momentum yield

$$\vec{v} = -k_i(p)\nabla p, \qquad -\nabla \cdot \vec{v} = f \qquad \text{in } Q_i(t) \, (i = 1, 2), \tag{7.1}$$

$$p_1 = p_2 \qquad \text{on } \mathcal{S}, \tag{7.2}$$

$$\vec{v} \cdot \vec{\nu} = -k_1(p_1)\nabla p_1 \cdot \vec{\nu} = -k_2(p_2)\nabla p_2 \cdot \vec{\nu} \qquad \text{on } \mathcal{S}. \tag{7.3}$$

The initial fluid configuration must also be specified:

$$\Omega_1(0) = \Omega_1^0 \quad \text{(prescribed subset of } \Omega\text{)}, \tag{7.4}$$

whence obviously $\Omega_2(0) = \Omega \setminus \overline{\Omega_1^0}$. On the other hand, there is no reason for specifying the initial pressure distribution, since no time derivative of this variable occurs in the problem.

The equations (7.1) through (7.4) must be coupled with appropriate boundary conditions, for instance, prescribing either the value p or the normal flux. This setting constitutes the *strong formulation* of the *Muskat problem*. [24] This is a two-phase free boundary problem, and in general is ill-posed. In fact, when the interface advances towards the less viscous fluid, it loses regularity, can develop *fingers,* and even break up into several connected components.

---

[24] This problem was studied by Muskat [411, 412] in the 1930s.

Let us set $\chi = 1$ in Q_1 and $\chi = -1$ in Q_2, as in (3.2), and define $k(p, \chi)$ as in (4.1)$_2$. By the procedure of Proposition 4.1, it is easy to see that (7.1) through (7.3) can be written in a weak form as

$$-\nabla \cdot [k(p,\chi)\nabla p] = f \qquad \text{in } \mathcal{D}'(Q). \tag{7.5}$$

However, here it is not clear how the fluids might be characterized in terms of the pressure, to complete the weak formulation of the problem.

The Verigin Problem. If the medium is unsaturated, the quasi-steady equation (7.1)$_2$ must be replaced by

$$\varphi \frac{\partial s}{\partial t} - \nabla \cdot \vec{v} = f \qquad \text{in } Q_i(t) \, (i = 1, 2), \tag{7.6}$$

where φ is the *porosity* (i.e., the average ratio between the volume of the pores and that of the medium), and s is the saturation (i.e., the ratio between the volume of the fluid and that of the pores). Neglecting hysteresis, the saturation can be regarded as a nondecreasing function of p: $s = \hat{s}(p)$ (we neglect the dependence of s from the fluid). This is known as the *Verigin problem*. [25]

In place of (7.5), here one gets the weak equation

$$\varphi \frac{\partial}{\partial t} \hat{s}(p) - \nabla \cdot [k(p,\chi)\nabla p] = f \qquad \text{in } \mathcal{D}'(Q). \tag{7.7}$$

The Hele-Shaw Problem. Let us consider two immiscible viscous fluids as previously, but now assume that they occupy a bounded nonporous region located between two slightly separated parallel plates, a so-called *Hele-Shaw cell*. If the interplate distance is small enough, we can represent this region by a two-dimensional set, which we still denote by Ω. Let us define S, $\Omega_i(t)$ and Q_i ($i = 1, 2$) as previously.

We assume that the viscosity of the fluid labelled by $i = 2$ (air, for instance) is so small that we can set $p = 0$ in Q_2; this yields a one-phase problem. Now let either some fluid be injected into the cell with a syringe, or the interplate distance be slightly reduced. This corresponds to $f \geq 0$; hence

$$p \geq 0 \qquad \text{in } Q_1, \tag{7.8}$$

and the more viscous fluid expands. As for the Muskat problem, this has the effect of regularizing the moving interface.

A suitable approximation of the equations of fluid-dynamics yields (7.1). Omitting the index 1, (7.2) and (7.3) then read

$$p = 0, \qquad \vec{v} \cdot \vec{\nu} = -k(p)\nabla p \cdot \vec{\nu} \qquad \text{on } S. \tag{7.9}$$

---

[25] In the one-dimensional setting, this problem was studied by Verigin [548] in 1954; existence of a classical solution was then proved by Kaminin [312].

The equations (7.1), (7.4), and (7.9) coupled with appropriate boundary conditions constitute the *strong formulation* of the *Hele-Shaw problem,* which is a one-phase free boundary problem. Defining the phase field χ as previously, (7.8) yields

$$\chi \in \text{sign}(p) \qquad \text{in } Q. \tag{7.10}$$

By the procedure of Proposition 4.1, it is easy to see that (7.1) and the second condition in (7.9) can be written in weak form as

$$\frac{1}{2}\frac{\partial \chi}{\partial t} - \nabla \cdot [k(p)\nabla p] = f \qquad \text{in } \mathcal{D}'(Q). \tag{7.11}$$

Setting $\chi^0 = 1$ in $\Omega_1(0)$ and $\chi^0 = -1$ outside, (7.4) is equivalent to

$$\chi(\cdot, 0) = \chi^0 \qquad \text{in } \Omega. \tag{7.12}$$

The equations (7.10) through (7.12) coupled with a boundary condition constitute the *weak formulation of the Hele-Shaw problem.* By the results of Chap. II, this problem is well-posed; in particular Theorems II.1.2, II.2.1, II.2.2, and II.2.3 can be applied.

If either some fluid is extracted from the Hele-Shaw cell or the two plates are slightly moved apart, then $f \leq 0$. In this case

$$p \leq 0 \qquad \text{in } Q_1, \tag{7.13}$$

the more viscous fluid contracts, and the moving interface loses regularity. Here we still have (7.11), but in the weak formulation (7.10) is replaced by

$$\chi \in \text{sign}(-p) \qquad \text{in } Q. \tag{7.14}$$

This problem is known as the *inverse Hele-Shaw problem,* since it is equivalent to a *backward* Hele-Shaw problem, and turns out to be ill-posed [26]

Surface Tension. The loss of regularity of the boundary when the less viscous fluid expands is contrasted by surface tension. At a mesoscopic length scale this can be accounted for by replacing (7.8) with a condition analogous to the Gibbs-Thomson law (2.2):

$$p = -\sigma \kappa \qquad \text{on } \mathcal{S}, \tag{7.15}$$

where σ is the surface tension coefficient, and κ the mean curvature of $\mathcal{S}$, which is assumed to be negative if the fluid boundary is convex. This also must be coupled with a contact angle condition like (2.4).

The equations (7.4), (7.11), and (7.15) coupled with boundary conditions constitute the *Hele-Shaw problem with surface tension.* This problem was proposed by

---

[26] See, e.g., DiBenedetto and Friedman [189].

Mullins and Sekerka as a model of solute diffusion in phase transitions in heterogeneous systems, [27] and is also known as the *Mullins-Sekerka problem*. Surface tension can also be introduced into the Muskat problem.

The Quasi-Steady Stefan Problem. The Hele-Shaw problem can be derived as a limit of the one-phase Stefan problem. If the heat capacity C_V is very small, one can replace the heat equation by the quasi-stationary equation

$$-\nabla \cdot [k(\theta, \chi)\nabla \theta] = f \qquad \text{in } Q_i \ (i = 1, 2). \tag{7.16}$$

In the weak formulation of either the two- or one-phase Stefan problem, one then gets (7.11) (with p replaced by θ, and different coefficients) in place of (4.5). As an initial condition one must then specify $\chi(\cdot, 0)$ instead of $u(\cdot, 0)$, cf. (7.12).

Exercise.

7.1 Which problem is obtained by assuming a *long* time scale in the Stefan problem?

Hint. Let t represent a reference time scale, set $\tau := t/\varepsilon$, and consider a Stefan problem on the *long* time scale τ. In view of the passage to the limit as $\varepsilon \to 0$, note that $\partial/\partial \tau = \varepsilon \partial/\partial t$.

IV.8 A Stefan-Type Problem Arising in Ferromagnetism

Phase transitions in solid–liquid systems and magnetic processes exhibit several interesting analogies. Here we outline a macroscopic model of ferromagnetism without hysteresis.

Macroscopic Ferromagnetism. In Sect. I.3, we already introduced the electromagnetic fields $\vec{H}$, $\vec{M}$, and $\vec{B} := \vec{H} + 4\pi \vec{M}$, and the system of *Maxwell's equations* (I.3.3) and (I.3.4), written using *Gauss units*. Let us assume the constitutive laws (I.3.5). Dealing with an electrically conducting material (e.g. a metal) at usual frequencies, the ϵ-term is much smaller than the σ-term. Hence one can neglect the *displacement current* term $\epsilon \partial \vec{D}/\partial t$ in (I.3.3). [28] Let us denote by Ω the region occupied by the system, and set $Q := \Omega \times]0, T[$. The *Ohm's law* (I.3.5)$_2$ then yields (I.3.8):

$$4\pi\sigma \frac{\partial \vec{B}}{\partial t} + c^2 \nabla \times \nabla \times \vec{H} = 4\pi c \sigma \nabla \times \vec{g} \qquad \text{in } Q. \tag{8.1}$$

---

[27] See Mullins and Sekerka [407, 408]. In this case θ represents the concentration of a component. See Sect. V.2 for a presentation of the model, and Sect. VIII.5 for the analysis of the problem.

[28] Although this statement is usually assumed in the physical literature (see, e.g., Landau and Lifshitz [339]), it would deserve a rigourous mathematical derivation. In the linear case it is easily proven. But if $\vec{B}$ and $\vec{H}$ are related by a nonlinear constitutive law, the argument does not seem obvious.

Besides, we have the law (I.3.4)$_2$: [29]

$$\nabla \cdot \vec{B} = 0 \quad \text{in } Q. \tag{8.2}$$

Let us assume that the fields $\vec{B}$ and $\vec{H}$ are related by a constitutive law of the form

$$\vec{B} \in \vec{\mathcal{F}}(\vec{H}) \quad \text{in } Q, \tag{8.3}$$

where $\vec{\mathcal{F}} : \mathbf{R}^3 \to 2^{\mathbf{R}^3}$ is a (possibly multi-valued) maximal monotone graph. For instance,

$$\vec{B} \in \vec{H} + 4\pi \mathcal{M} \vec{\alpha}(\vec{H}), \quad \text{in } Q, \tag{8.4}$$

where $\mathcal{M}$ is a positive constant and $\vec{\alpha}$ is the subdifferential of the modulus function:

$$\vec{\alpha}(\vec{v}) := \begin{cases} \{\frac{\vec{v}}{|\vec{v}|}\} & \text{if } \vec{v} \neq \vec{0}, \\ \{\vec{v} \in \mathbf{R}^3 : |\vec{v}| \leq 1\} & \text{if } \vec{v} = \vec{0}, \end{cases} \quad \forall \vec{v} \in \mathbf{R}^3; \tag{8.5}$$

cf. Fig. 4. For instance, this may represent the behaviour of soft iron for high field saturation. In this case, the unmagnetized and magnetically saturated phases are respectively characterized by $\vec{B} = \vec{0}$ and $|\vec{B}| \geq 4\pi \mathcal{M}$. In general, the occurrence of a mixed phase characterized by $0 < |\vec{B}| < 4\pi \mathcal{M}$ (a sort of *magnetical mushy region*) is not a priori excluded.

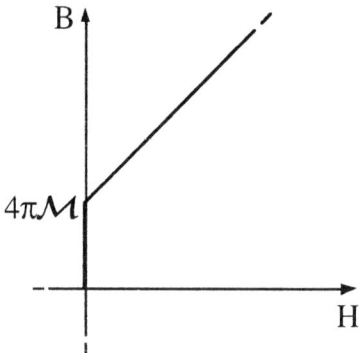

Figure 4. Constitutive relation between the moduli of the colinear vectors $\vec{H}$ and $\vec{B}$, for an (isotropic) ferromagnetic material with negligible hysteresis.

A Weak Formulation. By coupling (8.1) through (8.3), one obtains a *vectorial* quasilinear parabolic equation, which exhibits some analogies with the weak formulation of the Stefan problem, cf. (4.14). Here $\vec{H}$, $\vec{M}$, and $\vec{B}$ play similar roles

---

[29] This law obviously follows from (8.1), if it is assumed to hold at the initial instant.

to those of the scalar variables θ, χ and u, respectively, in the weak formulation of the Stefan problem.

If the system has planar symmetry, that is, all variables only depend on two space coordinates (x, y, say), and if the fields $\vec{H}$ and $\vec{B}$ are parallel to the z-axis, then they can be represented by scalars. In this case, the operator $\nabla \times \nabla \times$ equals $-\Delta$, the system (8.1) through (8.3) is reduced to (II.1.2), and the results of Chap. II can be applied.

In the general vectorial setting, one can also prove well-posedness of corresponding initial and boundary value problems, as well as several complementary results, by the techniques of Chap. II. Here a major difference from the scalar setting stays in the fact that the injection of the space $\{v \in L^2(\Omega; \mathbf{R}^3) : \nabla \times v \in L^2(\Omega; \mathbf{R}^3)\}$ into $L^2(\Omega; \mathbf{R}^3)$ is not compact. [30] Therefore standard compactness does not yield

$$\iint_Q \vec{B}_m \cdot \vec{H}_m \, dx \, dt \to \iint_Q \vec{B} \cdot \vec{H} \, dx \, dt, \qquad (8.6)$$

cf. (II.1.37), which is the basic property to pass to the limit in the nonlinear term, once L^2-estimates have been derived; cf. Lemma XI.5.1. However, by (8.2) one can derive (8.6) via *compensated compactness,* by applying the *div-curl lemma;* cf. Theorem XI.3.13. [31]

A Free Boundary Problem. If $\mathcal{F}$ is multi-valued, *formally* the system (8.1) through (8.3) is the weak formulation of a free boundary problem. In general the existence of an interface between the magnetically saturated and unsaturated phases is not obvious a priori, even under regularity hypotheses. However, under appropriate symmetry assumptions (e.g., planar symmetry) such a surface seems to exist.

Proposition 8.1 *(Discontinuity Conditions) Assume that:*

(i) $\mathcal{S} \subset Q$ *is a smooth 3-dimensional manifold, and* $\mathcal{S}_t := \mathcal{S} \cap (\Omega \times \{t\})$ *is a (possibly disconnected) smooth surface, for any* $t \in]0, T[$;

(ii) $\partial \vec{B}/\partial t, \nabla \times \nabla \times \vec{H} \in L^1(Q \setminus \mathcal{S}; \mathbf{R}^3)$;

(iii) the traces of $\vec{B}$ *and* $\nabla \times \vec{H}$ *exist on both sides of* $\mathcal{S}$.

Let $\vec{\nu} \in \mathbf{R}^3$ *be a unit vector field normal to* $\mathcal{S}_t$, *and* $v := \vec{v} \cdot \vec{\nu}$ *be the (normal) speed of* $\mathcal{S}_t$, *for any* $t \in [0, T]$. *Denote by* $[\![\cdot]\!]$ *the difference between the traces on the two sides of* $\mathcal{S}_t$. *Assume also that* [32]

(iv) $\vec{\nu} \times [\![\vec{H}]\!] = \vec{0}$ *a.e. on* $\mathcal{S}$.

Then the system (8.1), (8.2) in the sense of distributions is (formally) equivalent to the same equations pointwise in $Q \setminus \mathcal{S}$, *coupled with the discontinuity condition*

$$4\pi\sigma v [\![\vec{B}]\!] = c^2 \vec{\nu} \times [\![\nabla \times \vec{H}]\!] \quad \text{a.e. on } \mathcal{S}. \qquad (8.7)$$

---

[30] The former space is naturally associated with the operator $\nabla \times \nabla \times$, just as $H^1(\Omega)$ with $-\Delta$.

[31] See Damlamian [165].

[32] By the Ampère law (I.3.3)$_1$, the assumption (iv) is equivalent to the absence of any surface current.

116   The Stefan Problem

Proof. This argument can be compared with that of Proposition 4.1.

Let us denote by $\vec{n} := (\vec{n}_x, n_t) \in \mathbf{R}^4$ a unit vector field normal to $\mathcal{S}$. [33] As $n_t = -\vec{v} \cdot \vec{n}_x$ and $\vec{n}_x$ is parallel to $\vec{v}$, the condition (8.7) can be written in the form

$$4\pi\sigma n_t [\![\vec{B}]\!] + c^2 \vec{n}_x \times [\![\nabla \times \vec{H}]\!] = \vec{0} \qquad \text{a.e. on } \mathcal{S}. \tag{8.8}$$

Denoting by $\langle \cdot, \cdot \rangle$ the duality pairing between $\mathcal{D}'(Q; \mathbf{R}^3)$ and $\mathcal{D}(Q; \mathbf{R}^3)$, we have

$$\begin{aligned}
&\left\langle 4\pi\sigma \frac{\partial \vec{B}}{\partial t} + c^2 \nabla \times \nabla \times \vec{H}, \vec{\varphi} \right\rangle \\
&= \iint_Q \left\{ -4\pi\sigma \vec{B} \cdot \frac{\partial \vec{\varphi}}{\partial t} + c^2 \nabla \times \vec{H} \cdot \nabla \times \vec{\varphi} \right\} dx\, dt \\
&= \iint_{Q \setminus \mathcal{S}} \left(4\pi\sigma \frac{\partial \vec{B}}{\partial t} + c^2 \nabla \times \nabla \times \vec{H} \right) \cdot \vec{\varphi}\, dx\, dt \\
&\pm \int_{\mathcal{S}} \left\{ 4\pi\sigma n_t [\![\vec{B}]\!] + c^2 \vec{n}_x \times [\![\nabla \times \vec{H}]\!] \right\} \cdot \vec{\varphi}\, d\mathcal{S} \qquad \forall \vec{\varphi} \in \mathcal{D}(Q; \mathbf{R}^3);
\end{aligned} \tag{8.9}$$

the choice of the sign of the last term depends on the orientation of $\vec{n}$. Notice that $\nabla \times \vec{H}$ cannot exhibit any *Dirac-type* measure on $\mathcal{S}$, by the assumption (iv).

Therefore the equation (8.1) in the sense of distributions is equivalent to the same equations pointwise in $Q \setminus \mathcal{S}$ iff the last integral of (8.9) vanishes for any test function $\vec{\varphi}$, that is, iff (8.7) holds. □

A different model of ferromagnetism is outlined in Sect. IX.4.

Remarks. (i) By a similar procedure, from (8.2) one gets

$$\vec{v} \cdot [\![\vec{B}]\!] = 0, \qquad \vec{v} \cdot [\![\nabla \times \vec{H}]\!] = 0 \qquad \text{a.e. on } \mathcal{S} \tag{8.10}$$

(the former condition can also be derived from (8.7)). If $\vec{\mathcal{F}}$ is defined as in (8.4), then (8.7) yields

$$16\pi^2 \mathcal{M}\sigma v = c^2 \left| [\![\nabla \times \vec{H}]\!] \right| \qquad \text{a.e. on } \mathcal{S}. \tag{8.11}$$

(ii) The previous discussion can be extended to a constitutive relation of the form of Fig. 5, which is here proposed for (low temperature) *superconductivity*. Here H_c is the *critical field;* $|\vec{B}| = 0$, $0 < |\vec{B}| < H_c$, and $|\vec{B}| \geq H_c$, respectively, correspond to the superconducting, intermediate, and normal states. [34] □

---

[33] We use the arrow to denote vectors of $\mathbf{R}^3$ as well as $\mathbf{R}^4$.

[34] See, e.g., London [360], and Landau and Lifshitz [339; Chap. VI].

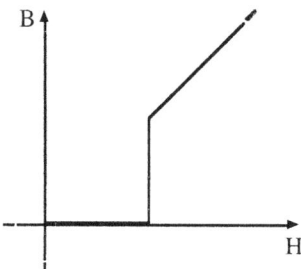

Figure 5. Constitutive relation between the moduli of the colinear vectors $\vec{H}$ and $\vec{B}$, for a low temperature superconducting material.

IV.9 On the History of the Stefan Problem

The mathematical literature dealing with phase transitions is so large as to discourage any attempt to review it exhaustively. Here we only select some landmarks and express some comments. For more detailed historical accounts, we refer to the monographs of Rubinstein [479], and Meirmanov [388], and to the surveys of Primicerio [456, 457], Danilyuk [173], Niezgódka [420], Magenes [373], Fasano [221], Tarzia [525], Rodrigues [472], and Oleĭnik, Primicerio and Radkevich [436]. Wilson, Solomon and Trent [579], Cannon [120], and Tarzia [527] are rich sources of references. The F.B.P. News [475] offer a capillary updating.

We confine ourselves to mathematical works, and neglect the large literature that deals with the physics of phase transitions. Actually, while physicists and material scientists have studied this large class of phenomena in much detail, for a long time mathematicians have been mainly concerned with the traditional basic Stefan formulation, whose analysis indeed exhibits several difficult and interesting features. However, in the last years this gap has been getting smaller.

In order to organize this presentation, rather artificially we distinguish few approaches, which originated in different periods but were then investigated in parallel.

The Pioneers. The first model of phase transition seems to be due to Lamé and Clayperon [337], and dates back to 1831. F. Neumann studied this problem in the 1860s. In 1889 the Austrian physicist Josef Stefan proposed a model for melting of the polar ices; in a series of papers [516], he dealt with several aspects of the one- and two-phase problems in a single dimension of space. This early analysis witnesses the importance Stefan recognized in this model, and explains why the problem has been named after him.

Rubinstein [480; p. 4] points out that "from 1889 to 1931 there was no serious publication devoted to the [Stefan] problem." Starting with the 1930s, several works were devoted to phase transitions. In particular in 1931 Leĭbenzon [346] proposed the replacement of the heat equation by the corresponding quasi-steady elliptic equation. [35] Actually, in several cases this approach offers an acceptable approximation, and is still often used for some applications. In 1939, Huber [304] proposed a method of solution of the one-phase one-dimensional problem; this was based on approximation of the free boundary by a piecewise linear function of time, determined by the Stefan condition at nodal instants.

So far explicit solution had been derived for special settings, but no result of existence of a solution for *large* classes of data had been proved.

Rigourous Analysis of the One-Dimensional Problem. In 1947, Rubinstein [479] formulated the one-dimensional two-phase Stefan problem in terms of a system of integral equations, and proved existence and uniqueness of a solution in a small time interval. He also studied several related questions. Other formulations of the one- and two-phase problems via various integral equations were then considered by several authors, G.W. Evans [214], Sestini [499], Friedman [247], Kolodner [328], Jiang [310], and others. Different techniques were based on the Huber method (see Fasano and Primicerio [223]), on approximation via a time delay in the Stefan condition (see Cannon and Hill [123]), and so on.

The well-posedness for large time of the two-phase problem in a single dimension of space was proved in several ways, jointly with approximation, regularity results, information on the asymptotic behaviour, and several other properties. See, for example, Cannon and Hill [123], Friedman [248; Chap. 8], [249, 250, 252], Schaeffer [496], Fasano, Primicerio and Kamin [229], and Rubinstein, Fasano and Primicerio [485]. Physically motivated generalizations of the problem were also studied in the one-dimensional setting; see, for example, Fasano and Primicerio [224] for a rather general result.

Weak Formulation in Several Dimensions. Around 1960, Kamenomostskaya [311] and Oleĭnik [435] proposed the weak formulation of the Stefan problem for multi- and one-dimensional systems, respectively, and proved existence and uniqueness of the solution. Of course this was also related to the parallel development of research on weak solutions of P.D.Es, to which the second of these authors contributed much. This approach is equivalent to the so-called *enthalpy formulation,* which was already well-known to engineers and physicists. [36] Related models had been considered by Tikhonov and Samarskiĭ [534], and by Albasiny [5].

Despite its physical foundation, this new approach needed quite some time to be fully accepted by a number of mathematicians, who regarded the strong solution as the only *legitimate* one. Studies on the strong formulation of the one-dimensional problem progressed in parallel to those on the weak formulation in several dimensions of space. The weak formulation was also considered in a one-dimensional

---

[35] As we saw, this corresponds to the *Hele-Shaw problem.*

[36] Indeed, as we saw in Sect. IV.4, the weak formulation can be directly derived from basic physical laws.

setting, and its relation with the strong Stefan problem was investigated; see, for example, Douglas, Cannon and Hill [201], Cannon, Henry and Kotlow [122], and Friedman [249, 250].

In the early 1970s, a free boundary problem issued from porous medium filtration, the so-called *dam problem,* was also attracting attention. This problem had been known for a long time to engineers, but it was only in 1971 that Baiocchi [44, 45] proposed his integral transformation; this allowed him to reformulate the problem as a variational inequality, a subject whose study was flourishing in that period. In this way he could prove existence and uniqueness of the solution. [37]

In 1973-74, Duvaut [203, 204] and Frémond [244] applied an analogous transformation to the Stefan problem: integrating the temperature in time, they formulated the multi-dimensional one- and two-phase problems as variational inequalities. By the discussion of Sect. II.5, the reader already knows that for the two-phase problem this approach is essentially equivalent to the enthalpy formulation.

Analysis of Regularity. The onset of weak formulations stimulated investigations about the regularity of the weak solution of the multi-dimensional Stefan problem, especially in the late 1970s and early 1980s. Results on the regularity of the free boundary for the multi-dimensional one-phase Stefan problem were obtained by using the variational inequality formulation by Friedman and Kinderlehrer [255] and Caffarelli [106, 107]. Eventually Kinderlehrer and Nirenberg [321, 322] were able to prove that the weak solution is also strong, under appropriate restrictions. Continuity of the temperature was proved by Caffarelli and Friedman [109] for the one-phase problem, and by DiBenedetto [187, 188], Ziemer [596], and Caffarelli and L.C. Evans [108] for the two-phase problem.

In 1979, Meirmanov [385, 386] proved existence of the strong solution of the multi-dimensional two-phase Stefan problem in a small time interval. An analogous result was then shown by Hanzawa [289] for the one-phase problem by using a different technique, based on the Nash-Moser regularity theory.

Since the early 1980s, these results stimulated the study of *mushy regions;* see Meirmanov [387], Primicerio [458], and Showalter [503]. In 1974-75, Atthey [34, 35] had already shown that regions of that sort may appear in the presence of a distributed heat source. On the other hand, Berger and Rogers [71] proved that no mushy region is formed in several dimensions, in the absence of any distributed heat source and if the coefficients are constants. Götz and Zaltzman [271] showed that the mushy region does not expand, under natural conditions. Other results were then obtained, for example, by Fasano and Primicerio [226, 227, 228]; see also Primicerio and Ughi [459], and the survey [221] of Fasano.

Recent studies on the regularity are due to DiBenedetto and Vespri [192], Athanassopoulos, Caffarelli and Salsa [31, 32, 33], and others.

---

[37] See also Baiocchi and Capelo [47], as well as other monographs on variational inequalities quoted in the *Book Selection,* and the proceedings of the meeting on free boundary problems.

Numerical Aspects. The analysis of the numerical methods of solution of the various formulations of the Stefan problem has been developed parallel to that of more theoretic aspects. In particular, it is of high theoretical and practical importance to compute efficiently the motion of the interface, and to estimate the approximation error in the temperature approximation.

Finite element methods have been used by Elliott [209], Nochetto and Verdi [429, 430], Paolini, Sacchi and Verdi [442], Verdi [544], and others. Adaptive finite element methods have been studied, for example, by Nochetto, Paolini and Verdi [425 – 428].

For more information we refer to the surveys of Furzeland [258], Magenes [373], Meyer [390, 391], Nochetto [423, 424], and Verdi [546].

Beyond the Stefan Model. One should also mention that a series of conferences devoted to free boundary problems have been regularly organized since the mid-1970s. [38] In the last ten years or so, activity on phase transition problems has also been organized in the framework of international projects.

Those meetings gathered mathematicians, material scientists, engineers, and other researchers in applied physics. On those occasions, much concern was devoted to phase transitions and related problems. Interaction with applied scientists was especially fruitful, and drove the attention of mathematicians into several physically justified generalizations of the Stefan model. [39]

Since about the middle of the 1980s, attention has also been attracted by new models, such as kinetic laws, phase relaxation, the Cahn-Hilliard equation for phase separation, the phase-field model, surface tension effects, mean curvature flow, and so on. The exigence also emerged of dealing with *free surfaces,* namely, unknown surfaces that are not set boundaries, and with lower dimensional free manifolds as well.

Special concern has been devoted to undercooling and to *microstructural* aspects. This also reflects an increasing collaboration between mathematicians and applied scientists, as well as the development of new tools, for example, *viscosity solutions* and *Young measures,* in other areas of applied analysis.

This considerably broadens the horizon of research on phase transitions, and we conclude here this brief account, emphasizing that the landscape of mathematical research on phase transition models is much broader than it may appear from this brief survey.

---

[38] See the *Book Selection* for the corresponding proceedings.

[39] In this respect, see also Friedman [254].

IV.10 Comments

Phase transitions are of major importance in physics and engineering. For instance, a large engineering literature deals with metal solidification and crystal growth.

The model of phase transition that we outlined in this chapter is quite simplified (like other models that we consider in this book). For instance, stress and deformation in the solid are neglected, as well as convection in the liquid, change of density, and so on. [40] In the next chapter, this model is amended to account for nonequilibrium at the solid–liquid interface, and is extended to heterogeneous systems, such as binary alloys. In Chaps. VI through IX we deal with surface tension and related phenomena like phase nucleation.

In this chapter and in most of the mathematical literature on phase transitions, solidification and melting are described as qualitatively similar phenomena, and are represented in a *symmetrical* form. This is contradicted by the experience: melting appears a rather regularizing process, whereas solidification may yield a variety of morphologies. This asymmetry is especially evident at the *mesoscopic* length scale, and is largely related to *metastability;* on the contrary, the traditional Stefan model deals with the macroscopic scale, and assumes local stability. [41] We briefly discuss this issue in Sect. VII.2, dealing with nucleation.

Phase Transition Phenomena. Phase transitions occur in several processes of physical and industrial interest, and can be represented by models that are closely related to the Stefan problem. For instance, they include:

(i) Monocrystal growth (see, e.g., Almgren, Taylor and Wang [9], Almgren and Wang [10], Crank and Ockendon [155], Langer [341], and Rubinstein [482, 483, 484]).

(ii) Continuous casting (see, e.g., Rodrigues [470], [471; Sect. 9.4]).

(iii) Soil freezing (see, e.g., Frémond [244], Wheeler [576]).

(iv) Solid-solid phase transitions (see, e.g., Brokate and Sprekels [98; Chap. 8], Hömberg [299], Verdi and V. [547], and V. [557]).

(v) Phase transitions in polymers (see, e.g., Andreucci et al. [20, 21], Astarita [28], and Fasano [222]).

(vi) Phase transitions in systems with *concentrated capacities;* (see, e.g., Fasano, Primicerio and Rubinstein [230], Magenes [374, 375, 376], Rubinstein [481], Rubinstein, H. Geiman, M. Shachaf [486], and Shillor [502]).

(vii) Phase transitions coupled with mass diffusion (which we consider in the next chapter) or with other diffusion processes (see, e.g., Bossavit [77, 78], V. [555]).

Stefan type problems also arise in modelling *thermal welding, ablation, thermal switches, food conservation,* and so on. There is a large technical literature on these

---

[40] Convection in phase transitions has been studied, e.g., by Cannon and DiBenedetto [121], DiBenedetto and Friedman [190], and Rodrigues [474].

[41] However, the one-phase Stefan problem with undercooling has been dealt with, e.g., by DiBenedetto and Friedman [189].

problems.

The *Hele-Shaw cell* was devised by Hele-Shaw in 1897 to model two-dimensional flow through porous media. This setting was then studied, for example, by Saffman and Taylor [493]. The problem of injection of fluid into a Hele-Shaw cell, which we outlined in Sect. IV.6, was considered, for example, by Richardson [465, 466], Elliott and Janovski [211], and Čižek and Janovski [159]. Injection moulding is used as an industrial process.

A metal body can be either machined or formed, by using it as an anode in an electrolytic cell. A simplified model of this process is equivalent to the Hele-Shaw problem; see, for example, McGeough [382], McGeough and Rasmussen [383], Elliott [208], and Rodrigues [471; Sect. 9.5]. Accounts of these and other free boundary problems of industrial interest can be found, for example, in Elliott and Ockendon [212], Crank [154; Sect. 2.12], Friedman [254], and in the proceedings listed in the *Book Selection*.

Apparently the model of ferromagnetism without hysteresis which we outlined in Sect. IV.8 has not yet been systematically investigated, despite its obvious relevance for applications. In particular, this author is not aware of any result concerning the existence of an interface between the magnetically saturated and unsaturated phases. For related problems, see, for example, Bossavit [77 – 80], Bossavit and Damlamian [81], Bossavit and Vérité [83], Damlamian [165], and V. [555, 551].

Chapter V. Generalizations of the Stefan Problem

Outline

Two features that determine the onset of *undercooling* and *superheating* in phase transitions are studied here:

(i) Nonequilibrium at solid–liquid interfaces. Here this is described by so-called *phase relaxation,* which also provides a model for glass formation.

(ii) Nonhomogeneity, that is, the presence of a second chemical component. This is described at first by coupling the equations of thermal and material diffusion, and then in the framework of the theory of *nonequilibrium thermodynamics.* The latter approach is also outlined in a more general setting, including the presence of several components and chemical reactions.

These problems can be treated by applying results and techniques of Chaps. II and III.

The general role of free energy in thermodynamical evolution is briefly examined on the basis of the second principle.

Prerequisites. Calculus and basic notions of thermodynamics are applied. Some simple variational inequalities are considered in Sect. V.1.

V.1 Kinetic Undercooling and Phase Relaxation

Like the previous chapter, the present one is devoted to introducing some models of phase transitions. In this section we amend the classical Stefan problem, and replace the local equilibrium condition (IV.1.8) by a dynamical law.

First Mode: Directional Solidification (or Columnar Growth). What drives phase transition? Looking at the classical Stefan problem (see Sect. IV.1), one might guess that phase transition is due to absorption or release of latent heat at the solid–liquid interface. This interpretation is not physically correct, and here we intend to modify the Stefan model accordingly.

The local equilibrium condition (IV.1.8) is not precise, since phase transition is driven either by *undercooling* or by *superheating*. "If the interface is not at the equilibrium temperature, then either melting or solidification occurs, at a rate which increases with the difference between the actual temperature and the equilibrium temperature". [1] This implication can also be inverted: phase transition occurs only if the solid–liquid interface is not at the equilibrium temperature. So phase transition is triggered by deviation from the equilibrium temperature, and exchange of latent heat at the interface is the effect (not the source) of phase transition.

Let us consider a one-dimensional system, that is, a system in which space symmetries allow us to reduce the number of dimensions to a single one. On account of the statement we previously quoted, we replace (IV.1.12) by the *kinetic law*

$$\dot{s}(t) + \gamma(\theta(s(t), t)) = 0; \qquad (1.1)$$

cf. Fig. 1. The *kinetic function* γ depends on the material. In several cases one can assume that

$$\gamma : \mathbf{R} \to \mathbf{R} \text{ is continuous and strictly increasing, } \gamma(0) := 0. \qquad (1.2)$$

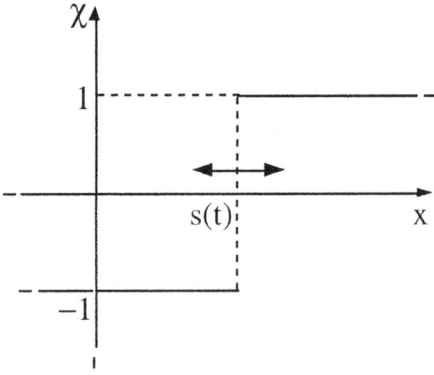

Figure 1. *Directional solidification* for a one-dimensional system.

Examples of Kinetic Laws. We mainly refer to solidification, which is physically more interesting than melting, and more important in applications. The form of the function γ depends on the interface structure; [2] here are some examples:

$$\gamma_1(\theta) = a\theta, \qquad \gamma_2(\theta) = -b \exp\left(\frac{c}{\theta}\right), \qquad \gamma_3(\theta) = -d\theta^2 \qquad \text{on } \mathcal{S}; \qquad (1.3)$$

here it is assumed that $\theta < 0$, and a, b, c, d are positive constants that depend on the material.

Under the action of undercooling, an atomically smooth solid interface grows by formation of successive monoatomic (or monomolecular) layers. Each of these layers is at first nucleated as a small *island* lying over the previous one. This process of *two-dimensional nucleation* exhibits several analogies with three-dimensional

---

[1] From Chalmers [129; p. 91].

[2] See, e.g., Chalmers [129; Chap. 2], and Flemings [240; Chap. 9].

nucleation, which is discussed in Sect. VII.2. Then the layer spreads over the interface by *lateral growth*.

At low nucleation rates, the normal velocity of the interface is proportional to the undercooling (namely, $-\theta$). At higher nucleation rates, the velocity is a nonlinear, rapidly increasing function of the undercooling. For instance, if the solid interface exhibits *screw dislocations*, the normal velocity is proportional to the square of the undercooling; on the other hand, for rough interfaces and small undercooling, the velocity is proportional to the undercooling. [3]

If $\gamma'(0) \neq 0$ and the deviation from the equilibrium temperature is small, one can use the corresponding linearized law

$$\alpha \dot{s}(t) + \theta(s(t), t) = 0; \quad (1.4)$$

here $\alpha := 1/\gamma'(0)$ is a positive relaxation coefficient, and its order of magnitude depends on the material and on the time scale. Often α is so small that the local equilibrium condition (IV.1.8) can be assumed.

So one can deal with Problem IV.1.2, with (IV.1.12) replaced either by (1.1) or by the linearized law (1.4). This system represents a *one-dimensional two-phase Stefan problem with kinetic law*. In the metallurgical literature, this mode of solidification is called *directional solidification*, and the corresponding undercooling is often referred to as *kinetic undercooling*.

Second Mode: Equiaxed Solidification (or Phase Relaxation). As we saw, in three space dimensions the weak formulation of the Stefan problem is well-posed, unlike the strong formulation. This suggests replacing (IV.1.8) by a nonequilibrium condition written in terms of the phase function χ.

At first, note that the condition (IV.4.11) can be equivalently written in the form

$$\text{sign}^{-1}(\chi) \ni \theta \qquad \text{in } Q. \quad (1.5)$$

It is then just natural to consider the relaxation law

$$a \frac{\partial \chi}{\partial t} + \text{sign}^{-1}(\chi) \ni \theta \qquad \text{in } Q, \quad (1.6)$$

where a is a positive coefficient; cf. Fig. 2. More generally, we can choose a function γ fulfilling (1.2), and consider the equation

$$\frac{\partial \chi}{\partial t} + \text{sign}^{-1}(\chi) \ni \gamma(\theta) \qquad \text{in } Q. \quad (1.7)$$

Here we have not written the constant a, since it can be included in the function γ by dividing the inclusion by a. Indeed $(1/a)\text{sign}^{-1} = \text{sign}^{-1}$ for any $a > 0$.

---

[3] See Chalmers [129; Chap. 2], Flemings [240; Chap. 9], and Woodruff [583; Chap. 8]. Most of the physical statements of this section are based upon these texts.

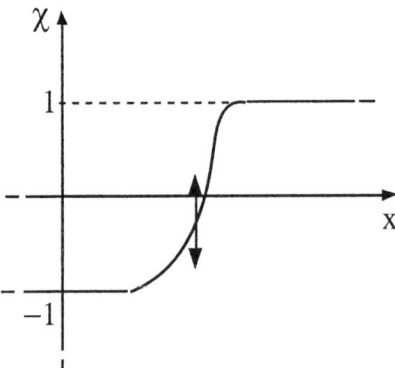

Figure 2. *Phase relaxation*, or *equiaxed solidification*, for a one-dimensional system.

The inclusion (1.7) is equivalent to the following variational inequality:

$$\begin{cases} -1 \leq \chi \leq 1 & \text{in } Q, \\ \left(\dfrac{\partial \chi}{\partial t} - \gamma(\theta)\right)(\chi - v) \leq 0 & \forall v \in [-1, 1], \text{ in } Q, \end{cases} \quad (1.8)$$

that is, [4]

$$\begin{cases} -1 \leq \chi \leq 1 & \text{in } Q, \\ \dfrac{\partial \chi}{\partial t} = \begin{cases} \gamma(\theta)^+ & \text{where } \chi = -1, \\ \gamma(\theta) & \text{where } -1 < \chi < 1, \\ -\gamma(\theta)^- & \text{where } \chi = 1, \end{cases} & \text{in } Q. \end{cases} \quad (1.9)$$

In the metallurgical literature, this mode of phase transition is called *equiaxed solidification*. Here phase transition is independent of the temperature gradient, and solidification occurs isotropically.

A Weak Formulation. Let us assume that (1.2) is fulfilled, that C_V, k are positive continuous functions, that $\theta^0, \chi^0 : \Omega \to \mathbf{R}$, $|\chi^0| \leq 1$, and $\hat{\theta} : \partial\Omega \times]0, T[\to \mathbf{R}$ are (sufficiently regular) given fields. Then we can state a *formal* problem.

Problem 1.1 *(Stefan Problem with Phase Relaxation)* To find $\theta, \chi : Q \to \mathbf{R}$ such that

$$C_V(\theta, \chi)\dfrac{\partial \theta}{\partial t} + \dfrac{L(\theta)}{2}\dfrac{\partial \chi}{\partial t} - \nabla \cdot [k(\theta, \chi)\nabla \theta] = f \quad \text{in } \mathcal{D}'(Q), \quad (1.10)$$

---

[4] The conditions (1.6), (1.7), and (1.8), are written only *formally* (we refer to Sect. I.2 for an explanation of the meaning of this term). In (1.9), $\partial\chi/\partial t$ is the time derivative from the right, that is, the limit of the incremental ratio for *positive* increments. In fact, χ may be piecewise of class C^1.

$$\frac{\partial \chi}{\partial t} + \text{sign}^{-1}(\chi) \ni \gamma(\theta) \quad \text{in } Q, \tag{1.11}$$

$$\theta = \hat{\theta} \quad \text{on } \partial\Omega \times]0, T[, \tag{1.12}$$

$$\theta(\cdot, 0) = \theta^0, \quad \chi(\cdot, 0) = \chi^0 \quad \text{in } \Omega. \tag{1.13}$$

This is not a free boundary problem, since (1.11) provides some regularity for $\partial \chi/\partial t$, and then (1.10) can be assumed to hold pointwise. For instance, if C_V, k, L are constant, then well-posedness can be proved by means of the techniques of Chap. II. [5] In particular, if γ has affine growth at infinity, a priori estimates can be *formally* derived by multiplying (1.10) by θ, (1.11) by $\partial \chi/\partial t$, and then summing these identities. If γ is Lipschitz continuous, further estimates can be obtained by multiplying (1.10) by $\partial \theta/\partial t$, the time derivative of (1.11) by $\partial \chi/\partial t$, and then summing these identities. In either case, further regularity can be derived for θ by comparing the terms of (1.10).

Comparison of the First and Second Mode. The laws (1.1) and (1.7) describe different evolution modes, although both represent relaxation towards (local) equilibrium. The equation (1.1) describes motion of the interface separating two pure phases, without formation of any mushy region. On the other hand, the second mode represents phase transition by formation of a mushy region, and (1.7) describes the evolution of the liquid concentration in that zone. Therefore these two modes are, respectively, associated with the strong and weak formulations of the Stefan problem in a rather natural way.

Third Mode. Directional and equiaxed growth are the basic modes of solidification of a pure material. For instance, in casting metal at first an equiaxed zone is formed in contact with the wall of the mould. Then a columnar region moves towards the interior, while in the remainder of the liquid, nucleation occurs and an equiaxed solid phase grows, until the two phases impinge on and eventually occupy the whole volume. [6] This is schematically represented in Fig. 3.

On the basis of this experimental evidence, we propose another mode that synthesizes the two previous ones. Like directional solidification, this mode is easily written for one-dimensional systems.

Let us consider a two-phase system: a solid phase $]a, s(t)[$ and a liquid phase $]s(t), b[$, where $a < s(t) < b$ in $]0, T[$. If the liquid phase is undercooled, the interface $x = s(t)$ moves to the right. Moreover, we assume that a mushy region is formed ahead of the interface, and χ evolves by a relaxation law such as (1.7).

Therefore the interface actually moves through a mushy region, where the the density of latent heat equals $[1 + \chi(s(t) + 0, t)]L/2$ ($\in [0, L]$). Then the Stefan equation (IV.1.11) and the equilibrium condition (IV.1.12) must be respectively replaced by

---

[5] See, e.g., V. [550], and Kenmochi [316].

[6] See, e.g., Flemings [240; Chap. 5], and Kurz and Fisher [331; Sect. 1.1.2].

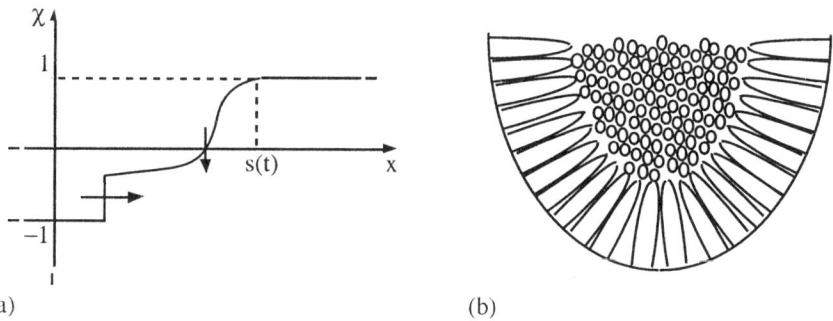

Figure 3. Combined *columnar* and *equiaxed solidification* for a one-dimensional system in (**a**). Schematic representation of crystal grown from an undercooled liquid in a vessel in (**b**): the solid columns advanced from the border, and impinged on the equiaxed grains which formed in the bulk.

$$\left(k_1(\theta)\frac{\partial\theta}{\partial x}\right)(s(t)+0,t) - \left(k_2(\theta)\frac{\partial\theta}{\partial x}\right)(s(t)-0,t)$$
$$= -\frac{1+\chi(s(t)+0,t)}{2}L\frac{ds}{dt}(t) \qquad \text{for } 0 < t < T, \tag{1.14}$$

and by the kinetic law

$$\frac{1+\chi(s(t)+0,t)}{2}\dot{s}(t) + \gamma(\theta(s(t),t)) = 0 \qquad \text{for } 0 < t < T. \tag{1.15}$$

In fact, we know that the mushy region consists of a very fine mixture of the two phases. As the moving front meets a liquid part, we can assume that it moves according to the law (1.1). On the other hand, as it encounters a solid part, the latter simply coalesces with the expanding solid phase; so there the interface moves instantaneously. As the liquid fraction equals $[1 + \chi(s(t)+0,t)]/2$, one gets (1.15). (The precise derivation of (1.14) and (1.15) is proposed as an exercise.) [7]

Incidentally, note that χ can be eliminated from (1.14) and (1.15); this yields

$$\left(k_1(\theta)\frac{\partial\theta}{\partial x}\right)(s(t)+0,t) - \left(k_2(\theta)\frac{\partial\theta}{\partial x}\right)(s(t)-0,t) = -L\gamma(\theta(s(t),t))$$
$$\text{for } 0 < t < T. \tag{1.16}$$

Glass Formation. A glass is an undercooled solid phase, which retains a large part of the latent heat of phase transition. The solid behaviour is caused by high viscosity, which in turn is due to the undercooling. A glass is *amorphous:* despite the low temperature, its crystal structure is not complete, since the viscosity reduces the mobility of particles in their migration to reach the crystal sites. Therefore

---

[7] See V. [553] for the analysis of this problem.

glasses contain only a part (although usually a large one) of the latent heat of phase transition. "At extremely rapid cooling rates, say $10^5 - 10^6$ Ks^{-1}, rather than forming solid crystal, a glass is produced. An important consequence of this is that under these conditions almost no latent heat of solidification needs to be absorbed". [8] Even metal glasses can be produced in this way. However, polymers are more prone to form a glass, as we see in Sect. VII.2.

Glass formation can be represented by means of a nonlinear kinetic law of the form (1.1), where the function γ is as follows:

$$\gamma : \mathbf{R} \to \mathbf{R} \text{ is continuous, and such that, for a certain } \tilde{\theta} < 0,$$
$$\gamma = 0 \quad \text{in }]-\infty, \tilde{\theta}], \quad \gamma < 0 \quad \text{in }]\tilde{\theta}, 0[, \quad \gamma > 0 \quad \text{in }]0, +\infty[; \quad (1.17)$$

see Fig. 4.

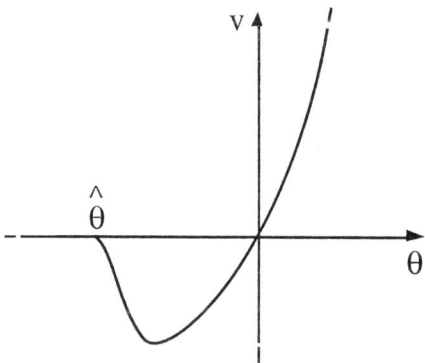

Figure 4. Kinetic law for glasses.

Let us apply (1.1), with γ as in (1.17), in a *quenched* (i.e., rapidly undercooled) one-dimensional system. The phase interface S advances through the liquid, but it stops as θ drops below $\tilde{\theta}$. A glassy phase is then formed in the undercooled zone ahead of S; there equiaxed growth does not occur, since nucleation is inhibited by high viscosity. In this way we get a crystal in contact with a glass.

For a three-dimensional system, we can consider (1.7), still with γ as in (1.17). Here after quenching a mushy region appears, in which crystals are formed and grow; but as θ drops below $\tilde{\theta}$ the undercooled liquid is transformed into a glass. This leaves a fine mixture of crystal and glass.

Radial Case with Surface Tension. For large velocities of the interface S_t, the Gibbs-Thomson law (IV.2.2), which represents (local) equilibrium condition, should be replaced by a *relaxation dynamics*.

---

[8] From Szekely [518].

Let us denote by v_S the normal velocity of S_t, assumed positive for melting, and consider the following *(mean) curvature flow* (with forcing term): [9]

$$av_S = \theta + \frac{2\sigma T_E}{L}\kappa \qquad \text{on } S, \tag{1.18}$$

where a is a positive relaxation constant. For instance, let the system be radially symmetric, $S := \{(x,t) \in Q : |x| = s(t)\}$ with $s : [0,T] \to \mathbf{R}^+$, and the solid phase coincide with $\{(x,t) \subset Q : |x| < s(t)\}$. Then (1.18) becomes

$$a\dot{s}(t) + \theta(s(t),t) = -\frac{2\sigma T_E}{Ls(t)} \qquad \text{for } 0 < t < T. \tag{1.19}$$

In more general geometries, (1.19) should be written in a weak form in terms of the phase function χ, and coupled with the energy conservation equation (IV.4.5). [10] In glasses, (1.18) should be replaced by

$$v_S = \gamma\left(\theta + \frac{2\sigma T_E}{L}\kappa\right) \qquad \text{on } S, \tag{1.20}$$

with γ as in (1.17).

Exercises.

1.1 On the model of Problem IV.1.2, provide a strong formulation of the one-dimensional Stefan problem with kinetic undercooling, namely, with (1.1) in place of (IV.1.12).

1.2 Provide a weak formulation of Problem 1.1 in Sobolev spaces, on the model of Problem II.1.1. Then assume that C_V, k, L are constant, and prove existence of a solution, along the lines outlined in the text.

1.3 Discuss the extension of the previous results to cases in which either C_V or k or L are not constant.

1.4 Derive equations (1.14) and (1.15).

1.5 Provide a precise formulation for a strong Stefan problem in radial symmetry, cf. Sect. IV.1.2, with the equation (1.19) of mean curvature flow in place of the equilibrium condition at the interface.

---

[9] See Sects. VII.6 and VII.7.
[10] The problem of coupling phase transition with mean curvature flow of the interface (with forcing term) is under current investigation; see, e.g., V. [567, 568].

V.2 Phase Transition in Two-Component Systems

In this section we extend the Stefan model to phase transitions in heterogeneous materials.

The Mass Diffusion Equation. We consider a *binary alloy;* that is, a homogeneous mixture of two components, which are soluble in each other in all proportions in both phases, outside a critical range of temperature. [11] Here *homogeneity* means that the constituents are intermixed on the atomic length scale to form a single phase, either solid or liquid. We regard one of the two components as the *solute,* for instance that with the lower solid–liquid equilibrium temperature. [12]

We label by 1 and 2 quantities relative to the liquid and solid phases, respectively, and use the following notation:

Q_i: (open) subset of Q corresponding to the phase i,
$S := \partial Q_1 \cap \partial Q_2$: (possibly disconnected) space-time manifold that separates the phases,
$S_t := S \cap (\Omega \times \{t\})$: configuration of S at an instant $t \in [0, T]$,
u: density of internal energy,
θ: relative temperature,
$\vec{q}$: heat flux (per unit surface),
$C_{V_i}(\theta)$: heat capacity per unit volume (namely, the heat needed to increase the temperature of a unit volume by one degree),
$k_i(\theta)$: thermal conductivity,
$L(\theta)$: density of latent heat of phase transition (namely, the heat exchanged by phase transition of a unit volume),
c: concentration of the solute,
$\vec{j}$: flux of solute (per unit surface),
$D_i(c)$: *mass diffusivity* in the phase i.

In the solid the latter coefficient is rather small, but not vanishing.
By the principle of mass conservation and the *Fick law,* we have

$$\frac{\partial c}{\partial t} = -\nabla \cdot \vec{j} \quad \text{in } Q_i, \tag{2.1}$$

$$\vec{j} = -D_i(c)\nabla c \quad \text{in } Q_i, \tag{2.2}$$

for $i = 1, 2$. These laws yield the *mass diffusion equation* in the interior of each phase:

$$\frac{\partial c}{\partial t} - \nabla \cdot [D_i(c)\nabla c] = 0 \quad \text{in } Q_i \ (i = 1, 2). \tag{2.3}$$

---

[11] Besides *completely miscible* substances, there exist compounds that exhibit a *miscibility gap.* See, e.g., Astarita [29; Sect. 9.3], Lupis [369; Sect. 3.4], and Kittel and Krömer [325; Chap. 11].

[12] However, in applications it is usual to choose the material with smaller concentration as the solute.

We use the following additional notation, for $i = 1, 2$:
$\vec{j}_i$: mass flux (per unit surface) through S contributed by the phase i,
c_i: limit of c on S from the phase i,
$\vec{v}$: (normal) velocity of S_t,
$\vec{\nu} \in \mathbf{R}^3$: a unit vector normal to S_t.
On S, we have $\vec{j}_i \cdot \vec{\nu} = c_i \vec{v} \cdot \vec{\nu}$, for $i = 1, 2$; hence

$$\vec{j}_2 \cdot \vec{\nu} - \vec{j}_1 \cdot \vec{\nu} = (c_2 - c_1)\vec{v} \cdot \vec{\nu} \qquad \text{on } S. \tag{2.4}$$

By (2.2), denoting by $\partial c_i / \partial \nu$ the normal derivative of c taken from the phase i ($i = 1, 2$), (2.4) yields the following discontinuity condition:

$$D_1(c_1)\frac{\partial c_1}{\partial \nu} - D_2(c_2)\frac{\partial c_2}{\partial \nu} = (c_2 - c_1)\vec{v} \cdot \vec{\nu} \qquad \text{on } S. \tag{2.5}$$

Temperature versus Concentration Diagrams. As in Sect. IV.1, the temperature θ must fulfill the following conditions

$$C_{Vi}(\theta)\frac{\partial \theta}{\partial t} - \nabla \cdot [k_i(\theta)\nabla \theta] = 0 \qquad \text{in } Q_i \ (i = 1, 2), \tag{2.6}$$

$$k_1(\theta)\frac{\partial \theta_1}{\partial \nu} - k_2(\theta)\frac{\partial \theta_2}{\partial \nu} = -L(\theta)\vec{v} \cdot \vec{\nu} \qquad \text{on } S, \tag{2.7}$$

$$\theta \text{ is continuous across } S. \tag{2.8}$$

Notice the analogy between the balance laws (2.3) and (2.6) in the interior of the phases, and between the discontinuity conditions (2.5) and (2.7) at the interface. The one-dimensional formulation of (2.3) and (2.5) is also analogous to (IV.1.10) and (IV.1.11).

We assume that the two components have different phase transition temperatures. Then the transition temperature of the mixture depends on the concentration, and (IV.1.8) is not fulfilled.

At equilibrium, the temperature is continuous across the interface, whereas the concentration has a discontinuity; see Fig. 5. On S, the temperature and the two limits c_1 and c_2 of the concentration from either phase are related as follows

$$\theta = \eta_1(c_1) = \eta_2(c_2) \qquad \text{on } S, \tag{2.9}$$

where η_1 and η_2 are known functions such that

$$\begin{gathered} \eta_i \in C^1([0,1]), \quad \eta_i' < 0 \ (i = 1, 2), \quad \eta_2 \leq \eta_1, \\ \eta_1(0) = \eta_2(0) = 0, \quad \eta_1(1) = \eta_2(1) = \hat{\theta}, \end{gathered} \tag{2.10}$$

where $\hat{\theta}$ is a negative constant; see Fig. 5.

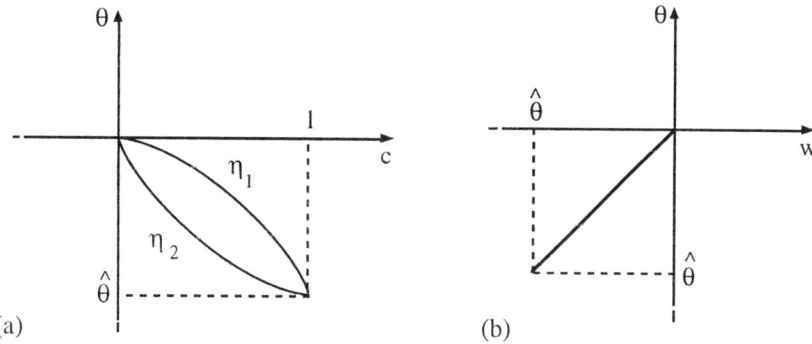

Figure 5. Constitutive law relating temperature and concentration at the interface at equilibrium in (a); analogous law relating temperature and the transformed variable w (defined in (2.12)) in (b). The graphs of η_1 and η_2 are traditionally called *liquidus* and *solidus*, respectively.

The temperature is continuous across $\mathcal{S}$, whereas there the concentration has a jump. The region between the two graphs represents either metastable or unstable states. In this model these states are not accessible, because of the assumption of (stable) equilibrium.

At equilibrium, we have

$$\theta \geq \eta_1(c) \quad \text{in } Q_1, \qquad \theta \leq \eta_2(c) \quad \text{in } Q_2. \tag{2.11}$$

The states characterized by $\eta_2(c) \leq \theta \leq \eta_1(c)$ are not stable. [13]

As previously, we assume that the process occurs in *local equilibrium*, in the sense that we mentioned in Sect. IV.1. [14]

Problem 2.1 (*Strong Formulation of the Three-Dimensional Problem of Phase Transitions in Binary Mixtures*) *To find $\theta, c \in C^0(\bar{Q})$ and a partition $\{Q_1, Q_2, \mathcal{S}\}$ of Q such that:*

(i) Q_1 *and* Q_2 *are open sets;*

(ii) $\mathcal{S} \subset Q$ *is a smooth 3-dimensional manifold, and* $\mathcal{S}_t := \mathcal{S} \cap (\Omega \times \{t\})$ *is a (possibly disconnected) smooth surface, for any* $t \in {]0, T[}$;

---

[13] If in a single phase system the variables are forced to attain such values (e.g., by a rapid change of the temperature), then a secondary phase nucleates and grows.

Asymptotically, then the two phases reach the respective concentrations $c_i = \eta_i^{-1}(\theta)$ $(i = 1, 2)$. The proportion between the phases is such that the weighted average of the concentrations equals the original one, so that the mass is conserved. However, due to difference of densities, the solid phase migrates, and precipitates to the bottom of the container in the form of crystals.

This process of *phase separation* in binary alloys is also known as *spinodal decomposition*, and is described by the classical *Cahn-Hilliard equation* (VII.6.2); see, e.g., Cahn [116], and Cahn and Hilliard [117].

[14] Several other simplifications are implicitly introduced. "We neglect all cross effects in diffusion and heat transfer as well as all thermal effects of mechanical origin. Finally, we neglect the change of heat flux terms when one passes from the system of the centre of mass to the system of the average volume velocity." (from Rubinstein [481; p. 261])

(iii) θ and c are smooth in Q_1 and Q_2;
(iv) the equations (2.3), (2.5) through (2.9) are fulfilled;
(v) θ and c attain given values on $\Omega \times \{0\}$ and on $\partial\Omega \times]0, T[$, and $\overline{S} \cap (\Omega \times \{0\})$ is also prescribed.

A Transformation of Variable. On account of the discontinuity of c across the interface, we introduce the new variable

$$w := \eta_i(c) \, (\in [\hat{\theta}, 0]) \qquad \text{in } Q_i \ (i = 1, 2). \tag{2.12}$$

In the linearized formulation that we consider later on, w is proportional to the *chemical potential*. [15] By (2.9) and (2.11),

$$w \text{ is continuous across } \mathcal{S}, \qquad w = \theta \quad \text{on } \mathcal{S}, \tag{2.13}$$

$$\theta \geq w \quad \text{in } Q_1, \qquad \theta \leq w \quad \text{in } Q_2. \tag{2.14}$$

Setting $\zeta_i := \eta_i^{-1}$, for $i = 1, 2$, we have

$$c = \zeta_i(w) \qquad \text{in } Q_i \ (i = 1, 2), \tag{2.15}$$

$$\nabla c = \zeta_i'(w) \nabla w \qquad \text{in } Q_i \ (i = 1, 2). \tag{2.16}$$

As $\zeta_i' < 0$, we set

$$\hat{D}_i(w) := -D_i(w)\zeta_i'(w) \, (> 0) \qquad \forall w \in [\hat{\theta}, 0] \, (i = 1, 2). \tag{2.17}$$

Weak Formulation. We set $\chi := -1$ in Q_2 and $\chi := 1$ in Q_1, as in (IV.3.2). By Proposition IV.4.1, under the assumption (2.8), the system (2.6), (2.7) is formally equivalent to

$$C_V(\theta, \chi)\frac{\partial \theta}{\partial t} + \frac{L(\theta)}{2}\frac{\partial \chi}{\partial t} - \nabla \cdot [k(\theta, \chi)\nabla \theta] = 0 \qquad \text{in } \mathcal{D}'(Q). \tag{2.18}$$

Here in place of (IV.4.11) we have

$$\chi \in \text{sign}(\theta - w) \qquad \text{in } Q. \tag{2.19}$$

The equation (2.12) can also be written in the form

$$c = \zeta_1(w)\frac{\chi + 1}{2} + \zeta_2(w)\frac{1 - \chi}{2} \qquad \text{in } Q. \tag{2.20}$$

---

[15] See Crank [154; Sect. 6.2.7].

Let us define the specific heat $C_V(\theta, \chi)$ and the thermal conductivity $k(\theta, \chi)$ as in (IV.4.1), and similarly set

$$\hat{D}(w, \chi) := \hat{D}_1(w)\frac{\chi+1}{2} + \hat{D}_2(w)\frac{1-\chi}{2} \qquad \forall w \in [\hat{\theta}, 0], \forall \chi \in [-1, 1]. \quad (2.21)$$

Proposition 2.1 *(Weak Formulation) Assume that:*
  (i) $S \subset Q$ is a smooth 3-dimensional manifold, and $S_t := S \cap (\Omega \times \{t\})$ is a (possibly disconnected) smooth surface, for any $t \in]0, T[$;
  (ii) $\partial c/\partial t, \partial^2 c/\partial x_i \partial x_j \in C^0(\bar{Q} \setminus S)$, for $i, j = 1, 2, 3$;
  (iii) the trace of $\partial c/\partial \nu$ exists on both sides of S.
  Define w as in (2.12), and set $\chi := -1$ in Q_2, $\chi := 1$ in Q_1. Then the system (2.3), (2.5) is equivalent to the following equation in the sense of distributions [16]

$$-\frac{\partial}{\partial t}\left(\zeta_1(w)\frac{\chi+1}{2} + \zeta_2(w)\frac{1-\chi}{2}\right) - \nabla \cdot [\hat{D}(w, \chi)\nabla w] = 0 \quad \text{in } \mathcal{D}'(Q). \quad (2.22)$$

The argument is analogous to that of Proposition IV.4.1. Notice that the equation (2.22) is *forward parabolic*, as $\zeta_1', \zeta_2' < 0$.

The equations (2.18), (2.19), and (2.22) (for the unknown functions θ, w, χ) coupled with appropriate initial and boundary conditions, constitute the *weak formulation* of the *problem of phase transition in binary mixtures*.

Linearized Constitutive Laws. If the solute concentration c is *small* (as often occurs in practice), it is possible to linearize the η_is, that is, to replace (2.9) by

$$\theta = \eta_i'(0)c_i =: -\frac{1}{r_i}c_i \qquad \text{on } S \ (i = 1, 2), \quad (2.23)$$

with $0 \leq r_2 < r_1$; cf. Fig. 6. Then we set

$$w := -\frac{1}{r_i}c_i \ (\leq 0) \qquad \text{in } Q_i \ (i = 1, 2), \quad (2.24)$$

and from (2.20) get

$$c = -r_1 w \frac{\chi+1}{2} - r_2 w \frac{1-\chi}{2} \qquad \text{in } Q. \quad (2.25)$$

Although the linearization only applies for *small* values of c, here the range of c is assumed to be the whole $\mathbf{R}^+$, which corresponds to $w \leq 0$.

---

[16] We refer to Sect. XI.1 for the definition of *distributions* and of the corresponding derivatives.

136   V. Generalizations of the Stefan Problem

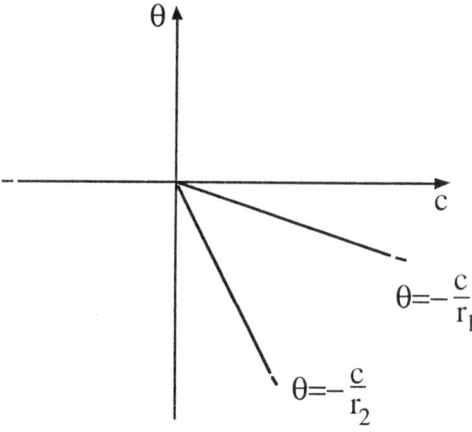

Figure 6. Linearized θ versus c constitutive laws.

The *thermal equations* (2.18) and (2.19) (for prescribed w), and the *mass equations* (2.19) and (2.22) (for prescribed θ) separately have a *monotone structure;* but as a system they miss the properties needed to apply the monotonicity techniques of Chaps. II and III. Indeed, as far as this author knows, existence of a weak solution has not been proved for this formulation. Moreover, the preceding equations do not account for *cross effects,* such as heat flux induced by chemical potential gradient, and mass flux due to temperature gradient. These drawbacks are overcome by the theory of *nonequilibrium thermodynamics,* which we outline in the next section.

Exercises.

2.1 Provide a strong formulation of the problem of phase transition in binary mixtures in one space dimension, cf. Problem 2.1.

2.2 Insert the kinetic undercooling into the problem of the previous exercise.

2.3 Provide a weak formulation of the problem of phase transition in binary mixtures in three space dimensions, cf. Problem 2.1.

2.4 Insert the phase relaxation into the problem of the previous exercise.

V.3 Approach via Nonequilibrium Thermodynamics

> *The customer cannot win at this game;*
> *this is the first law.*
> *In fact, the customer is likely to lose;*
> *this is the second law.*
> (From a book of instructions
> for croupiers at the roulette table) (17)

In this section we outline another model of phase transition in two component systems, in the framework of the so-called theory of *nonequilibrium thermodynamics*. In Sect. V.5 this model is extended to multi-component systems.

Balance Laws and the Gibbs Formula. We label the two components by $i = 1, 2$, and use the following notation:

u: density of internal energy,
τ: absolute temperature,
$\vec{\omega}$: energy flux (per unit surface) due to heat and mass transport,
h: intensity of an energy source or sink per unit volume,
c_i: concentration of the component i,
$\vec{j}_i$: flux (per unit surface) of the component i,
μ_i: *chemical potential* of the component i,
s: *specific entropy*, namely entropy per unit volume. This is regarded as a function of the state variables (u, c_1, c_2).

In the absence of chemical reactions, the principles of energy and mass conservation yield

$$\frac{\partial u}{\partial t} = -\nabla \cdot \vec{\omega} + h \qquad \text{in } Q, \tag{3.1}$$

$$\frac{\partial c_i}{\partial t} = -\nabla \cdot \vec{j}_i \qquad \text{in } Q \ (i = 1, 2). \tag{3.2}$$

At equilibrium the following classical *Gibbs formula* is fulfilled:

$$ds = \frac{1}{\tau}du - \frac{\mu_1}{\tau}dc_1 - \frac{\mu_2}{\tau}dc_2 \qquad \forall (u, c_1, c_2) \in \text{Dom}(s). \tag{3.3}$$

By this we mean that there exists a differentiable function $\hat{s} :]0, +\infty[\times [0, 1]^2 \to \mathbf{R}$ such that, for any process evolving through equilibrium states,

$$s(x, t) = \hat{s}(u(x, t), c_1(x, t), c_2(x, t)) \qquad \forall (x, t) \in Q,$$

$$\frac{\partial \hat{s}}{\partial u} = \frac{1}{\tau}, \qquad \frac{\partial \hat{s}}{\partial c_1} = -\frac{\mu_1}{\tau}, \qquad \frac{\partial \hat{s}}{\partial c_2} = -\frac{\mu_2}{\tau} \qquad \forall (u, c_1, c_2) \in \text{Dom}(\hat{s}).$$

---

[17] Astarita [29] acutely inserted this quotation at the beginning of the section titled *Irreversibility and Dissipation*.

$\hat{s}$ is a concave function of u, c_1, c_2; henceforth we identify $\hat{s}$ with s.

A basic postulate of *nonequilibrium thermodynamics* assumes that (3.3) also applies to systems that are not in *local equilibrium*. Actually, the limits of validity of the whole theory strongly depend on those of this formula.

Entropy Balance. Since $c_1 + c_2 = 1$ and $\vec{j}_1 + \vec{j}_2 = \vec{0}$, setting $\mu := \mu_1 - \mu_2$, $c := c_1$ and $\vec{j} := \vec{j}_1$, (3.2) and (3.3) are equivalent to

$$\frac{\partial c}{\partial t} = -\nabla \cdot \vec{j} \qquad \text{in } Q, \tag{3.4}$$

$$ds = \frac{1}{\tau} du - \frac{\mu}{\tau} dc \qquad \forall u > 0, \forall c \in [0,1]. \tag{3.5}$$

$s :]0, +\infty[\times [0,1] \to \mathbf{R}$ is also a concave function. Multiplying (3.1) by $1/\tau$, (3.4) by $-\mu/\tau$ and using (3.5), we get the *entropy balance equation*

$$\begin{aligned}
\frac{\partial s}{\partial t} &= -\frac{1}{\tau} \nabla \cdot \vec{\omega} + \frac{\mu}{\tau} \nabla \cdot \vec{j} + \frac{h}{\tau} \\
&= -\nabla \cdot \frac{\vec{\omega} - \mu \vec{j}}{\tau} + \vec{\omega} \cdot \nabla \frac{1}{\tau} - \vec{j} \cdot \nabla \frac{\mu}{\tau} + \frac{h}{\tau} \\
&=: -\nabla \cdot \vec{j}_s + \pi + \frac{h}{\tau} \qquad \text{in } Q,
\end{aligned} \tag{3.6}$$

where we set

$$\vec{j}_s := \frac{\vec{\omega} - \mu \vec{j}}{\tau} : \text{entropy flux (per unit surface),}$$

$$\pi := \vec{\omega} \cdot \nabla \frac{1}{\tau} - \vec{j} \cdot \nabla \frac{\mu}{\tau} : \text{entropy production rate (per unit volume).}$$

h/τ is the rate of entropy provided by an external source or sink, per unit volume. Incidentally, we note that $\vec{q} := \tau \vec{j}_s$ is the heat flux (per unit surface).

The *local formulation of the second principle of thermodynamics* [18] states that entropy production is nonnegative and vanishes only at equilibrium, that is,

$$\pi \geq 0, \qquad \pi = 0 \quad \text{only if} \quad \nabla \frac{1}{\tau} = -\nabla \frac{\mu}{\tau} = \vec{0} \qquad \text{in } Q; \tag{3.7}$$

$\pi = 0$ ($\pi > 0$, resp.) corresponds to a reversible (irreversible, resp.) process.

Phenomenological Laws. Let us assume that the *generalized fluxes* $\vec{J} := (\vec{\omega}, \vec{j})$ are functions of the state variables $z := (1/\tau, -\mu/\tau)$ and of the *generalized forces* $\nabla z = (\nabla(1/\tau), -\nabla(\mu/\tau))$. This yields the so-called *phenomenological laws* [19]

$$\vec{J}_i = \vec{F}_i(z, \nabla z) \qquad \forall z \in \text{Dom}(s^*) \, (i=1,2). \tag{3.8}$$

---

[18] See, e.g., Glansdorff and Prigogine [269], and Prigogine [455].

[19] Here $s^*(z) := \inf_{w \in \text{Dom}(s)} \left(-\sum_{i=0}^{M} z_i w_i - s(w) \right) (\leq +\infty)$ for any $z \in \text{Dom}(s^*)$. s^* is called the *concave conjugate* function of s, and coincides with the negative of the *convex conjugate* function of $-s$, which is defined in Sect. XI.4.

By (3.7), $\vec{F}_1$ and $\vec{F}_2$ must be such that

$$\sum_{i=1,2} \vec{F}_i(z,\vec{\xi}) \cdot \vec{\xi}_i \geq 0 \quad \forall z, \forall \vec{\xi} := (\vec{\xi}_1, \vec{\xi}_2),$$
$$\sum_{i=1,2} \vec{F}_i(z,\vec{\xi}) \cdot \vec{\xi}_i = 0 \quad \text{only if} \quad \vec{\xi} = (\vec{0},\vec{0}). \tag{3.9}$$

In the absence of phase transition, the $\vec{F}_i$s are continuous with respect to z.

In a neighbourhood of equilibrium, namely, for *small* forces, one can assume that the fluxes depend linearly on the forces:

$$\vec{J} = L(z)\nabla z \quad \forall z \in \text{Dom}(s^*), \tag{3.10}$$

that is,

$$\begin{pmatrix} \vec{\omega} \\ \vec{j} \end{pmatrix} = L\left(\frac{1}{\tau}, -\frac{\mu}{\tau}\right) \begin{pmatrix} \nabla\frac{1}{\tau} \\ -\nabla\frac{\mu}{\tau} \end{pmatrix} \quad \text{in } Q. \tag{3.11}$$

By (3.7), no affine term occurs in this expression, and $L = L(z)$ is a positive definite 2×2-tensor for any z. In a neighbourhood of a fixed state, L can be assumed to be constant. A fundamental result of the theory of nonequilibrium thermodynamics states that, in the absence of applied magnetic field, L is symmetric:

$$L_{12} = L_{21} \quad \text{(Onsager reciprocity relation)}. \tag{3.12}$$

In conclusion, we have derived the system (3.1), (3.4), (3.5), (3.11), which must be coupled with appropriate initial and boundary conditions.

Relation with the Model of Sect. V.2. Let us consider a single-phase system. The Fourier and Fick laws respectively read

$$\vec{q} = -k\nabla\tau, \quad \vec{j} = -D\nabla c \quad \text{in } Q, \tag{3.13}$$

where $k = k(\tau, c)$ and $D = D(\tau, c)$ are positive coefficients. The energy flux (per unit surface) equals

$$\vec{\omega} = \vec{q} + \frac{\partial u}{\partial c}\vec{j} = -k\nabla\tau - D\frac{\partial u}{\partial c}\nabla c \quad \text{in } Q. \tag{3.14}$$

Let us define the *free energy* density f as the negative of the *Legendre transform* [20] of the energy density u with respect to the variable τ; that is, $f := u - \tau s$. (3.5) yields

$$df = -sd\tau + \mu dc, \quad \text{i.e.,} \quad \frac{\partial f}{\partial \tau} = -s\left(=\frac{f-u}{\tau}\right), \quad \frac{\partial f}{\partial c} = \mu,$$

---

[20] See Sect. XI.4.

whence $\partial u/\partial c = \mu - \tau \partial \mu/\partial \tau$. A simple calculation yields

$$J := \frac{\partial\left(\frac{1}{\tau}, -\frac{\mu}{\tau}\right)}{\partial(\tau, c)} = \begin{pmatrix} -\frac{1}{\tau^2} & 0 \\ \frac{\mu}{\tau^2} - \frac{1}{\tau}\frac{\partial \mu}{\partial \tau} & -\frac{1}{\tau}\frac{\partial \mu}{\partial c} \end{pmatrix} = \begin{pmatrix} -\frac{1}{\tau^2} & 0 \\ \frac{1}{\tau^2}\frac{\partial u}{\partial c} & -\frac{1}{\tau}\frac{\partial^2 f}{\partial c^2} \end{pmatrix}.$$

Therefore, by comparing (3.11) with (3.13) and (3.14), we can relate the coefficients $k = k(\tau, c)$ and $D = D(\tau, c)$ with $\{L_{ij}(z)\}$:

$$L = -\begin{pmatrix} k & D\frac{\partial u}{\partial c} \\ 0 & D \end{pmatrix} J^{-1} = \begin{pmatrix} k & D\frac{\partial u}{\partial c} \\ 0 & D \end{pmatrix} \begin{pmatrix} \tau^2 & 0 \\ \tau\frac{\partial u}{\partial c}\left(\frac{\partial^2 f}{\partial c^2}\right)^{-1} & \tau\left(\frac{\partial^2 f}{\partial c^2}\right)^{-1} \end{pmatrix},$$

that is,

$$\begin{cases} L_{11} = k\tau^2 + D\tau\left(\frac{\partial u}{\partial c}\right)^2\left(\frac{\partial^2 f}{\partial c^2}\right)^{-1}, \\ L_{12} = L_{21} = D\tau\frac{\partial u}{\partial c}\left(\frac{\partial^2 f}{\partial c^2}\right)^{-1}, \\ L_{22} = D\tau\left(\frac{\partial^2 f}{\partial c^2}\right)^{-1}. \end{cases} \quad (3.15)$$

Therefore the (nonlinear) Fourier and Ficks laws are equivalent to the Gibbs law (3.3) and the phenomenological laws (3.11). However, the equivalence fails if k, D, and $\{L_{ij}\}$ are assumed to be constant, since the two approaches correspond to different linearizations. This also holds for two-phase systems.

Exercises.

3.1 Provide a weak formulation of the system (3.1), (3.4), (3.5), (3.11), coupled with suitable initial and boundary conditions, in three space dimensions. Then extend this to two-phase systems.

3.2 Provide a strong formulation (based on the approach of nonequilibrium thermodynamics) of the problem of phase transition in binary mixtures, in one space dimension.

3.3 Insert the phase relaxation into the problem of the exercise 3.1.

3.4 Insert the kinetic undercooling into the problem of the exercise 3.2.

3.5 Guess a generalization of the law (1.18) of mean curvature flow for binary alloys. Then provide a corresponding strong formulation in a radially symmetrical system.

V.4 Analysis of the Model of Section V.3

In this section we discuss some mathematical aspects of the model of coupled heat and mass diffusion, which we outlined in the previous section. These developments can be extended to two-phase systems.

A General Formulation. Let us set $w := (u, c)$, and denote by s^* the *concave conjugate* function of s. For the sake of simplicity, here we assume that both s and s^* are differentiable. [21] As $z := (1/\tau, -\mu/\tau)$, the Gibbs formula (3.5) reads $w = \partial s^* / \partial z$, and is equivalent to $z = \partial s / \partial w$.

Let us set $H := (h, 0)$ and $\vec{\mathcal{F}} := (\vec{F}_1, \vec{F}_2)$. The balance laws (3.1), (3.4), the Gibbs formula (3.5), and the phenomenological laws (3.8) yield the following system:

$$\begin{cases} \dfrac{\partial w}{\partial t} + \nabla \cdot \vec{\mathcal{F}}(z, \nabla z) = H \\ w = \dfrac{\partial s^*}{\partial z}(z) \end{cases} \quad \text{in } Q. \tag{4.1}$$

Let us assume that $\vec{F}_1$ and $\vec{F}_2$ are globally continuous, and uniformly strongly monotone with respect to the second argument, that is,

$$\exists c > 0 : \forall z \in \text{Dom}(s^*), \forall \vec{X}_1, \vec{X}_2 \in \mathbf{R}^3, \text{ for } i = 1, 2, \\ \left[\vec{F}_i(z, \vec{X}_1) - \vec{F}_i(z, \vec{X}_2)\right] \cdot \left(\vec{X}_1 - \vec{X}_2\right) \geq c \left|\vec{X}_1 - \vec{X}_2\right|^2. \tag{4.2}$$

The latter condition is a natural analytical assumption. However, the second principle only prescribes (3.9), which is weaker than (4.2) in general, and equivalent to it only in the linear case.

The vector function $z \mapsto \partial s^* / \partial z$ is antimonotone, hence (4.1) is *forward parabolic*. Under further assumptions, a vectorial result analogous to Theorem III.4.2 yields the existence of a solution of the corresponding initial and boundary value problem, cf. Problem III.4.2.

Remark. At first sight, difficulties might be expected to arise because of the physical constraint $z_1 := 1/\tau > 0$. However, when studying (4.1) at room temperature, it is reasonable to assume that $\text{Dom}(s^*) = \mathbf{R}^2$. Similarly, dealing with heat diffusion problems at room temperature, usually one does not care to prove that the absolute temperature is nonnegative.

In any case, our constitutive laws cannot be extrapolated to absolute temperatures close to $\tau = 0$. □

---

[21] The assumption of differentiability can be dropped, provided that $\partial s^* / \partial z$ is replaced by the *superdifferential* of s^*. The latter coincides with the negative of the *subdifferential* of $-s^*$, in the sense of *convex analysis;* see Sect. XI.4.

142   V. Generalizations of the Stefan Problem

Linearization. Henceforth we assume that the system stays *close* to equilibrium, that is, ∇z is *small*. Then $\vec{\mathcal{F}}$ can be linearized with respect to the second argument, cf. (3.10), so that (4.1) reads

$$\begin{cases} \dfrac{\partial w}{\partial t} + \nabla \cdot [L(z)\nabla z] = H \\ w = \dfrac{\partial s^*}{\partial z}(z) \end{cases} \quad \text{in } Q. \tag{4.3}$$

As we saw, by the second principle of thermodynamics, the matrix [22] $L(z)$ is positive definite for any z.

If L does not depend on z, one can prove that the weak formulation of (4.3) coupled with appropriate initial and boundary conditions is well-posed; see the discussion about systems of P.D.E.s at the end of Sect. II.3. In particular, a priori estimates can be derived by multiplying the first equation by z. If L is symmetric, the elliptic operator is self-adjoint, and one can multiply the equation by $\partial z / \partial t$; this yields further regularity. Thus the Onsager relation (3.12) has a tangible analytical counterpart.

A Kirchhoff-Type Transformation. The *Kirchhoff transformation* $v \mapsto \tilde{v} := \int_0^v k(\xi)d\xi$ allows us to replace the nonlinear term $\nabla \cdot [k(v)\nabla v]$ by $\Delta\tilde{v}$, and is often helpful in the analysis of parabolic equations. It is not always possible to extend this transformation to systems of equations. There exist functions $G_1, G_2 :$ Dom$(s^*) \to \mathbf{R}$ such that

$$\vec{J}_i = \sum_{j=1,2} L_{ij}(z)\nabla z_j = \nabla G_i(z) \quad \text{in } Q\ (i=1,2) \tag{4.4}$$

only if

$$\nabla \times \vec{J}_i = \vec{0} \quad \text{in } Q\ (i=1,2), \tag{4.5}$$

that is, energy and matter flow irrotationally — which appears to be a reasonable physical condition. (4.5) is also sufficient for (4.4) to hold, since Dom(s^*) is simply connected. Note that (4.5) is equivalent to the (local) integrability of the differential forms $\sum_{j=1,2} L_{ij}(z)dz_j$, that is,

$$\frac{\partial}{\partial z_2}L_{i1}(z) = \frac{\partial}{\partial z_1}L_{i2}(z) \quad \forall z \in \text{Dom}(s^*)\ (i=1,2). \tag{4.6}$$

Let us set $G := (G_1, G_2)$. Under the assumption (4.4), (4.3) reads

$$\begin{cases} \dfrac{\partial w}{\partial t} + \Delta G(z) = H \\ w = \dfrac{\partial s^*}{\partial z}(z) \end{cases} \quad \text{in } Q. \tag{4.7}$$

---

[22] Having fixed the coordinate system, we do not distinguish between tensors and matrices.

This system has the form of Problem III.4.1; a result for systems analogous to Theorem III.4.1 yields existence of a solution of the corresponding initial and boundary value problem.

If the tensor L is symmetric, $\vec{G}$ is a gradient. But, if L is nonconstant, this does not seem to yield further regularity of the solution in an obvious way.

Two-Phase Systems. So far we considered a single phase. Two-phase systems are characterized by discontinuities in the constitutive relation between w and z. This setting can be treated just by replacing the gradient $\partial s^*/\partial z$ with the *superdifferential,* and interpreting (4.1)$_1$ in the sense of distributions, as in Proposition 2.1. [23] Here condition (4.4) means that the generalized fluxes are irrotational in the interior of each phase, and that their tangential components are continuous across the moving interface. From the physical viewpoint this looks natural.

Exercises.

4.1 Provide a precise formulation of the system (4.3), coupled with suitable initial and boundary conditions, in three space dimensions. Then assume that the matrix L does not depend on z, and prove that the problem is well-posed.

4.2 Extend the results of the previous exercise to two-phase systems.

4.3 Insert phase relaxation into the problem of the previous exercise.

V.5 General Nonequilibrium Thermodynamics

The developments of Sect. V.3 are just an example of a more general theory, known as *nonequilibrium thermodynamics.* This model is essentially based on a local formulation of the second principle of thermodynamics, and can be applied to many processes, including chemical reactions. At the basis of this approach there is a simple derivation which we outline here.

Balance Laws and the Gibbs Formula. Let us consider a system of $M+1$ (≥ 2) components, in which P chemical reactions occur. We use the following notation:

H_ℓ: heat absorbed in the ℓth reaction,

$\nu_{i\ell}$: *stoichiometric coefficient* of the ith component in the ℓth reaction,

v_ℓ: rate of the ℓth reaction.

The laws of energy and mass conservation yield [24]

$$\frac{\partial u}{\partial t} = -\nabla \cdot \vec{\omega} - \sum_{\ell=1}^{P} H_\ell v_\ell \quad \text{in } Q, \quad (5.1)$$

---

[23] See Donnelly [198], and Luckhaus and V. [367].
[24] These laws implicitly illustrate the meaning of $H_\ell, v_\ell, \nu_{i\ell}$.

$$\frac{\partial c_i}{\partial t} = -\nabla \cdot \vec{j}_i + \sum_{\ell=1}^{P} \nu_{i\ell} v_\ell \quad \text{in } Q \ (i = 1, \ldots, M+1). \tag{5.2}$$

Assuming that the space density of entropy s depends on u and $c_1, \ldots, c_{M+1}$, we have the *Gibbs formula*

$$ds = \frac{1}{T} du - \sum_{i=1}^{M+1} \frac{\mu_i}{T} dc_i. \tag{5.3}$$

This law must be interpreted similarly to (3.3), and is also assumed to hold *close to equilibrium*.

Some Notation. Note that

$$\sum_{i=1}^{M+1} c_i = 1, \quad \sum_{i=1}^{M+1} \vec{j}_i = \vec{0}, \quad \sum_{i=1}^{M+1} \nu_{i\ell} = 0 \quad \forall \ell.$$

Hence equation (5.2) for $i = M+1$ follows from the other equations. Moreover, setting $\tilde{\mu}_i := \mu_i - \mu_{M+1}$ for $i = 1, \cdots, M$, (5.3) is equivalent to

$$ds = \frac{1}{T} du - \sum_{i=1}^{M} \frac{\tilde{\mu}_i}{T} dc_i. \tag{5.4}$$

Let us set $\nu_{0\ell} := -H_\ell$ for any ℓ, and define:
(i) the *intensive variables* $z_0 := 1/T$, $z_i := -\tilde{\mu}_i/T$ for $i = 1, \ldots, M$;
(ii) the *extensive variables* $w_0 := u$, $w_i := c_i$ for $i = 1, \ldots, M$;
(iii) the *generalized fluxes* $\vec{J}_0 := \vec{\omega}$, $\vec{J}_i := \vec{j}_i$ for $i = 1, \ldots, M$;
(iv) the *generalized forces* $\vec{X}_i := \nabla z_i$ for $i = 0, \ldots, M$.

The balance laws (5.1), (5.2), and the *Gibbs formula* then read

$$\frac{\partial w_i}{\partial t} = -\nabla \cdot \vec{J}_i + \sum_{\ell=1}^{P} \nu_{i\ell} v_\ell \quad \text{in } Q \ (i = 0, \ldots, M), \tag{5.5}$$

$$ds = \sum_{i=0}^{M} z_i dw_i. \tag{5.6}$$

Entropy Balance. By multiplying (5.5) by z_i and summing in i, we get

$$\frac{\partial s}{\partial t} = \sum_{i=0}^{M} z_i \frac{\partial w_i}{\partial t} = -\sum_{i=0}^{M} z_i \nabla \cdot \vec{J}_i + \sum_{i=0}^{M} \sum_{\ell=1}^{P} \nu_{i\ell} z_i v_\ell$$

$$= -\nabla \cdot \sum_{i=0}^{M} z_i \vec{J}_i + \sum_{i=0}^{M} \vec{J}_i \cdot \nabla z_i + \sum_{i=0}^{M} \sum_{\ell=1}^{P} \nu_{i\ell} z_i v_\ell \tag{5.7}$$

$$=: -\nabla \cdot \vec{j}_s + \pi \quad \text{in } Q,$$

V.5 General Nonequilibrium Thermodynamics

where we set

$$\vec{j}_s := \sum_{i=0}^{M} z_i \vec{J}_i = \frac{\vec{\omega}}{\tau} - \sum_{i=1}^{M} \frac{\tilde{\mu}_i}{\tau} \vec{j}_i : \text{entropy flux (per unit surface)},$$

$$B_\ell := \sum_{i=0}^{M} \nu_{i\ell} z_i \qquad (\ell = 1, \ldots, P),$$

$$\pi := \sum_{i=0}^{M} \vec{J}_i \cdot \nabla z_i + \sum_{\ell=1}^{P} B_\ell v_\ell : \text{entropy production rate (per unit volume)}.$$

$\vec{q} := \tau \vec{j}_s$ is the heat flux (per unit surface), and τB_ℓ is the *chemical affinity* of the ℓth reaction. [25]

As we saw, the local formulation of the *second principle of thermodynamics* states that the entropy production is nonnegative and vanishes only at equilibrium. Here this reads

$$\begin{aligned} &\pi \geq 0; \quad \pi = 0 \text{ only if} \\ &\nabla z_i = \vec{0} \ (i = 0, \ldots, M), \quad B_\ell = 0 \ (\ell = 1, \ldots, P) \end{aligned} \quad \text{in } Q. \quad (5.8)$$

$\pi = 0$ ($\pi > 0$, resp.) corresponds to a reversible (irreversible, resp.) process.

Phenomenological Laws. Let us set

$$z := (z_0, \ldots, z_M) = \left(\frac{1}{\tau}, -\frac{\tilde{\mu}_1}{\tau}, \ldots, -\frac{\tilde{\mu}_M}{\tau}\right),$$

$$w := (w_0, \ldots, w_M) = (u, c_1, \ldots, c_M),$$

$$\vec{X} := (\nabla z_0, \ldots, \nabla z_M) = \left(\nabla \frac{1}{\tau}, -\nabla \frac{\tilde{\mu}_1}{\tau}, \ldots, -\nabla \frac{\tilde{\mu}_M}{\tau}\right),$$

$$\vec{J} := (\vec{J}_0, \ldots, \vec{J}_M) = (\vec{\omega}, \vec{j}_1, \ldots, \vec{j}_M),$$

$$B := (B_1, \ldots, B_P), \qquad V := (v_1, \ldots, v_P).$$

One can assume that a set of constitutive relations (called *phenomenological laws*) of the following form holds

$$(\vec{J}, V) = \vec{F}(z, \vec{X}, B) \qquad \forall z \in \text{Dom}(s^*), \forall \vec{X} \in \left(\mathbf{R}^3\right)^{M+1}, \forall B \in \mathbf{R}^P. \quad (5.9)$$

By (5.8), these laws are subject to the following restriction

$$\begin{aligned} &\vec{F}(z, \vec{X}, B) \cdot (\vec{X}, B) \geq 0 \qquad \forall z, \vec{X}, B, \\ &\vec{F}(z, \vec{X}, B) \cdot (\vec{X}, B) = 0 \quad \text{only if} \quad \vec{X} = \vec{0}, B = 0, \end{aligned} \quad (5.10)$$

---

[25] See Prigogine [455].

where by the dot we denote the scalar product in $(\mathbf{R}^3)^{M+1} \times \mathbf{R}^P$.

Linearized Phenomenological Laws. In a neighbourhood of equilibrium (i.e. for small $\vec{X}$ and B), one can linearize (5.9). [26] Within the limits of validity of this linearization, the vectorial variables $\vec{J}, \vec{X}$ are uncoupled from the scalars V, B, because of the so-called *Curie principle:* generalized forces cannot have more elements of symmetry than the generalized fluxes they produce. [27] This yields

$$\vec{J}_i = \sum_{j=0}^{M} L_{ij}(z)\vec{X}_j \quad \forall i, \qquad v_\ell = \sum_{m=1}^{P} \hat{L}_{\ell m}(z)B_m \quad \forall \ell. \qquad (5.11)$$

By (5.8), L and $\hat{L}$ are positive definite. A classical result of thermodynamics states that, in the absence of applied magnetic field, L and $\hat{L}$ are symmetric in a neighbourhood of equilibrium, namely, for *small* generalized forces:

$$L_{ij}(z) = L_{ji}(z), \quad \hat{L}_{\ell m}(z) = \hat{L}_{m\ell}(z) \qquad \forall z, \forall (i,j), \forall (\ell, m) \qquad (5.12)$$

(Onsager's reciprocity relations). Therefore, setting

$$\Phi(z, \vec{X}) := \frac{1}{2}\sum_{i,j=0}^{M} L_{ij}(z)\vec{X}_i \cdot \vec{X}_j, \qquad \Psi(z, B) := \frac{1}{2}\sum_{\ell,m=1}^{P} \hat{L}_{\ell m}(z)B_\ell B_m, \quad (5.13)$$

the constitutive relation (5.9) also reads

$$\vec{J}_i := \frac{\partial \Phi(z, \vec{X})}{\partial \vec{X}_i} \quad \forall i, \forall (z, \vec{X}), \qquad v_\ell := \frac{\partial \Psi(z, B)}{\partial B_\ell} \quad \forall \ell, \forall (z, B). \qquad (5.14)$$

These formulae must be fulfilled for any $\vec{X} \in (\mathbf{R}^3)^{M+1}$, any $B \in \mathbf{R}^P$, any $z \in \mathrm{Dom}(s^*)$.

If s^* is differentiable, then $w_i = \partial s^*/\partial z_i(z)$ for any i, and (5.5) can be written in the form

$$\begin{cases} \dfrac{\partial w_i}{\partial t} + \nabla \cdot \dfrac{\partial \Phi(z, \vec{X})}{\partial \vec{X}_i} + \sum_{\ell=1}^{P} \nu_{i\ell} \dfrac{\partial \Psi(z, B)}{\partial B_\ell} = 0 \quad (i = 0, \cdots, M) \\ w = \dfrac{\partial s^*}{\partial z}(z) \end{cases} \quad \text{in } Q. \quad (5.15)$$

---

[26] However, the range of validity of the linearization is fairly small for most chemical reaction laws.

[27] See, e.g., Glansdorff and Prigogine [269] or Prigogine [455]. Outside the *linear region* this is no longer true, and *symmetry breaking* may occur.

For any z, $\Phi(z, \cdot)$ and $\Psi(z, \cdot)$ are convex functions, and the same holds for $-s^*$. We discussed problems of this sort in Sect. III.4.

Remarks. (i) This approach can be extended to general nonlinear relations of the form (5.9), provided that $\vec{F}(z, \cdot)$ and $G(z, \cdot)$ are monotone for any z.

(ii) So far the functions s^*, $\Phi(z, \cdot)$ and $\Psi(z, \cdot)$ have been assumed to be differentiable for any z. If this does not hold, it suffices to replace partial derivatives by *subdifferentials,* or *superdifferentials,* [28] The systems (4.1) and (5.15) have similar analytical structures, hence the analysis of Sect. V.4 can easily be extended to the more general setting of this section.

(iii) Here we considered one-phase systems. Different phases are characterized by discontinuities *(jumps)* of the variables w_is and of the tensors $L(z)$, $\hat{L}(z)$. This can be discussed along the lines of the previous section. [29] □

V.6 The Evolution of the Free Energy

In this section we use the approach of nonequilibrium thermodynamics to show that a large class of processes is driven by the tendency to decrease the free energy, and derive some *gradient flow* dynamics.

Let us consider a system whose state is characterized by the specific entropy s and by a scalar variable Y, so that at equilibrium the specific internal energy u can be regarded as a function $u(s, Y)$. The form of this constitutive relation depends on the specific material, and we assume that it also holds outside equilibrium, within a range to be specified. For instance, Y might represent the concentration of a component in a binary alloy; in that case $\partial u / \partial Y$ would represent the chemical potential. Under regularity conditions, setting $\tau := \partial u / \partial s$ (which is the thermodynamical definition of the absolute temperature), obviously we have

$$du = \tau ds + \frac{\partial u}{\partial Y} dY. \tag{6.1}$$

The density of *free energy* f is defined as the *Legendre transform* [30] of u with respect to s: $f = f(\tau, Y) := u - \tau s$; hence $s = -\partial f / \partial \tau$. Setting $\mu := \partial f / \partial Y$, we have

$$\mu dY := \frac{\partial f}{\partial Y}(\tau, Y) dY = df + s d\tau = du - \tau ds = \frac{\partial u}{\partial Y}(s, Y) dY. \tag{6.2}$$

---

[28] In the sense of *convex analysis,* cf. Sect. XI.4.
[29] See Luckhaus and V. [367].
[30] See Sect. XI.4.

148   V. Generalizations of the Stefan Problem

Now let us consider a space-distributed system. We assume that it is incompressible, [31] and that there is no exterior source or sink of Y. Let us denote the energy flux by $\vec{\omega}$, the heat flux by $\vec{q}$, the flux of Y by $\vec{j}$, and define the entropy flux $\vec{j}_s := \vec{q}/\tau$ (all these fluxes are per unit surface). (6.1) and (6.2) yield

$$\vec{\omega} = \tau \vec{j}_s + \frac{\partial u}{\partial Y}(s, Y)\vec{j} = \vec{q} + \mu(\tau, Y)\vec{j}. \tag{6.3}$$

Let us denote by h the intensity of an energy source or sink, and by π the entropy production rate (both per unit volume). The energy conservation principle and the entropy balance read

$$\frac{\partial u}{\partial t} = -\nabla \cdot \vec{\omega} + h \quad \text{in } Q, \tag{6.4}$$

$$\frac{\partial s}{\partial t} = -\nabla \cdot \vec{j}_s + \frac{h}{\tau} + \pi \quad \text{in } Q. \tag{6.5}$$

By the local formulation of the second principle of thermodynamics, we have the classical *Clausius-Duhem inequality* (for an incompressible material):

$$\pi \geq 0, \quad \pi = 0 \text{ only at equilibrium,} \quad \text{in } Q. \tag{6.6}$$

The equations (6.3) and (6.4) yield

$$\pi = \frac{\partial s}{\partial t} - \frac{1}{\tau}\frac{\partial u}{\partial t} + \nabla \cdot \left(\vec{j}_s - \frac{\vec{\omega}}{\tau}\right) + \vec{\omega} \cdot \nabla \frac{1}{\tau} \quad \text{in } Q, \tag{6.7}$$

that is, by (6.2) and (6.3),

$$\pi = -\frac{\mu}{\tau}\frac{\partial Y}{\partial t} - \nabla \cdot \frac{\mu \vec{j}}{\tau} + \vec{\omega} \cdot \nabla \frac{1}{\tau} \quad \text{in } Q. \tag{6.8}$$

We distinguish two special cases, which are alternative to each other.

First Case: Absence of Flux of Y. In this setting $\vec{j} = \vec{0}$, $\vec{\omega} = \vec{q}$, and (6.8) reads

$$\pi = -\frac{\mu}{\tau}\frac{\partial Y}{\partial t} + \vec{q} \cdot \nabla \frac{1}{\tau} \quad \text{in } Q. \tag{6.9}$$

Here one can regard $-\mu/\tau$ and $\nabla(1/\tau)$ as *generalized forces*, $\partial Y/\partial t$ and $\vec{q}$ as *generalized displacements,* and assume *phenomenological laws* of the form

$$\left(\frac{\partial Y}{\partial t}, \vec{q}\right) = \Phi\left(\tau, Y, -\frac{\mu}{\tau}, \nabla \frac{1}{\tau}\right), \tag{6.10}$$

---

[31] By this assumption we exclude any mechanical work; however, our discussion might be extended to include that, too.

with Φ positive definite. By (6.8), $\pi = \vec{q} \cdot \nabla(1/\tau)$ whenever $\mu = 0$; on the other hand, if τ is uniform, $\pi = -(\mu/\tau)\partial Y/\partial t$. If we assume that the dependence from $-\mu/\tau$ and $\nabla(1/\tau)$ uncouples, the phenomenological laws take the form

$$\frac{\partial Y}{\partial t} = \varphi\left(\tau, Y, -\frac{\mu}{\tau}\right), \qquad \vec{q} = \vec{\psi}\left(\tau, Y, \nabla\frac{1}{\tau}\right), \qquad (6.11)$$

and (6.6) yields the following restrictions

$$\frac{\partial Y}{\partial t}\frac{\mu}{\tau} \leq 0, \qquad \frac{\partial Y}{\partial t} = 0 \quad \text{iff} \quad \mu = 0$$
$$\vec{q} \cdot \nabla\frac{1}{\tau} \geq 0, \qquad \vec{q} = \vec{0} \quad \text{iff} \quad \nabla\tau = \vec{0} \qquad \text{in } Q. \qquad (6.12)$$

The linearized form of (6.11) then reads

$$a\frac{\partial Y}{\partial t} = -\frac{1}{\tau}\frac{\partial f}{\partial Y}\left(=: -\frac{\mu}{\tau}\right), \qquad \vec{q} = k\nabla\frac{1}{\tau} \qquad \text{in } Q, \qquad (6.13)$$

where a and k are positive coefficients which may depend on τ, Y. The first of these equations is an example of what in Sect. VII.6 we call a *gradient flow*.

Therefore, in a system with state variables (τ, Y) in which energy only flows by heat conduction, the evolution of Y tends to decrease the free energy of the system. Notice that here Y is not assumed to be a conserved quantity; that is, the conservation law (6.16) (see the following) is not imposed.

These results have been derived without assuming that the temperature field is either uniform or constant. If τ is stationary (but not necessarily uniform), we have

$$\frac{df}{dt}(\tau(x), Y(x, t)) = \frac{\partial f}{\partial Y}(\tau(x), Y(x, t))\frac{\partial Y}{\partial t}(x, t) \left(=: \mu\frac{\partial Y}{\partial t}\right), \qquad (6.14)$$

and the first part of (6.12) yields

if $\tau = \tau(x)$ in Q, then $\quad$ (i) $\dfrac{df}{dt} \leq 0, \quad$ (ii) $\dfrac{df}{dt} = 0 \quad$ iff $\quad \dfrac{\partial f}{\partial Y} = 0. \qquad (6.15)$

In general, in the absence of flux of Y, a closed mathematical system is obtained by coupling the equations (6.4) (with $\vec{\omega} = \vec{q}$), (6.10) (or (6.11)), and specifying appropriate initial and boundary conditions.

Second Case: Y is a Conserved Quantity. This setting is analogous to that developed in Sect. V.3 for mass diffusion. As we excluded any exterior source or sink of Y, the conservation law reads

$$\frac{\partial Y}{\partial t} = -\nabla \cdot \vec{j} \qquad \text{in } Q. \qquad (6.16)$$

By multiplying this equation by μ/τ, we get

$$\frac{\mu}{\tau}\frac{\partial Y}{\partial t} + \nabla \cdot \frac{\mu \vec{j}}{\tau} = \vec{j} \cdot \nabla \frac{\mu}{\tau} \quad \text{in } Q; \tag{6.17}$$

hence (6.8) reads

$$\pi = -\vec{j} \cdot \nabla \frac{\mu}{\tau} + \vec{\omega} \cdot \nabla \frac{1}{\tau} \quad \text{in } Q. \tag{6.18}$$

Here one can assume *phenomenological laws* of the form

$$(\vec{j}, \vec{\omega}) = \Phi\left(\tau, Y, -\nabla \frac{\mu}{\tau}, \nabla \frac{1}{\tau}\right), \tag{6.19}$$

with Φ positive definite. If we assume that the dependence from $-\nabla(\mu/\tau)$ and $\nabla(1/\tau)$ uncouples, these laws take the form

$$\vec{j} = \vec{\varphi}\left(\tau, Y, -\nabla \frac{\mu}{\tau}\right), \quad \vec{\omega} = \vec{\psi}\left(\tau, Y, \nabla \frac{1}{\tau}\right), \tag{6.20}$$

and (6.6) yields the following restrictions

$$\begin{aligned}\vec{j} \cdot \nabla \frac{\mu}{\tau} &\leq 0, \quad \vec{j} = \vec{0} \quad \text{iff} \quad \nabla \frac{\mu}{\tau} = \vec{0} \\ \vec{\omega} \cdot \nabla \frac{1}{\tau} &\geq 0, \quad \vec{\omega} = \vec{0} \quad \text{iff} \quad \nabla \tau = \vec{0}\end{aligned} \quad \text{in } Q. \tag{6.21}$$

The linearized form of (6.20) then reads

$$\vec{j} = -c\nabla \frac{\mu}{\tau}, \quad \vec{\omega} = k\nabla \frac{1}{\tau} \quad \text{in } Q, \tag{6.22}$$

where c and k are positive coefficients that may depend on τ, Y. If c is constant, (6.16) and the first of these laws yield (as $\mu := \partial f/\partial Y$)

$$\frac{\partial Y}{\partial t} = c\Delta\left(\frac{1}{\tau}\frac{\partial f}{\partial Y}\right) \quad \text{in } Q, \tag{6.23}$$

or equivalently,

$$\Delta^{-1}\frac{\partial Y}{\partial t} = \frac{c}{\tau}\frac{\partial f}{\partial Y} \quad \text{in } Q, \tag{6.24}$$

which can be compared with the *Cahn-Hilliard equation*, cf. (VII.6.2).

This is a *gradient flow* for the potential $f(\tau, \cdot)$, like (6.13)$_1$, but it corresponds to a different choice of the relaxation dynamics. [32]

---

[32] See also Sect. VII.6 for the concept of *gradient flow*.

If τ is stationary, we have (6.14); (6.17) then yields

$$\frac{1}{\tau}\frac{df}{dt} + \nabla \cdot \frac{\mu \vec{j}}{\tau} = \frac{\mu}{\tau}\frac{\partial Y}{\partial t} + \nabla \cdot \frac{\mu \vec{j}}{\tau} = \vec{j} \cdot \nabla \frac{\mu}{\tau} \quad \text{in } Q, \qquad (6.25)$$

whose meaning becomes more transparent if the temperature is constant. Let us define the *free energy production*

$$\xi := \frac{df}{dt} + \nabla \cdot (\mu \vec{j}). \qquad (6.26)$$

By (6.25) and the first part of (6.21), we have

if τ is constant in Q, then (i) $\xi \leq 0$, (ii) $\xi = 0$ iff $\nabla \frac{\partial f}{\partial Y} = \vec{0}$. (6.27)

In general, if Y is a conserved quantity, a closed mathematical system is obtained by coupling the equations (6.4), (6.16), (6.19) (or (6.20), and specifying appropriate initial and boundary conditions.

Conclusions. [33] We can draw the following conclusions for an incompressible system with state variables (τ, Y) and with no exterior source or sink of Y. By assuming that the two phenomenological laws are uncoupled, on the basis of the second principle of thermodynamics, we have derived the following properties.

(i) If the energy only flows by heat conduction, the evolution of Y can be described by the gradient flow dynamics $(6.11)_1$. If τ is stationary (but not necessarily uniform), then the free energy of the system decreases.

(ii) If Y is a conserved quantity, then we have (6.16). In this case the evolution of Y can be described by the gradient flow dynamics (6.23). If τ is constant (in space and time), then the free energy production ξ (defined in (6.26)) is nonpositive, and only vanishes at equilibrium. Thus in this case the free energy can diffuse, but its space average does not increase if there is no supply from the exterior.

These results extend a well-known principle of classical thermodynamics to the dynamical setting: at constant temperature, any system in equilibrium minimizes its free energy. They have been derived under the assumption that the phenomenological laws uncouple, which in general is not strictly true. However, this gives us some insight into the dynamics of the free energy, and anyway holds for isothermal phenomena.

This discussion can easily be extended to cases in which Y is a vector, or a set of variables.

---

[33] The discussion of Penrose and Fife [451; p. 47] can be compared with these conclusions.

Exercise.

6.1 Extend the previous discussion to the case in which there is an exterior source or sink r of Y; that is,

$$\frac{\partial Y}{\partial t} = -\nabla \cdot \vec{j} + r \qquad \text{in } Q. \tag{6.28}$$

Notice that if r is prescribed, then it does not appear in the expression of the entropy production π. On the other hand, if r depends on the state variables, then it can be treated as we did in Sect. VI.5 for chemical reactions.

V.7 Comments

In this chapter we only described two important generalizations of the classical Stefan model. Many other physically justified models of phase transitions can be considered. For more information we refer the reader to the proceedings of the meetings on *free boundary problems*, see the *Book Selection*.

The Hyperbolic Stefan Problem. One can replace the Fourier conduction law (IV.1.1) by the *Cattaneo law* $a\partial \vec{q}/\partial t + \vec{q} = -k(\theta,\chi)\nabla\theta$, where a is a positive relaxation constant. In this case the parabolic problem (IV.4.14) is replaced by the quasilinear hyperbolic system

$$\begin{cases} a\dfrac{\partial^2 \tilde{u}}{\partial t^2} + \dfrac{\partial \tilde{u}}{\partial t} - \nabla \cdot [k(\theta,\chi)\nabla\theta]\tilde{\theta} = f & \text{in } \mathcal{D}'(Q), \\ \tilde{u} \in \alpha(\tilde{\theta}) & \text{in } Q. \end{cases} \tag{7.1}$$

with α as in (IV.4.14)$_2$. The analysis of the corresponding initial and boundary value problem is still open, even for k constant, as far as this author knows. However, usually a is very small, [34] and in most applications the Fourier law (IV.1.2) is an acceptable approximation, although it represents instantaneous heat diffusion. See also Showalter and Walkington [505] for a related model.

Mixtures. The analysis of phase transition in heterogeneous systems is especially relevant for applications; see the books on the physics of phase transitions quoted in the *Book Selection*. The model that we outlined in Sect. V.2 has been studied by Tayler [528], Fix [236], Crowley [156], Crowley and Ockendon [157], Bermudez and Saguez [73], Wilson, Solomon and Alexiades [577], and others; see, for example, Crank [154; Sects. 1.3.7, 6.2.7], and Rubinstein [481 – 484].

---

[34] Here we refer to the scaling normally used for macroscopic experiments.

As far as this author knows, no result of existence of a solution has (yet) been proved for this model; nevertheless, the numerical simulations provide fairly acceptable results, and indeed the model is often used in engineering. This may be ascribed to the fact that this approach is not physically well-justified, since it neglects cross effects between heat and mass diffusion; on the other hand, in typical situations the omitted terms do not seem to be quantitatively very significant.

The approach based on nonequilibrium thermodynamics looks physically and analytically competitive. It has been applied to phase transitions in multi-component systems by Donnelly [198], and Luckhaus and V. [367]. In V. [555] it has been used to model a free boundary problem in ferromagnetism without hysteresis, representing the coupling between electromagnetic evolution and heat diffusion of the Joule heat.

About Nonequilibrium Thermodynamics. This theory is also known as *irreversible thermodynamics*. It can be applied to a multitude of processes; for a detailed account we refer the reader to the specific texts quoted in the *Book Selection*. Here we list the basic elements of this theory, and point out the corresponding analytical interpretation:

(i) A choice of the relevant state variables, $\vec{z}$, including the internal energy.

(ii) The *Gibbs formula*, that is, a constitutive relation between the entropy density, s, and those variables. For physical reasons, s must be a concave function of $\vec{z}$; this yields a *cyclically monotone* [35] relation between $-\vec{w} := \partial s / \partial \vec{z}$ and $\vec{z}$. This formula is assumed to hold outside equilibrium as well.

(iii) A system of balance equations, including the first principle. If chemical reactions are neglected, this reads $\partial \vec{w} / \partial t + \nabla \cdot \vec{J} = \vec{G}$, where $\vec{J}$ represents the generalized fluxes and $\vec{G}$ is a given source term.

(iv) A local formulation of the second principle.

(v) A set of *phenomenological laws* $\vec{J} = \vec{F}(\vec{z}, \nabla \vec{z})$, which are assumed to be consistent with the second principle. This provides the parabolicity of the differential system.

(vi) The linearization of the phenomenological laws in a neighbourhood of equilibrium, and the *Onsager relations*. The latter conditions are tantamount to the integrability of those laws, and consequently allow us to derive a priori estimates by multiplying the system by $\partial \vec{w} / \partial t$.

This approach can be applied to two-phase systems, provided that:

(i) the balance equations are assumed to hold in the sense of distributions,

(ii) the derivatives of the entropy and of its conjugate function are replaced by (multi-valued) superdifferentials,

(iii) the constitutive relations (i.e., the Gibbs formula and the phenomenological laws) are replaced by inclusions.

---

[35] See Sect. XI.5 for the definition of *cyclical monotonicity*.

The resulting system is of the form (III.4.7), or (III.1.1) if $\vec{F}$ does not depend on $\vec{z}$. This provides a unifying analytical treatment for a number of dissipative processes, and allows us to apply the results and techniques of Chap. III.

The *Curie principle* is slightly controversial. To conclude with a light note, we borrow another quotation from Astarita [29; p. 167].

> *Prigogine and Mazur:* "... all coupling between quantities of different tensorial character being forbidden..."
> *Kirkwood and Crawford:* "We must treat scalars, vectors and tensors separately, for entities of different tensorial character cannot interact (Curie theorem)."
> "What this theorem is, we may have some difficulty in divining, since the terms 'interact' and 'couple' are not found in books of algebra, although they do appear frequently in the The Arabian Nights." (C. Truesdell)

Chapter VI. The Gibbs-Thomson Law

Outline

The *free energy* and the *entropy* are the relevant thermodynamical potentials at constant temperature and at constant energy, respectively. Expressions are here derived for these potentials in solid–liquid systems, including interface contributions.

It is then shown that relative minimizers of the free energy (as well as relative maximizers of the entropy) fulfill the *Gibbs-Thomson law* at phase interfaces, and the associated contact angle condition on the boundary.

Finally, the so-called *phase field* model is outlined.

Prerequisites. Calculus and basic notions of thermodynamics are used in Sects. VI.1 and VI.2. Nonlinear functional analysis is used in Sect. VI.3 (whose reading however is not essential for understanding the remainder of the book). Simple elements of functional analysis and differential geometry are applied in Sect. VI.4.

Reference is made to material of Chap. IV.

VI.1 Free Energy

The Role of Free Energy. In classical thermodynamics, the free energy [1] is a *convex* functional of the state variables; at prescribed uniform temperature, equilibrium states minimize this functional. In Sect. V.6 we saw that the system tends to minimize its free energy also under the action of a nonuniform stationary temperature, if the state variables other than the temperature do not flow.

In the framework of a *mesoscopic* representation of solid–liquid systems, the interface between the phases and the boundary of the system also contribute to the free energy. In that setting the free energy is a *nonconvex* functional (see Sect. VI.4); hence it can also admit *relative* minimizers, which represent states of *metastable* equilibrium. In that framework, we say that a state is of *local equilibrium* if it is *close* (in a sense to be specified) to an (either absolute or relative) minimum point of the free energy.

---

[1] We refer to the *Helmholtz* free energy, since we assume that the system evolves at constant volume. If instead the pressure were prescribed, the same functional should be interpreted as the *Gibbs* free energy (or *free enthalpy*).

In this section we derive the expression of the free energy functional for stationary solid–liquid systems. In Sect. VII.3 we discuss stable and metastable states.

The Stationary Heat Equation. Under stationary conditions, the heat equation (IV.1.4) reduces to [2]

$$-\nabla \cdot [k(\theta, \chi)\nabla\theta] = h \quad \text{in } \mathcal{D}'(\Omega). \tag{1.1}$$

For the moment we assume that k does not depend on χ, so that the temperature is uncoupled from the phase. This hypothesis is rather restrictive, and in Sect. VI.3 we drop it; however, at thermal equilibrium (i.e., uniform temperature) obviously there is no point in making this restriction.

Under the preceding assumption on k, the dependence of k on the temperature can easily be treated by means of the *Kirchhoff transformation:* $\theta \mapsto \alpha(\theta) := \int_0^\theta k(\xi)d\xi$. This reduces (1.1) to the equation $-\Delta\alpha(\theta) = h$ in Ω. So, once appropriate boundary conditions have been specified, the field $\alpha(\theta)$ is uniquely determined. If $k > 0$, α can be inverted and θ is determined.

For instance, we can assume that our system is in contact with one or more reservoirs, and then prescribe θ (nonnecessarily constant) on the boundary $\partial\Omega$. In principle, any temperature field θ can then be maintained in Ω by an appropriate source distribution h. In this way the system can be stationary outside (global) thermodynamical equilibrium.

When not otherwise specified, we assume that the system is in *local* thermodynamical equilibrium; by this we mean that fields are *close* to values of (either stable or metastable) equilibrium, and vary slowly. We anticipate that this does not apply to nucleation, which we deal with in the next chapter.

Henceforth θ is just any function of $L^1(\Omega)$.

Bulk Contribution to the Free Energy. The free energy F is a functional of the temperature and phase fields: $F = F(\theta, \chi)$. It includes three phase-dependent contributions: one of them is proportional to the volume, and depends on temperature; a second one is due to contact between phases, and is proportional to the interface area; a third one is due to contact with some external material at the boundary of the system, and is proportional to the boundary area.

Let us denote by f the space density of free energy, by u the space density of internal energy, and by s the space density of entropy. These quantities are regarded as functions of the absolute temperature τ ($= \tau_E + \theta$) and of the *phase function* χ. In a neighbourhood of $\tau = \tau_E$, we can assume that C_V and L are constant; by (IV.4.7), we then have $u = C_V(\tau - \tau_E) + (L/2)\chi$, but for an unessential additive constant. The classical thermodynamical relations

$$s = -\frac{\partial f}{\partial \tau}, \qquad u = f + \tau s \tag{1.2}$$

---

[2] In this chapter we denote the source term by h in place of f, and reserve the latter notation for the density of free energy.

then yield $f - \tau \partial f/\partial \tau = C_V(\tau - \tau_E) + (L/2)\chi$. This equation has the general solution

$$f = -\frac{L}{2\tau_E}(\tau - \tau_E)\chi - C_V(\tau \log \tau + \tau_E) + \alpha\tau \qquad \forall \alpha \in \mathbf{R}.$$

Here no pressure or volume term occurs, since we assumed the material to be incompressible.

By integrating f over Ω, we get the following expression for the *linearized* bulk contribution to the free energy

$$\int_\Omega f\,dx = -\frac{L}{2\tau_E}\int_\Omega \theta\chi\,dx + \text{term independent of } \chi. \qquad (1.3)$$

In view of latter use, we note that for the (space) *density of entropy* s we have

$$s := -\frac{\partial f}{\partial \tau} = \frac{L}{2\tau_E}\chi + C_V \log \tau + \text{Constant}. \qquad (1.4)$$

Notice that this expression has been derived under the assumption that τ is in a neighbourhood of τ_E, since this is implicit in the assumption that u depends linearly on τ and χ.

Interface Contribution to the Free Energy. Henceforth we assume that Ω is a bounded domain of $\mathbf{R}^3$ of Lipschitz class, and define the *perimeter functional* $P: L^1(\Omega) \to \mathbf{R} \cup \{+\infty\}$:

$$P(\chi) := \begin{cases} \frac{1}{2}\int_\Omega |\nabla\chi| \ (\leq +\infty) & \text{if } |\chi| = 1 \text{ a.e. in } \Omega, \\ +\infty & \text{otherwise}; \end{cases} \qquad (1.5)$$

$\chi \mapsto \int_\Omega |\nabla\chi|$ is the total variation functional, cf. (XI.1.5).

We denote by σ the surface density of free energy [3] at the solid–liquid interface. We still assume that our system consists of a homogeneous and isotropic material. In this case σ is constant in Ω, and the total interface contribution equals $\sigma P(\chi)$.

Crystals are anisotropic, and indeed the isotropy assumption is more appropriate for liquid–vapour than for solid–liquid system. However, our discussion might be extended to heterogeneous and anisotropic materials, by replacing $\sigma \int_\Omega |\nabla\chi|$ by $\int_\Omega |A(x) \cdot \nabla\chi|$, where $A(x)$ is a positive definite 3×3-tensor, which depends continuously on $x \in \Omega$. [4]

---

[3] σ is also called the *surface tension coefficient*.

[4] In this respect, see Dobrushin, Kotecký and Shlosman [197], Taylor [529], Cahn and Taylor [118], Taylor and Cahn [530].

Boundary Contribution to the Free Energy. Our system is assumed to be in contact with a container and/or a gas at the boundary $\Gamma := \partial\Omega$. Let us denote by σ_L (σ_S, resp.) the surface density of free energy relative to a surface separating the liquid phase (solid phase, resp.) from an external material. This yields the following contribution to the free energy [5]

$$\int_\Gamma (\sigma_L \chi^+ + \sigma_S \chi^-) d\Gamma = \frac{\sigma_L - \sigma_S}{2} \int_\Gamma \chi d\Gamma + \frac{\sigma_L + \sigma_S}{2} |\Gamma| \quad (1.6)$$
$$\forall \chi \in BV(\Omega);$$

here $\chi^+ := \frac{1}{2}(\chi + |\chi|)$, $\chi^- := \frac{1}{2}(|\chi| - \chi)$, and we denote by $d\Gamma$ the surface element on Γ (recall that $|\chi| = 1$ a.e. in Ω).

Total Free Energy. By summing the three preceding contributions, we get the total free energy. For any prescribed temperature field $\theta \in L^1(\Omega)$, we deal with variations of the functional

$$F_\theta(\chi) := \begin{cases} \sigma P(\chi) + \dfrac{\sigma_L - \sigma_S}{2} \displaystyle\int_\Gamma \chi d\Gamma - \dfrac{L}{2\tau_E} \displaystyle\int_\Omega \theta \chi dx \\ \qquad\qquad\qquad\qquad \forall \chi \in \mathrm{Dom}(P), \qquad (1.7) \\ +\infty \qquad\qquad\quad \forall \chi \in L^1(\Omega) \setminus \mathrm{Dom}(P), \end{cases}$$

which represents the total phase contribution to the free energy, but for an additive constant that is independent of χ. Notice that

$$F_\theta(\chi) < +\infty \quad \text{iff} \quad \chi \in BV(\Omega) \text{ and } |\chi| = 1 \text{ a.e. in } \Omega; \quad (1.8)$$

hence, by minimizing F_θ, the *pure phase* constraint "$\chi = \pm 1$ a.e. in Ω" is automatically imposed. By (1.8), F_θ may also be regarded as a functional on $BV(\Omega)$.

The functional P is lower semicontinuous with respect to the strong topology of $L^1(\Omega)$. [6] To get the same property for F_θ, we need a further assumption.

Proposition 1.1 [7] *(Lower Semicontinuity) For any $\theta \in L^1(\Omega)$, the functional F_θ is lower semicontinuous with respect to the strong topology of $L^1(\Omega)$ iff*

$$|\sigma_L - \sigma_S| \leq \sigma. \quad (1.9)$$

We only provide a heuristic justification for this statement.

---

[5] See Theorem XI.2.6 for traces in $BV(\Omega)$. Henceforth we do not display the trace operator γ_0.

[6] See, e.g., Giusti [268; pp. 7], Evans and Gariepy [218; pp. 183], Ziemer [596; Sect. 5.2].

[7] See Massari and Pepe [379].

For a one-dimensional system $\Omega =]a, b[$, obviously we replace (1.7) by

$$F_\theta(\chi) := \sigma P(\chi) + \frac{\sigma_L - \sigma_S}{2}[\chi(a) + \chi(b)] - \frac{L}{2\tau_E}\int_a^b \theta\chi dx \quad \forall \chi \in \text{Dom}(P).$$

It is easy to see that in this case the condition (1.9) is needed for the lower semicontinuity to hold. For instance, if $\sigma_L - \sigma_S > \sigma$, let $\theta > 0$ in $]a, b[$ and set $\chi_n = -1$ in $]a, a + 1/n[$, $\chi_n = 1$ in $]a + 1/n, b[$ for any $n \in \mathbf{N}$. $\{\chi_n\}$ is a minimizing sequence that does not converge to any minimizer. On the other hand, if $\sigma_L - \sigma_S \leq \sigma$, then $\{\chi_n\}$ is not a minimizing sequence. This construction can be extended to more dimensions of space.

The condition (1.9) is not always fulfilled; for instance, for gold in contact with its vapour it does not hold. This means that solid and vapour should be always separated by a monoatomic liquid layer, and no superheating should be needed for liquid nucleation. [8]

This discussion can be extended to the case in which σ_L and σ_S vary on Γ.

Theorem 1.2 *(Existence) For any $\theta \in L^1(\Omega)$, if (1.9) is fulfilled, then the functional F_θ has an (absolute) minimizer.*

Proof. By Proposition 1.1, it suffices to apply Theorem XI.7.4 with $B = BV(\Omega)$ and $B_0 = L^1(\Omega)$. □

In general the (absolute) minimizer of F_θ is not unique since this functional is nonconvex.

In Sect. VI.4 we see that any relative minimizer of this functional fulfills the Gibbs-Thomson law (IV.2.2) and the contact angle condition (IV.2.4) with $\cos \omega = \sigma_S - \sigma_L \leq \sigma$. Of course this makes sense only if (1.9) is fulfilled.

Limit as $\sigma \to 0$. On the macroscopic length scale, $\sigma = 0$. It is not difficult to show that, as $\sigma \to 0$, F_θ Γ-converges [9] to

$$F_\theta^0(\chi) := \begin{cases} -\dfrac{L}{2\tau_E}\int_\Omega \theta\chi dx & \text{if } |\chi| \leq 1 \text{ a.e. in } \Omega, \\ +\infty & \text{otherwise.} \end{cases} \quad (1.10)$$

This functional is convex and lower semicontinuous in $L^1(\Omega)$. χ is a minimum point of F_θ^0 iff $\partial F_\theta^0(\chi) \ni 0$ in $L^\infty(\Omega)$, which is equivalent to the temperature-phase rule (IV.4.7).

---

[8] However, surprisingly enough, there is no evidence of this; see Chalmers [129; p. 85].

[9] In the sense of De Giorgi, see, e.g., De Giorgi and Franzoni [180], De Giorgi [179], and Dal Maso [163]. See also Sect. VI.5.

We conclude that the stationary macroscopic convex model can be retrieved as the limit of the mesoscopic nonconvex model as $\sigma \to 0$.

VI.2 Entropy

The Role of Entropy. In isolated systems the entropy is the relevant potential; its role is similar to that of the free energy in systems maintained at a prescribed temperature.

Isolated systems tend to maximize this potential. In classical thermodynamics, the entropy is a *concave* functional of the state variables. In the framework of a *mesoscopic* representation of solid–liquid systems, this functional is *nonconcave*; hence it can admit *relative* maximizers, which represent states of *metastable* equilibrium.

The Entropy Functional. We denote the surface densities of entropy, free energy and internal energy at phase interfaces by $\hat{s}$, $\hat{f}$ and $\hat{u}$, respectively. The following classical thermodynamical relations hold, cf. (1.2),

$$\hat{s} = -\frac{\partial \hat{f}}{\partial \tau}, \qquad \hat{f} = \sigma, \qquad \hat{u} = \hat{f} + \tau \hat{s}; \tag{2.1}$$

the second one is just the definition of σ.

So far we assumed the surface tension coefficient σ to be constant, as it is often considered in the literature; this yields $\hat{s} = 0$ and $\hat{u} = \hat{f}$. At variance with this hypothesis, here we assume that the surface density of entropy does not vanish, and that the surface density of internal energy is constant: $\hat{u} = u_0$. [10] (2.1) then yields

$$\hat{s} = \frac{\mu}{\tau_E}, \qquad \sigma = u_0 - \frac{\mu \tau}{\tau_E}, \tag{2.2}$$

where μ is a positive constant.

For the sake of simplicity, in this section we assume that $\sigma_L = \sigma_S$; however our developments can be extended to the general setting. By (1.4), for the total entropy S we get

$$S(\theta, \chi) = \frac{\mu}{\tau_E} P(\chi) + \int_\Omega \left(C_V \log(\tau_E + \theta) + \frac{L}{2\tau_E}\chi \right) dx, \tag{2.3}$$

but for an unessential additive constant.

---

[10] This is consistent with the analysis of Wollkind, Notestine and Maurer [582]; see also Wollkind and Notestine [581].

Isolated System. Let us now consider an isolated system without any internal heat source or sink. In this case the temperature is constant, and we can prescribe the total internal energy:

$$\theta = \bar{\theta} : \text{constant in } \Omega, \quad u_0 P(\chi) + \int_\Omega \left(C_V \bar{\theta} + \frac{L}{2}\chi \right) dx = U, \quad (2.4)$$

where U is a given positive constant.

Proposition 2.1 *The functional S has an (absolute) maximizer $(\bar{\theta}, \chi) \in \mathbf{R} \times L^1(\Omega)$ under the constraints (2.4).*

Proof. On account of the constraints, $(\bar{\theta}, \chi)$ maximizes the functional

$$\hat{S}(\bar{\theta}, \chi) := \frac{\mu}{u_0 \tau_E} \left[U - \int_\Omega \left(C_V \bar{\theta} + \frac{L}{2}\chi \right) dx \right] \\ + \int_\Omega \left(C_V \log(\tau_E + \bar{\theta}) + \frac{L}{2\tau_E}\chi \right) dx \quad (2.5)$$

in the set K of the $(\bar{\theta}, \chi) \in \mathbf{R} \times BV(\Omega)$ such that $|\chi| = 1$ a.e. in Ω and (2.4) holds. This constraint can be inserted into the functional $\hat{S}$, by using its indicator function I_K; let us denote by $\tilde{S}$ this modified functional, $\tilde{S} := \hat{S} + I_K$. To get the existence of the maximizer, then it suffices to apply Theorem XI.7.4 to $-\tilde{S}$, with $B := \mathbf{R} \times BV(\Omega)$ and $B_0 := \mathbf{R} \times L^1(\Omega)$. □

The Gibbs-Thomson Law. To circumvent the difficulties due to the irregularity of the constraint (2.4), we assume that, possibly after a rotation of the axes, the solid–liquid interface can be represented as the Cartesian graph of a function $\psi : G \to]a, b[$ of class C^1; here G is an open subset of $\mathbf{R}^2$, $a, b \in \mathbf{R}$ and $a < b$. This argument can be extended to the case in which such a representation can only be achieved locally.

Setting $y := (y_1, y_2)$, $\tilde{\nabla} := (\partial/\partial y_1, \partial/\partial y_2)$, defining the multi-valued function sign as in (XI.5.3), and assuming that the liquid phase stays below the graph of ψ, we have

$$\chi(y, \xi) = \text{sign}(\psi(y) - \xi) \quad \forall (y, \xi) \in G \times]a, b[,$$

$$P(\chi) = \int_G \sqrt{1 + |\tilde{\nabla}\psi(y)|^2} dy =: \tilde{P}(\psi).$$

Therefore $S(\bar{\theta}, \chi)$ can be represented as a *Fréchet differentiable* functional [11]

$$J(\bar{\theta}, \psi) := \frac{\mu}{\tau_E} \tilde{P}(\psi) + \int_G dy \int_a^b d\xi \left(C_V \log(\tau_E + \bar{\theta}) + \frac{L}{2\tau_E} \text{sign}(\psi(y) - \xi) \right).$$

---

[11] See Sect. XI.4 for the definition.

Hence there exists a *Lagrange multiplier* $\lambda \in \mathbf{R}$ such that $(\bar{\theta}, \chi, \lambda)$ is a stationary point of the *Lagrange functional*

$$\mathcal{L}(\bar{\theta}, \psi, \lambda)$$
$$:= J(\bar{\theta}, \psi) + \lambda \left[u_0 \tilde{P}(\psi) + \int_G dy \int_a^b d\xi \left(C_V \bar{\theta} + \frac{L}{2} \operatorname{sign}(\psi(y) - \xi) \right) - U \right].$$

By differentiating $\mathcal{L}(\bar{\theta}, \psi, \lambda)$ with respect to $\bar{\theta}$, we identify the Lagrange multiplier: $\lambda = -\bar{\tau}^{-1}$, where we set $\bar{\tau} := \tau_E + \bar{\theta}$. By (2.2), we get

$$\mathcal{L}\left(\bar{\theta}, \psi, -\frac{1}{\bar{\tau}}\right) = -\frac{\sigma}{\bar{\tau}} \tilde{P}(\psi) + \frac{U}{\bar{\tau}}$$
$$+ \int_G dy \int_a^b d\xi \left[\frac{L}{2} \operatorname{sign}(\psi(y) - \xi) \left(\frac{1}{\tau_E} - \frac{1}{\bar{\tau}} \right) + C_V \log(\tau_E + \bar{\theta}) - \frac{C_V}{\bar{\tau}} \bar{\theta} \right]$$

By taking the first variation of this functional with respect to ψ, we get the Gibbs-Thomson law (IV.2.2):

$$\kappa = -\frac{L\bar{\tau}}{2\sigma} \left(\frac{1}{\tau_E} - \frac{1}{\bar{\tau}} \right) = -\frac{L\bar{\theta}}{2\sigma\tau_E} \quad \text{on } \mathcal{S},$$

and the contact angle condition $\cos \omega = 0$, as we assumed $\sigma_L = \sigma_S$. This follows from Theorem 4.4, since the latter functional can be represented in terms of χ as

$$-\frac{\sigma}{\bar{\tau}} P(\chi) + \frac{U}{\bar{\tau}} + \int_\Omega \left[\frac{L\chi}{2} \left(\frac{1}{\tau_E} - \frac{1}{\bar{\tau}} \right) + C_V \log(\tau_E + \bar{\theta}) - \frac{C_V}{\bar{\tau}} \bar{\theta} \right] dx \quad (2.6)$$
$$\forall \chi \in L^1(\Omega).$$

Bulk Conditions. We distinguish two cases.

(i) *Coexisting Phases.* The surface tension coefficient is very small, and becomes significant only at a small length scale. Here we assume that $|\Omega|$ is not too small, and for the moment neglect the surface tension contribution. Setting $\bar{\chi} := |\Omega|^{-1} \int_\Omega \chi\, dx$, (2.4) then reads

$$|\Omega| \left(C_V \bar{\theta} + \frac{L}{2} \bar{\chi} \right) = U. \quad (2.7)$$

Hence the total entropy is a function of the scalars U and $\bar{\chi}$:

$$S = \tilde{S}(U, \bar{\chi}) := \frac{L|\Omega|}{2\tau_E} \bar{\chi} + C_V |\Omega| \log \left(\tau_E + \frac{U}{C_V |\Omega|} - \frac{L}{2C_V} \bar{\chi} \right). \quad (2.8)$$

Here $\bar{\chi}$ can be any element of $[-1, 1]$. If $\tilde{S}$ attains a maximum at $(U, \bar{\chi})$, then

$$\left(\frac{\partial \tilde{S}}{\partial \bar{\chi}}(U, \bar{\chi})\right)(\bar{\chi} - v) \geq 0 \qquad \forall v \in [-1, 1];$$

that is, as in (2.8) the argument of the logarithm equals the absolute temperature $\bar{\tau}$,

$$0 \leq \frac{L|\Omega|}{2}\left(\frac{1}{\tau_E} - \frac{1}{\bar{\tau}}\right)(\bar{\chi} - v) = \frac{L}{2C_V \bar{\tau} \tau_E}\left(U - \frac{L|\Omega|}{2}\bar{\chi}\right)(\bar{\chi} - v)$$
$$\forall v \in [-1, 1].$$

Hence at equilibrium

$$\begin{cases} \bar{\chi} = -1 & \text{if } U < -\frac{L}{2}|\Omega|, \\ \bar{\chi} = \frac{2U}{L|\Omega|} & \text{if } -\frac{L}{2}|\Omega| \leq U \leq \frac{L}{2}|\Omega|, \\ \bar{\chi} = 1 & \text{if } U > \frac{L}{2}|\Omega|; \end{cases} \qquad (2.9)$$

that is, by (2.7),

$$\begin{cases} \bar{\chi} = -1 & \text{if } \bar{\theta} < 0, \\ \bar{\chi} = \frac{2U}{L|\Omega|} & \text{if } \bar{\theta} = 0, \\ \bar{\chi} = 1 & \text{if } \bar{\theta} > 0, \end{cases} \qquad (2.10)$$

which yields the *temperature-phase rule* (IV.4.9). So undercooled and superheated states are here excluded.

(ii) *Single-Phase System*. Since there are no phase interfaces, the total entropy function $\tilde{S}$ still has the form (2.8), and must be maximized under the constraint $\bar{\chi} = $ constant $(= \pm 1)$. By (2.8) and (2.9), it is easy to see that there exists a value $\tilde{U} \in \,]-L|\Omega|/2, L|\Omega|/2[$ such that

$$\begin{cases} \tilde{S}(U, -1) > \tilde{S}(U, 1) & \text{if } U < \tilde{U}, \\ \tilde{S}(U, -1) = \tilde{S}(U, 1) & \text{if } U = \tilde{U}, \\ \tilde{S}(U, -1) < \tilde{S}(U, 1) & \text{if } U > \tilde{U}. \end{cases} \qquad (2.11)$$

Hence the absolute maximizers of S can be characterized as follows

$$\begin{cases} \chi = -1 & \text{if } U < \tilde{U}, \\ \chi = -1 \text{ or } \chi = 1 & \text{if } U = \tilde{U}, \\ \chi = 1 & \text{if } U > \tilde{U}. \end{cases} \qquad (2.12)$$

By (2.7), the conditions $\chi = -1$ in the whole Ω and $-L|\Omega|/2 < U \leq \tilde{U}$ correspond to a superheated solid; similarly, the conditions $\chi = 1$ in the whole Ω and $\tilde{U} \leq U < L|\Omega|/2$ correspond to an undercooled liquid. These states eventually decay by nucleation of a new phase, as we see in the next chapter. Notice that this requires *symmetry breaking:* the system, which is initially uniform in space, evolves into a nonuniform system.

By (2.8), expanding the logarithm in Taylor series, we have

$$\frac{\tilde{S}(U,\tilde{\chi})}{|\Omega|} = C_V \log \tau_E + \frac{U}{|\Omega|\tau_E} - \frac{1}{C_V \tau_E^2}\left(\frac{U}{|\Omega|} - \frac{L}{2}\tilde{\chi}\right)^2 \\ + C_V\, o\left(\frac{1}{C_V^2 \tau_E^2}\left(\frac{U}{|\Omega|} - \frac{L}{2}\tilde{\chi}\right)^2\right); \tag{2.13}$$

hence $\tilde{U} \simeq 0$.

This discussion also applies to small portions of systems near equilibrium, under the assumption of local equilibrium. In Chap. VIII dealing with evolution we take $\tilde{U} = 0$.

Conclusions. By the previous analysis, we can draw the following conclusions for an isolated system with no internal heat source. If surface tension is neglected, the states of *stable* equilibrium correspond to the standard temperature-phase rule (IV.4.9), without any superheating or undercooling. If surface contributions are introduced into the entropy functional, the Gibbs-Thomson law holds. However, either superheated or undercooled *metastable* states may occur prior to nucleation.[12] These states are consistent with the relation (2.12). For small systems (or subsystems) one can take $\tilde{U} = 0$.

About the Entropy Density. Now we derive an expression for the entropy density, in view of the study of the process of phase transition in Chap. VIII.

We neglect terms of the order of $(\theta/\tau_E)^3$ in the expression of the entropy density s, cf. (1.4), and denote an arbitrary constant by c. As $|\chi| = 1$, we have

$$\begin{aligned} s = -\frac{\partial f}{\partial \tau} &= \frac{L\chi}{2\tau_E} + C_V \log(\tau_E + \theta) + c \\ &\simeq \frac{L\chi}{2\tau_E} + \frac{C_V \theta}{\tau_E} - \frac{C_V \theta^2}{2\tau_E^2} + C_V \log \tau_E + c \\ &= \frac{u}{\tau_E} - \frac{1}{2\tau_E^2 C_V}\left(u - \frac{L}{2}\chi\right)^2 + C_V \log \tau_E + c \\ &= \frac{L}{2\tau_E^2 C_V}\left(u\chi - \frac{L\chi^2}{4}\right) + \gamma(u) = \frac{Lu\chi}{2\tau_E^2 C_V} + \tilde{\gamma}(u), \end{aligned} \tag{2.14}$$

---

[12] Stability and metastability are discussed in Sect. VII.3.

with $\gamma(u)$ and $\tilde{\gamma}(u)$ independent of χ. Notice that, by maximizing the last expression of $s = s_u(\chi)$ at constant u, one gets

$$\chi = \begin{cases} -1 & \text{where } u < 0, \\ -1 \text{ or } 1 & \text{where } u = 0, \\ 1 & \text{where } u > 0, \end{cases} \quad (2.15)$$

consistently with the previous discussion.

VI.3 Phase-Dependent Conductivity

So far we assumed that the thermal conductivity k did not depend on χ, so that the temperature was uncoupled from the phase. This allowed us to apply the Kirchhoff transformation, and this simplified our model very much. However, in general k does depend on the phase, and here we account for that dependence. [13]

It is convenient to extend k for values $|\chi| > 1$:

$$k(\theta, \chi) := \begin{cases} k(\theta, -1) & \text{if } \chi < -1, \\ k(\theta, 1) & \text{if } \chi > 1, \end{cases} \quad \forall \theta \in \mathbf{R}.$$

We assume that

$$k \in C^0\left(\mathbf{R}^2\right), \quad \exists k_1, k_2 > 0 : \forall (\theta, \chi) \in \mathbf{R}^2, \ k_1 \leq k(\theta, \chi) \leq k_2, \quad (3.1)$$

$$h \in H^{-1}(\Omega), \quad \chi \in L^1(\Omega), \quad g \in H^1(\Omega). \quad (3.2)$$

Let us now consider the problem of determining the temperature field that fulfills (1.1), coupled for instance with the nonhomogeneous Dirichlet boundary condition.

Problem 3.1 *To find $\theta \in H_0^1(\Omega) + g$ such that*

$$\int_\Omega k(\theta, \chi) \nabla \theta \cdot \nabla v \, dx = \langle h, v \rangle \quad \forall v \in H_0^1(\Omega). \quad (3.3)$$

(By $\langle \cdot, \cdot \rangle$ we denote the duality pairing between $H^{-1}(\Omega)$ and $H_0^1(\Omega)$.)

Lemma 3.1 [14] *Assume that h and g are as in (3.2), and that*

$$a \in C^0(\Omega \times \mathbf{R}), \quad \exists \nu > 0 : \forall (x, u) \in \Omega \times \mathbf{R}, \ a(x, u) \geq \nu, \quad (3.4)$$

---

[13] This section is an analytical island in a modelling sea, and can be skipped without compromising the understanding of the rest.

[14] This statement is a special case of Theorem 3.2 of Chipot and Michaille [133].

*the function $\xi \mapsto a(x,\xi)$ is Hölder continuous of index $\frac{1}{2}$,
uniformly with respect to $x \in \Omega$.* (3.5)

Then the solution of the problem

$$\begin{cases} u \in H_0^1(\Omega) + g, \\ \int_\Omega a(x,u)\nabla u \cdot \nabla v \, dx = \langle h, v \rangle & \forall v \in H_0^1(\Omega) \end{cases}$$ (3.6)

is unique, if it exists.

We can now prove the main result of this section.

Theorem 3.2 *(Existence and Continuous Dependence on the Data) Assume that (3.1), (3.2) are fulfilled, and that*

*the function $\theta \mapsto k(\theta, \xi)$ is Hölder continuous of index $\frac{1}{2}$,
uniformly with respect to $\xi \in \mathbf{R}$.* (3.7)

Then, for any $\chi \in L^1(\Omega)$, there exists one and only one solution $\theta = J(\chi)$ of Problem 3.1. Moreover, the operator J is continuous with respect to the strong topology of $L^1(\Omega)$ for its domain, and the weak topology of $H^1(\Omega)$ for its range.

Proof. Let us fix any $\chi \in L^1(\Omega)$. For any $\theta \in L^2(\Omega)$, there exists one and only one $\tilde{\theta} \in H_0^1(\Omega) + g$ such that

$$-\nabla \cdot \left[k(\theta, \chi) \nabla \tilde{\theta} \right] = h \quad \text{in } \mathcal{D}'(\Omega).$$ (3.8)

Thus an operator $\Lambda : L^2(\Omega) \to L^2(\Omega) : \theta \mapsto \tilde{\theta}$ is defined. We want to show that it has a fixed point.

We claim that Λ is strongly continuous. To show this, let $\theta_n \to \theta$ strongly in $L^2(\Omega)$, hence also in measure in Ω. Therefore $k(\theta_n, \chi) \to k(\theta, \chi)$ in measure in Ω, hence also strongly in $L^p(\Omega)$ for any $p \in [1, +\infty[$, by (3.1) and Proposition XI.3.10. By multiplying (3.8) by $\theta_n - g$, by (3.1) one gets that $\tilde{\theta}_n = \Lambda(\theta_n)$ is uniformly bounded in $H^1(\Omega)$. Hence there exists $\tilde{\theta}$ such that, possibly extracting a subsequence,

$$\Lambda(\theta_n) \to \tilde{\theta} \quad \text{weakly in } H^1(\Omega), \text{ hence strongly in } L^2(\Omega),$$

by the Rellich-Kondrachov theorem, cf. Theorem XI.3.4. By taking the limit in (3.8) written for θ_n and $\tilde{\theta}_n$, we conclude that θ and $\tilde{\theta}$ solve (3.8). Thus it is proved that Λ is strongly continuous.

The range of Λ is a closed and bounded subset of $H^1(\Omega)$, hence it is included in a strongly compact subset of $L^2(\Omega)$. Therefore, by the *Schauder fixed point theorem*,[15] Λ has a fixed point, and this solves (3.3).

By Lemma 3.1, under the assumption (3.7) the solution of Problem 3.1 is uniquely determined. Thus J is single-valued.

To show the continuity of J, let $\chi_n \to \chi$ strongly in $L^1(\Omega)$. Since J is uniformly bounded, there exists θ such that, possibly extracting a subsequence from $\{\theta_n := J(\chi_n)\}$,

$$\theta_n \to \theta \quad \text{weakly in } H^1(\Omega).$$

Passing to the limit in (3.3) written for θ_n and χ_n, one gets $\theta = J(\chi)$. Therefore the whole sequence $\{\theta_n\}$ converges. □

Now we consider the problem of minimizing the potential

$$\hat{F}: L^1(\Omega) \to \mathbf{R} \cup \{+\infty\} : v \mapsto F_{J(v)}(v),$$

with $F_{J(v)}$ defined as in (1.7). By the *direct method* of the calculus of variations, for example, by applying Theorem XI.7.4 with $B := BV(\Omega)$ and $B_0 := L^1(\Omega)$, one gets the following result.

Proposition 3.3 *(Existence) Under the assumptions of Proposition 3.1, there exists* $\chi \in L^1(\Omega)$ *such that*

$$\hat{F}(\chi) \leq \hat{F}(v) \quad \forall v \in L^1(\Omega). \tag{3.9}$$

In general the (absolute) minimizer of $\hat{F}$ is not unique, since this functional is nonconvex.

VI.4 The Gibbs-Thomson Law

In this section we derive the Gibbs-Thomson law and the contact angle condition by minimizing the functional F_θ, which represents the phase contribution to the free energy for a solid–liquid system at the relative temperature θ.

The functional F_θ was defined in (1.7); $L, \tau_E, \sigma, \sigma_L, \sigma_S$ are assumed to be positive constants. We require (1.9), and assume that $\theta \in L^1(\Omega)$ is prescribed.

Absolute and Relative Minimizers. We want to study absolute and relative minimizers of F_θ. Here relative minima are meant with respect to the strong topology of $L^1(\Omega)$, or equivalently of $L^p(\Omega)$ for any $p \in]1, +\infty[$, since we deal with bounded functions defined in a bounded set.

---

[15] See Theorem XI.9.4.

Definition. Any $\chi \in L^1(\Omega)$ is said to be a **relative minimizer** of F_θ iff there exists $K(\chi) > 0$ such that

$$F_\theta(\chi) \leq F_\theta(v) \qquad \forall v \in L^1(\Omega), \|v - \chi\|_{L^1(\Omega)} \leq K(\chi), \tag{4.1}$$

and χ is not an absolute minimizer.

Whenever we intend to include either relative and absolute minimizers, we simply speak of *minimizers*.

By Theorem 1.2, for any $\theta \in L^1(\Omega)$ there exists an (in general not unique) absolute minimizer of the functional F_θ.

The following result can be interpreted stating that the perimeter term *sustains* the occurrence of relative minima.

Proposition 4.1 (*Existence of Relative Minimizers*) *For any $\theta \in L^\infty(\Omega)$, if $\theta \leq 0$ in Ω, then "$\chi = 1$ in Ω" is a relative minimizer.*

However, if in (1.7) the σ-term were dropped, then F_θ would have no relative minimizer for any $\theta \in L^\infty(\Omega)$.

(Obviously an analogous result holds for $\theta \geq 0$ and $\chi = -1$.)

Proof. If we modify the field "$\chi = 1$ in Ω" by setting $\chi = -1$ in a ball $A \subset \Omega$, then the integral term of (1.7) decreases at most proportionally to $\|\theta\|_{L^\infty(\Omega)}|A|$, whereas the perimeter increases proportionally to the area of the surface of A. Hence, if the ball A is small enough, the surface contribution prevails over the bulk one. By the isoperimetric property of the sphere, the same holds *a fortiori* for shapes different from a ball. This yields the first claim.

To check the second statement, let A be any measurable subset of Ω in which the temperature-phase rule does not hold. If we modify χ by imposing that rule also in A, then F_θ decreases, as $\sigma = 0$. □

By anticipating developments of Sect. VII.3, we can say that at constant temperature (states represented by) absolute minimizers of the free energy persist for any time. On the other hand, relative minimizers may persist for some time, but eventually are destined to decay because of fluctuations. By the same token, maximizers and saddle points decays in an extremely short time.

The Gibbs-Thomson Law. By Proposition XI.8.3, the regularity of S which is required in the first part of the next theorem holds, whenever χ minimizes $F_\theta(\chi)$ and $\theta \in L^p(\Omega)$ for some $p > 3$.

Theorem 4.2 (*Gibbs-Thomson Law and Contact Angle Condition*) *Let $\theta, \chi \in L^1(\Omega)$ and* [16]

$$\liminf \frac{F_\theta(v) - F_\theta(\chi)}{\|v - \chi\|_{L^1(\Omega)}} \geq 0 \quad \text{as } v \to \chi \text{ strongly in } L^1(\Omega). \tag{4.2}$$

---

[16] (4.2) also reads $\partial^- F_\theta(\chi) \ni 0$ in $L^\infty(\Omega)$. See (XI.4.16) for the definition of the operator ∂^-. In particular this condition holds for any either relative or absolute minimizer of F_θ.

VI.4 The Gibbs-Thomson Law

Let us denote by S the boundary in Ω of the set $\Omega^+ := \{x \in \Omega : \chi(x) = 1\}$. Let N be an open subset of Ω; assume that $S \cap N$ is of class C^1 and $\theta \in W^{1,1}(N)$. Let us denote by $\vec{n}$ the unit normal vector to $S \cap N$ oriented towards Ω^+, and set $\kappa := \frac{1}{2} \nabla_S \cdot \vec{n}$. [17]

Then $\kappa \in L^1(S \cap N)$, and

$$\gamma_0 \theta = -\frac{2\sigma \tau_E}{L} \kappa \qquad \text{a.e. on } S \cap N. \tag{4.3}$$

Moreover, if S and the boundary Γ of Ω are of class C^1 in a neighbourhood $\hat{N}$ of a point of $S \cap \Gamma$, then [18]

$$\cos \omega = \frac{\sigma_S - \sigma_L}{\sigma} \qquad \text{a.e. on } S \cap \Gamma \cap \hat{N}, \tag{4.4}$$

where ω is the angle between $\vec{n}$ and the outward normal vector to Γ.

Proof. (i) By a standard procedure, we represent S locally in Cartesian form, and let the first variation of F_θ vanish for any local Cartesian perturbation of the interface.

By assumption, for any $\bar{x} \in S \cap N$ there exist an open set $G \subset \mathbf{R}^2$, $a, b \in \mathbf{R}$, and a function $\psi \in C^1(\bar{G}; [a, b])$, such that, possibly after a suitable rotation of the axes, $\bar{x} = (\bar{y}, \psi(\bar{y}))$ and

$$(\bar{G} \times [a, b]) \cap \Omega^+ = \{x = (y, z) \in \bar{G} \times [a, b] : z < \psi(y)\}.$$

Possibly replacing N and G by smaller sets, we can assume that the graph of ψ coincides with $S \cap N$. We denote by $\gamma_0 \theta$ the trace of θ onto $S \cap N$. By Theorem XI.2.3, we have $\gamma_0 \theta \in L^1(S \cap N)$, hence $\gamma_0 \theta \circ \psi \in L^1(G)$.

The contribution of $\bar{G} \times [a, b]$ to $F_\theta(\chi)$ equals

$$J(\psi) := \int_G \left(\sigma \sqrt{1 + |\tilde{\nabla} \psi(y)|^2} - \frac{L}{2\tau_E} \int_a^b \text{sign}(\psi(y) - \xi) \theta(y, \xi) d\xi \right) dy, \tag{4.5}$$

where $y := (y_1, y_2)$ and $\tilde{\nabla} := (\partial/\partial y_1, \partial/\partial y_2)$.

The functional $J : C^1(\bar{G}) \to \mathbf{R}$ is Fréchet differentiable, [19] and by (4.2) we have $J'(\psi) = 0$ in $C^1(\bar{G})'$. Hence, denoting by $\langle \cdot, \cdot \rangle$ the duality pairing between $\mathcal{D}'(G)$ and $\mathcal{D}(G)$, we have

$$\langle J'(\psi), v \rangle = \sigma \int_G \frac{\tilde{\nabla} \psi \cdot \tilde{\nabla} v}{\sqrt{1 + |\tilde{\nabla} \psi|^2}} dy - \frac{L}{\tau_E} \int_G (\gamma_0 \theta \circ \psi) v \, dy = 0 \qquad \forall v \in \mathcal{D}(G),$$

---

[17] That is, 2κ is the tangential divergence of $\vec{n}$ over $N \cap S$, in the sense of $H^{-1}(S \cap N)$, say. Thus κ represents the mean curvature of S.

$S \cap N$ is endowed with the two-dimensional Hausdorff measure. As it appears from the argument, $H^{-1}(S \cap N)$ can be defined via local charts; see also Aubin [40].

[18] $S \cap \Gamma \cap N$ is endowed with the one-dimensional Hausdorff measure.

[19] See Sect. XI.4 for the definition of the *Fréchet differential*, which is here denoted by J'.

that is

$$(2\kappa =) \sigma \tilde{\nabla} \cdot \frac{\tilde{\nabla}\psi}{\sqrt{1+|\tilde{\nabla}\psi|^2}} = -\frac{L}{\sigma T_E}\gamma_0 \theta \circ \psi \quad \text{in } \mathcal{D}'(G). \quad (4.6)$$

We conclude that $\kappa \in L^1(S \cap N)$ and (4.3) holds.

(ii) The contact angle condition can easily be derived if Γ has the shape of a cylinder in a neighbourhood of $x \in S \cap \Gamma$. In this case, possibly after a coordinate rotation, we can assume that x is of the form $x = (y, z)$, with the z-axis parallel to Γ. Then there exist an open set $G \subset \mathbf{R}^2$, $a, b \in \mathbf{R}$, and a function $\psi \in C^1(\bar{G})$ as in the preceding. Let us denote by E the maximal subset of ∂G such that $E \times]a, b[\subset \Gamma$.

The joint contribution of $\bar{G} \times [a, b]$ and $E \times [a, b]$ to $F_\theta(\chi)$ equals

$$\hat{J}(\psi) := J(\psi) + \frac{\sigma_L - \sigma_S}{2} \int_E [2\psi(y) - a - b] ds,$$

where $J(\psi)$ is defined as previously and ds represents the line element on E. This functional is *Fréchet differentiable,* and the condition $\hat{J}'(\psi) = 0$ reads

$$\langle \hat{J}'(\psi), v \rangle = \sigma \int_G \frac{\tilde{\nabla}\psi \cdot \tilde{\nabla}v}{\sqrt{1+|\tilde{\nabla}\psi|^2}} dy + (\sigma_L - \sigma_S)\int_E v ds - \frac{L}{T_E}\int_G (\gamma_0 \theta \circ \psi) v dy = 0$$

$$\forall v \in H^1(G).$$

After partial integration, by (4.6) we get

$$-\frac{\tilde{\nabla}\psi}{\sqrt{1+|\tilde{\nabla}\psi|^2}} \cdot \vec{\nu} = \frac{\sigma_L - \sigma_S}{\sigma} \quad \text{on } E, \quad (4.7)$$

where $\vec{\nu}$ denotes the bidimensional unit vector, normal to E and outward oriented. The latter condition is equivalent to the contact angle condition (4.4).

In the general case, locally one can reduce Γ to the shape of a cylinder by means of a nonlinear coordinate transformation, and then proceed as previously. (We do not detail this rather technical procedure.) □

Another Formulation of the Gibbs-Thomson Law. Although Theorem 4.2 suffices for our further developments, we indicate an alternative formulation of the Gibbs-Thomson law.

Proposition 4.3 (*Gibbs-Thomson Law in Integral Form*) [20] *Let $\theta \in W^{1,1}(\Omega)$, $\chi \in L^1(\Omega)$. Assume that the boundary S in Ω of the set $\Omega^+ := \{x \in \Omega : \chi(x) = 1\}$ is a surface of class C^1, and set*

$$\vec{n} := \frac{\nabla \chi}{|\nabla \chi|} \quad \text{(Radon-Nikodym derivative)} \quad \text{in } C_c^0(\Omega; \mathbf{R}^3)'. \quad (4.8)$$

---

[20] See Luckhaus [362].

Then the Gibbs-Thomson law (4.3) is equivalent to

$$\sigma \left\langle \sum_{i=1}^{3} \frac{\partial \xi_i}{\partial x_i} - \sum_{i,j=1}^{3} n_i n_j \frac{\partial \xi_i}{\partial x_j}, |\nabla \chi| \right\rangle \quad (4.9)$$

$$+ \frac{L}{\tau_E} \int_{\Omega} (\chi + 1) \sum_{i=1}^{3} \frac{\partial}{\partial x_i} (\theta \xi_i) dx = 0 \qquad \forall \xi \in C_0^{\infty}(\Omega; \mathbf{R}^3),$$

where by $\langle \cdot, \cdot \rangle$ we denote the duality pairing between $C_c^0(\Omega)$ and $C_c^0(\Omega)'$.

Proof. The measure $\vec{n}$ is concentrated on $\mathcal{S}$, where it equals the two-dimensional Hausdorff measure multiplied by the unit normal oriented towards Ω^+. Let us denote by γ_0 the trace operator $W^{1,1}(\Omega) \to L^1(\mathcal{S})$. [21] By the divergence theorem, (4.9) can be written in the form

$$2 \int_{\mathcal{S}} \left[\sigma \left(\sum_{i=1}^{3} \frac{\partial \xi_i}{\partial x_i} - \sum_{i,j=1}^{3} n_i n_j \frac{\partial \xi_i}{\partial x_j} \right) + \frac{L}{\tau_E} \gamma_0 \theta \sum_{i=1}^{3} \xi_i n_i \right] d\mathcal{S} = 0 \quad (4.10)$$

$$\forall \vec{\xi} \in C_0^{\infty}(\Omega; \mathbf{R}^3).$$

Note that the internal bracket equals $\nabla_{\mathcal{S}} \cdot \vec{\xi}$, the tangential divergence of $\vec{\xi}$ along $\mathcal{S}$.
 On $\mathcal{S}$ the field $\vec{\xi}$ can be decomposed into its tangential and normal components, which we denote by $\vec{\xi}_\tau$ and $\vec{\xi}_n$, respectively. So, setting $\xi_n := \vec{\xi} \cdot \vec{n}$, we have

$$\vec{\xi} = \vec{\xi}_\tau + \vec{\xi}_n = \vec{\xi}_\tau + \xi_n \vec{n} \qquad \text{on } \mathcal{S}.$$

By definition, $\nabla_{\mathcal{S}} \cdot \vec{n} = 2\kappa$ in the sense of $H^{-1}(\mathcal{S})$ (and also pointwise, if $\mathcal{S}$ is of class C^2). As $\nabla_{\mathcal{S}} \xi_n$ is orthogonal to $\vec{n}$, then we get

$$\nabla_{\mathcal{S}} \cdot \vec{\xi} = \nabla_{\mathcal{S}} \cdot \vec{\xi}_\tau + (\nabla_{\mathcal{S}} \xi_n) \cdot \vec{n} + \xi_n \nabla_{\mathcal{S}} \cdot \vec{n} = \nabla_{\mathcal{S}} \cdot \vec{\xi}_\tau + 2\xi_n \kappa \qquad \text{in } \mathcal{D}'(\mathcal{S}).$$

Moreover, $\int_{\mathcal{S}} \nabla_{\mathcal{S}} \cdot \vec{\xi}_\tau d\mathcal{S} = 0$ for any $\vec{\xi} \in C_0^{\infty}(\Omega; \mathbf{R}^3)$. Therefore, as $\eta = \xi_n$ is an arbitrary regular function $\mathcal{S} \to \mathbf{R}^3$, (4.10) is equivalent to

$$\left\langle 2\sigma\kappa + \frac{L}{\tau_E} \gamma_0 \theta, \eta \right\rangle_{\mathcal{S}} = 0 \qquad \forall \eta \in H_0^1(\mathcal{S}; \mathbf{R}^3). \quad (4.11)$$

Here we denote by $\langle \cdot, \cdot, \rangle_{\mathcal{S}}$ the duality pairing induced by the usual scalar product in $L^2(\mathcal{S}; \mathbf{R}^3)$. Finally, (4.11) is equivalent to the Gibbs-Thomson law (4.3). □

A Generalized Gibbs-Thomson Law. In Theorem 4.2 the assumption $\theta \in W^{1,1}(\Omega)$ entailed the existence of the trace of θ onto $\mathcal{S}$. Here we show that, if

---

[21] See Theorem XI.2.3.

we drop that hypothesis and assume that $\theta \in L^\infty(\Omega)$, then any absolute minimizer of F_θ fulfills the Gibbs-Thomson law in a generalized form.

At first, let us define the *approximate inferior/superior limit:* [22]

$$\operatorname{ap\,lim\,inf}_{z \to x} \theta(y) := \sup \left\{ \xi \in \mathbf{R} : \lim_{r \to 0} \frac{|\{z \in B_r(x) : \theta(z) < \xi\}|}{|B_r(x)|} = 0 \right\}, \quad (4.12)$$

$$\operatorname{ap\,lim\,sup}_{z \to x} \theta(y) := \inf \left\{ \xi \in \mathbf{R} : \lim_{r \to 0} \frac{|\{z \in B_r(x) : \theta(z) > \xi\}|}{|B_r(x)|} = 0 \right\}. \quad (4.13)$$

These approximate limits always exist, either finite or infinite.

Proposition 4.4 (*Generalized Gibbs-Thomson Law*) *Assume that Ω is of class C^1, $\theta \in L^\infty(\Omega)$, $\chi \in L^1(\Omega)$, and $F_\theta(\chi) = \inf F_\theta$. Then the boundary S in Ω of the set $\Omega^+ := \{x \in \Omega : \chi(x) = 1\}$ is a manifold of class $C^{1,\alpha}$, for any $\alpha \in {]}0, 1{[}$. Moreover, defining κ as in Theorem 4.2, $\kappa \in L^\infty(S)$,*

$$\operatorname{ap\,lim\,inf}_{z \to x} \theta(z) \leq -\frac{2\sigma\tau_E}{L}\kappa(x) \leq \operatorname{ap\,lim\,sup}_{z \to x} \theta(z) \qquad \text{for a.a. } x \in S, \quad (4.14)$$

and the contact angle condition (4.4) is fulfilled.

Proof. By Proposition XI.8.3, S is a manifold of class $C^{1,\alpha}$ for any $\alpha \in {]}0, \frac{1}{2}{[}$. Let the sequence $\{\theta_m\} \subset C^0(\bar{\Omega})$ be such that

$$\theta_m \to \theta \qquad \text{weakly star in } L^\infty(\Omega), \text{ and a.e. in } \Omega.$$

Such a sequence can be constructed by convolution with a sequence of smooth kernels converging to the *Dirac measure*, e.g., $\{k_m(x) := (\pi/m)^{-3/2} \exp\left(-m|x|^2\right)\}$. That is, $\theta_m := \int_{\mathbf{R}^3} \theta(x - \xi) k_m(\xi) d\xi$ for any $x \in \mathbf{R}^3$ and any m; here θ is set equal to 0 outside Ω.

For any m, let χ_m be a minimizer of the functional F_{θ_m}. The sequence $\{\chi_m\}$ is uniformly bounded in $BV(\Omega)$, therefore there exists $\chi \in BV(\Omega)$ such that, possibly extracting a subsequence, $\chi_m \to \chi$ weakly star in $BV(\Omega)$, hence also strongly in $L^1(\Omega)$.

For any m, let us denote by S_m the boundary in Ω of the set $\Omega_m^+ := \{x \in \Omega : \chi_m(x) = 1\}$. By Theorem XI.8.4, locally the S_ms are Cartesian graphs of functions ψ_ms defined on a same (smooth) set $G \subset \mathbf{R}^{N-1}$, possibly after a suitable rotation of the axes; moreover, these functions are uniformly bounded in $C^{1,\alpha}$, for any $\alpha < \frac{1}{2}$. Let us denote by $\vec{n}_m$ the unit normal vector to $S_m \cap N$ oriented towards Ω_m^+, and set $\kappa_m := \frac{1}{2}\nabla_{S_m} \cdot \vec{n}_m$.

---

[22] See, e.g., Evans and Gariepy [218; Sect. 5.9], and Ziemer [596; Sect. 5.9].

By setting $\vec{\varphi}(\vec{v}) := \vec{v}/\sqrt{1+|\vec{v}|^2}$ for any $\vec{v} \in \mathbf{R}^2$ and $\tilde{\nabla} := (\partial/\partial y_1, \partial/\partial y_2)$, (4.3) reads

$$-\frac{L}{\sigma \tau_E}\theta_m = 2\kappa_m = -\tilde{\nabla}\cdot\vec{\varphi}(\tilde{\nabla}\psi_m) = -\sum_{i,j=1,2}\frac{\partial\varphi_i}{\partial v_j}(\tilde{\nabla}\psi_m)\frac{\partial^2\psi_m}{\partial y_j \partial y_i} \quad (4.15)$$

a.e. in G.

The matrix $\{\partial\varphi_i/\partial v_j\}$ is positive definite, since $\vec{\varphi}$ is the gradient of the smooth convex function $\vec{v} \mapsto \sqrt{1+|\vec{v}|^2}$, is continuous and uniformly bounded with respect to y and m. As the sequence $\{\theta_m\}$ is bounded in $L^\infty(\Omega)$, the sequence $\{\psi_m\}$ is uniformly bounded in $C^{1,\alpha}_{\text{loc}}$, for any $\alpha \in]0, 1[$, by classical results of regularity for elliptic systems. [23] Hence ψ belongs to the same spaces. For any m, the contact angle condition (4.4) is fulfilled; hence it also holds for ψ, and $\mathcal{S}$ is of class $C^{1,\alpha}$ up to the boundary, for any $\alpha \in]0, 1[$.

Finally, as θ is essentially bounded, it is easy to see that for any $x \in \mathcal{S}$, for any m and any $\varepsilon > 0$, there exists a neighbourhood N_ε of x such that

$$\text{ap}\lim_{z\to y}\inf \theta(z) - \varepsilon \leq \theta_m(y) \leq \text{ap}\lim_{z\to y}\sup \theta(z) + \varepsilon \quad \forall y \in \mathcal{S}_m \cap N_\varepsilon.$$

By coupling these inequalities with (4.15) and passing to the limit at first as $m \to +\infty$ and then as $\varepsilon \to 0$, we get (4.14). □

Remarks. (i) The following example shows that (4.14) is optimal. Let $\theta \in C^0(\bar{\Omega})$, χ be an absolute minimizer of F_θ, and define $\mathcal{S}$ as in the preceding. By Theorem 4.2, the mean curvature of $\mathcal{S}$ equals $-L\theta/\sigma\tau_E$.

Now let $\tilde{\theta} \in L^\infty(\Omega)$ be any function such that $\tilde{\theta} > \theta$ in $\Omega^+ := \{x \in \Omega : \chi(x) = 1\}$, $\tilde{\theta} < \theta$ in $\Omega \setminus \Omega^+$. It is easy to see that χ minimizes $F_{\tilde{\theta}}$ as well.

(ii) Results similar to those of this section hold for the entropy functional S, defined in (2.3). In particular, any absolute maximizer of S subject to the constraint (2.4) is a stationary point of the Lagrange function (2.6), hence it fulfills the Gibbs-Thomson law (4.3). □

Exercises.

4.1 Discuss the generalization of Theorem 4.2 to the anisotropic setting, in which σ is replaced by a 3×3-tensor. Then discuss also the case in which σ depends on x.

4.2 Show that any $\chi \in L^1(\Omega)$ that fulfills (4.2) is necessarily an (either relative or absolute) minimum point of the functional F_θ.

Hint. In any neighbourhood of χ in $L^1(\Omega)$, the perimeter may have arbitrarily large variations.

---

[23] See, e.g., Gilbarg and Trudinger [266; Corollary 9.18].

VI.5 The Phase Field Model

In this section we outline a model that is set at a finer length scale than that we considered so far.

Double Wells. Let us consider a free energy functional of the form

$$J_\varepsilon(\chi) := \int_\Omega \left(\frac{\varepsilon a}{2} |\nabla \chi|^2 + \frac{1}{\varepsilon} W(\chi) - \frac{L}{2 T_E} \theta \chi \right) dx + \frac{\sigma_L - \sigma_S}{2} \int_\Gamma \chi \, d\Gamma \qquad (5.1)$$

$$\forall \chi \in H^1(\Omega).$$

Here a and ε are positive parameters, W is a so-called *double well* potential; for instance, [24]

$$W_1(v) := \left(1 - v^2\right)^2 \qquad \forall v \in \mathbf{R},$$

$$W_2(v) := 1 - v^2 + I_{[-1,1]}(v) = \begin{cases} 1 - v^2 & \text{if } |v| \le 1, \\ +\infty & \text{if } |v| > 1 \end{cases} \qquad \forall v \in \mathbf{R}.$$

W_1 is *Fréchet differentiable*, whereas W_2 is only *subdifferentiable* in the sense that $\partial^- W_2(v) \ne \emptyset$ for any $v \in \mathbf{R}$. [25] The corresponding functionals $J_{1\varepsilon}$ and $J_{2\varepsilon}$ exhibit several similar properties. By Theorem XI.7.4, either functional has an (in general not unique) absolute minimum point.

The terms $(\varepsilon a/2)|\nabla \chi|^2$ and $(1/\varepsilon)W(\chi)$ are in competition, if χ is nonuniform: the second one penalizes deviations from $|\chi| = 1$, whereas the first one penalizes the high gradients that are induced by sharp variations of χ. For small ε, any absolute minimum of $J_{i\varepsilon}$ ($i = 1, 2$) attains values close to ± 1 in the whole Ω, but for thin transition layers. In real systems, the coefficients a, ε are so small that the layer thickness is typically of the order of 10^{-7} cm. We regard this length scale as *microscopic*, since it is close to that of molecular phenomena. [26]

Dynamics. As the phase function χ is not a conserved quantity, we consider a relaxation dynamics of the form (V.6.13)$_1$: $c \partial \chi / \partial t + J'_{1\varepsilon}(\chi) = 0$ (the latter term is the *Fréchet differential* of $J_{1\varepsilon}$) yields the *Allen-Cahn* (or *Landau-Ginzburg*) equation [27]

$$c \frac{\partial \chi}{\partial t} - \varepsilon a \Delta \chi + \frac{4}{\varepsilon} \chi(\chi^2 - 1) = \frac{L\theta}{2 T_E} \quad \text{in } Q, \qquad (5.2)$$

where c is a positive coefficient.

Let us now consider $J_{2\varepsilon}$. This functional is of the form $J_{2\varepsilon} := \tilde{J}_{2\varepsilon} + I_{[-1,1]}$; $\tilde{J}_{2\varepsilon}$ is nonconvex and Fréchet differentiable, and $I_{[-1,1]}$ is convex lower semicontinuous.

---

[24] See Sect. XI.4 for the definition of $I_{[-1,1]}$.

[25] See (XI.4.11) and (XI.4.16) for these definitions.

[26] Actually, it is so close that the use of a continuous model might be questioned.

[27] See, e.g., Allen and Cahn [7].

Hence $\partial^- J_{2\varepsilon}(\chi) = \tilde{J}'_{2\varepsilon} + \partial I_{[-1,1]}$. Here the gradient flow dynamics reads $c\partial\chi/\partial t + \partial^- J_{2\varepsilon}(\chi) \ni 0$, that is,

$$c\frac{\partial \chi}{\partial t} - \varepsilon a \Delta \chi + \partial I_{[-1,1]}(\chi) - \frac{2\chi}{\varepsilon} \ni \frac{L\theta}{2\tau_E} \quad \text{in } Q; \tag{5.3}$$

this is equivalent to the following variational inequality

$$\begin{cases} |\chi| \leq 1 \text{ in } Q; \ \forall v \text{ such that } |v| \leq 1 \text{ in } Q, \\ \left(c\frac{\partial \chi}{\partial t} - \varepsilon a \Delta \chi - \frac{2\chi}{\varepsilon}\right)(\chi - v) \leq \frac{L\theta}{2\tau_E}(\chi - v) \quad \text{in } Q. \end{cases} \tag{5.4}$$

The so-called *phase field* model consists of coupling the energy balance equation (IV.4.5) with either (5.2) or (5.3). In either case sharp interfaces are replaced by thin transition layers, because of the regularity of χ due to the occurrence of the H^1-seminorm in the functional.

The functionals $J_{1\varepsilon}$ and F_θ (cf. (1.7)) are related by a Γ-limit operation. [28]

Theorem 5.1 (*Γ-limit*) [29] *Let $\theta \in L^1(\Omega)$ and $a := 9\sigma^2/128$. As $\varepsilon \to 0$, the family of functionals $\{J_{1\varepsilon}\}$ $\Gamma(L^1(\Omega)^-)$-converges to F_θ.*

The latter statement means that: [30]
(i) for any $u \in L^1(\Omega)$ and any sequence $\{u_\varepsilon\}$ such that $u_\varepsilon \to u$ strongly in $L^1(\Omega)$, $\liminf_{\varepsilon \to 0} J_{1\varepsilon}(u_\varepsilon) \geq F_\theta(u)$;
(ii) for any $u \in L^1(\Omega)$ there exists a sequence $\{u_\varepsilon\}$ such that $u_\varepsilon \to u$ strongly in $L^1(\Omega)$ and $\lim_{\varepsilon \to 0} J_{1\varepsilon}(u_\varepsilon) = F_\theta(u)$.
A similar result holds for $\{J_{2\varepsilon}\}$, with a different choice of a.

Macro-, Meso-, and Micro- Length Scales. We saw that solid–liquid systems can be described at three length scales; see Table 1.

(i) At a *macroscopic* scale the free energy is convex, cf. (1.10). Under several simplifying assumptions, the evolution is described by the weak formulation of the Stefan problem; this can represent either a sharp or a diffuse interface, depending on the occurrence of the mushy region.

(ii) At a *mesoscopic* scale the free energy is nonconvex, cf. (1.7), and the interface is sharp. The corresponding process is studied in Chap. VIII.

---

[28] In the sense of De Giorgi; see, e.g., De Giorgi and Franzoni [180], De Giorgi [179], and Dal Maso [163].

[29] See Modica [395, 396] and also Luckhaus and Modica [365] for related results. The argument rests on a well-known technique of Modica and Mortola [397]. See also Fonseca and Tartar [241] and Owen [438].

[30] Here we assume that ε vanishes along a sequence.

(iii) At a *microscopic* scale the free energy is nonconvex, cf. (5.1), and the interface is diffuse. The evolution can be described by the phase field model (IV.4.5), (5.2).

For instance, the mushy region is represented by $|\chi| < 1$ at a macroscopic length scale. At a mesoscopic scale, one distinguishes solid from liquid parts, hence $|\chi| = 1$. At a microscopic scale, interfaces are replaced by transition layers, across which χ varies smoothly; there $|\chi| \leq 1$. The process of *zooming out* from the microscopic to the mesoscopic scale is represented by the limit as $\varepsilon \to 0$, and from the mesoscopic to the macroscopic scale by the limit as $\sigma \to 0$.

Macroscopic scale	Mesoscopic scale	Microscopic scale
$\|\chi\| \leq 1$	$\|\chi\| = 1$	None (or $\|\chi\| \leq 1$)
Sharp/diffuse interface	Sharp interface	Diffuse interface
Convex free energy	Nonconvex free energy	Nonconvex free energy
No $\nabla \chi$ in free energy	$\sigma \int_\Omega \|\nabla \chi\|$	$\frac{\varepsilon a}{2} \int_\Omega \|\nabla \chi\|^2 dx$
$\chi \in L^\infty(\Omega)$	$\chi \in BV(\Omega)$	$\chi \in H^1(\Omega)$
$\theta = 0$ in $\{x : \|\chi(x)\| < 1\}$	$av_S = \theta + \frac{2\sigma\tau_E}{L}\kappa$ on S	$c\frac{\partial \chi}{\partial t} - \varepsilon a \Delta \chi + \frac{1}{\varepsilon} W'(\chi) = \frac{L\theta}{2\tau_E}$

Table 1. Comparison among some properties of
   (i) the macroscopic model of solid–liquid systems (i.e., the weak formulation of the Stefan problem, see Sect. IV.4 and (1.10));
   (ii) the mesoscopic model of surface tension of Sect. VI.1 (see Sect. VII.6 for the equation of mean curvature flow);
   (iii) the microscopic phase field model.
The lines from the second to the seventh, respectively, concern: the constraint acting on the phase function χ, the interface structure, the convexity of the free energy, the space interaction accounted for by the free energy, the space regularity of χ, the constitutive condition relating χ with θ.

VI.6 Comments

Here are the main issues of this chapter:
   (i) Construction of the (nonconvex) free energy and entropy functionals for two-phase systems, including surface tension contributions. These are the relevant potentials, respectively, at prescribed (nonuniform) temperature and at prescribed internal energy.
   (ii) Analysis of the stationary heat equation with phase-dependent conductivity.

(iii) Derivation of the Gibbs-Thomson law and of the contact angle condition for stationary points of the free energy and entropy functionals.

(iv) Outline of the microscopic phase field model.

Classical thermodynamics deals with convex potentials and with equilibrium; the theory of nonequilibrium thermodynamics overcomes the latter restriction. At a macroscopic length scale, two-phase systems can studied in the convex framework; but accounting for undercooling and phase nucleation requires the use of a mesoscopic length scale and of nonconvex potentials.

The thermodynamical foundations of surface tension in two-phase systems have been studied in particular by Gurtin [275 – 283].

The theory of hypersurfaces of prescribed mean curvature have been investigated by Allard, Almgren, Federer, Finn, Fleming, Nietsche, Reifenberg, and others; fundamental contributions have been given by the Italian school: Bombieri, Caccioppoli, De Giorgi, Emmer, Giaquinta, Giusti, Gonzalez, Massari, Miranda, Pepe, Tamanini, and others. See the *Book Selection*. Some results are reviewed in Sect. XI.8.

Chapter VII. Nucleation and Growth

Outline

For materials capable of attaining two phases, the free energy potential is a *nonconvex* functional of the state variables. Absolute and relative minimizers of this functional are related to *stable* and *metastable* states, respectively. Metastability explains the undercooling required for solid nucleation.

A model is proposed to select the metastable states that can be expected to persist on a prescribed time scale, at a given (possibly nonuniform) temperature.

The role of fluctuations in *nucleation* is briefly discussed.

Two modes of phase transition are distinguished: discontinuous evolution (e.g., nucleation), and continuous front motion. The latter is represented by *mean curvature flow* with forcing term, which is derived as *gradient flow* of the free energy functional. A modification is proposed to include nucleation.

Hysteresis in front motion is also outlined, on the basis of a model due to Cahn.

Prerequisites. Calculus and basic notions of thermodynamics are applied. Elements of functional analysis are used in Sects. VII.3 and VII.4.

Reference is made to material of Chap. IV and Sects. VI.1 and VI.4.

VII.1 Local and Global Minimizers

In view of the study of nucleation, in this section we deal with (either relative or absolute) minimizers of the free energy functional F_θ, cf. (VI.1.7).

Local Minimizers. We distinguish between compactly and noncompactly supported variations of the phase function χ.

Definition. We say that $v \in L^1(\Omega)$ is an (either relative or absolute) **local minimizer** of a functional $\Psi : L^1(\Omega) \to \mathbf{R}$, if v is an (either relative or absolute) minimizer of Ψ with respect to variations having compact support in (the open set) Ω. Local (either relative or absolute) maximizers are similarly defined. Minimizers (respectively, maximizers) with respect to any variation of the argument are said to be **global**. [1]

---

[1] Local and global minimizers should not be confused with relative and absolute minimizers, cf. Sect. VI.4. Indeed, relative minimizers refer to the strong topology of $L^1(\Omega)$, whereas local minimizers refer to the topology of Ω. Moreover this use of the term *local* has no relation with the physical concept of *local equilibrium*.

VII.1 Local and Global Minimizers

Let us consider a homogeneous liquid that occupies a ball Ω of radius $\tilde{R}$ at a uniform temperature $\theta < 0$. We want to study the conditions under which a solid phase is formed in some domain $A \subset \Omega$ that is not in contact with the boundary Γ of Ω (so-called *homogeneous nucleation*). To that aim, we investigate whether $\chi = 1$ is an (either relative or absolute) *local* minimizer of the free energy functional F_θ.

In our simplified analysis, the newborn phase is regarded as isotropic. This assumption is acceptable for liquid nucleation in a vapour, but not for crystallization in a liquid. Nevertheless, we still refer to solid nucleation, since most of our results can be extended to the anisotropic setting. [2]

By the isoperimetric property of the sphere, [3] if A is a ball, then the corresponding function χ ($\chi := 1$ in A, $\chi := -1$ outside) minimizes the potential F_θ among compact subsets of Ω of prescribed volume. Hence here we confine ourselves to varying the phase in balls.

Case of Radial Symmetry. We denote by φ_R the characteristic function of a solid ball of radius R contained in Ω; that is, $\varphi_R := 1$ in the ball, $\varphi_R := 0$ outside. The position of the center is immaterial, as we assumed the temperature $\theta(<0)$ to be uniform. Regarding the variation of the free energy functional as a function of R, we have [4]

$$\delta F_\theta(R) := F_\theta(\varphi_R) - F_\theta(\varphi_0) = 4\pi\sigma R^2 + \frac{4\pi L\theta}{3\tau_E} R^3 \quad \forall R \in [0, \tilde{R}], \quad (1.1)$$

whence

$$\delta F'_\theta(R_c) = 0 \quad \text{for} \quad R_c(=R_c(\theta)) := -\frac{2\sigma\tau_E}{L\theta}(>0), \quad (1.2)$$

$$\delta F_\theta(R_c) = \frac{16\pi\sigma^3\tau_E^2}{3L^2\theta^2} = \frac{4\pi\sigma R_c^2}{3} = \frac{L|\theta|}{2\tau_E}\frac{4\pi R_c^3}{3}. \quad (1.3)$$

The *critical radius* R_c coincides with the value prescribed by the Gibbs-Thomson law (IV.2.2). We assume that $\tilde{R} > 3\sigma\tau_E/L|\theta|$, so that $\delta F_\theta(\tilde{R}) < 0$. Then $R = \tilde{R}$, $R = R_c$, and $R = 0$ are an absolute minimizer, a relative maximizer, and a relative minimizer of δF_θ in $[0, \tilde{R}]$, respectively; cf. Fig. 1.

---

[2] Although it would be straightforward *to translate* our model from solid–liquid to liquid–vapour systems, in the latter setting several modifications should be introduced. For instance, one could not neglect variations of pressure and density.

Indeed, the classical nucleation theory was initially formulated for liquid nucleation, and then extended to solid nucleation.

[3] This result states that the sphere minimizes the surface area among all solids of prescribed volume (and, dually, it maximizes the volume among all the solids of prescribed surface area). This classical property plays an important role in the *geometric measure theory*.

[4] ΔF_θ would be a more customary notation than δF_θ for free energy variations, but we reserve the symbol Δ for the Laplace operator.

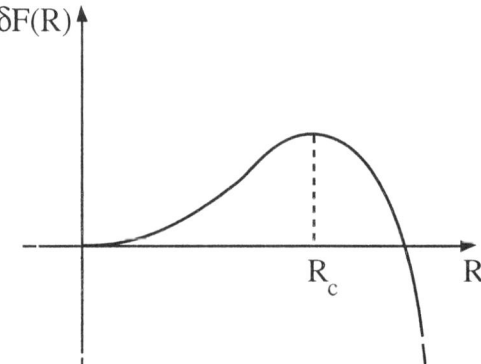

Figure 1. Graph of the function $\delta F_\theta(R) := 4\pi\sigma R^2 + (4\pi L\theta/3\tau_E)R^3$. $R = 0$ is in a *potential well*.

Let us set $\chi_R := 2\varphi_R - 1$ for any $R \geq 0$. $\chi_{\hat{R}}$ (corresponding to *all solid*) is a local absolute minimizer of F_θ, whereas χ_0 (corresponding to *all liquid*) is a local relative minimizer. On the other hand, χ_{R_c} maximizes F_θ with respect to radial variations of the solid phase, but not for all variation modes: χ_{R_c} is a *saddle point* for the potential F_θ (in a sense to be specified).

These conclusions can easily be extended to any set Ω that fulfills the following *internal ball condition:* for any $x \in \Omega$, there exists $y \in \Omega$ such that $x \in B_{R^*}(y) \subset \Omega$ (where $R^* := 3\sigma\tau_E/L|\theta|$).

Global Minimizers. Now we deal with arbitrary phase variations, still assuming that θ is uniform and negative. By Theorem VI.4.2, [5] if $\partial^- F_\theta(\chi) \ni 0$, cf. (VI.4.2), then the interface S has the critical mean curvature $\kappa_c(\theta) := 1/R_c(\theta)$, cf. (1.2), and the contact angle condition (IV.2.4) is fulfilled.

Let us assume that Ω is convex and consider the formation of a solid *cap* on part of Γ. By such a cap we mean the intersection of Ω with a ball; the radius and the (mean) curvature of that ball are referred to as the radius and the curvature of the cap. A *critical cap* (namely, a cap that has critical radius and fulfills the contact angle condition) represents a saddle point of the potential F_θ.

Let us now consider a system containing a particle of another material in its interior. This particle can be represented as a hole in the region Ω; for the sake of simplicity, let us assume that this hole has the shape of a ball of radius $\hat{R} < R_c$. If the material of the particle is such that $\sigma_L = \sigma_S$, then its surface gives no contribution to F_θ, and the previous discussion about the variation of F_θ by formation of a solid ball (here a solid *shell* that includes the particle) can be carried over to this setting. If instead $\sigma_L \neq \sigma_S$, the hole surface contributes to the free energy. The values of these parameters do not affect the critical radius, but modify the depth of the

---

[5] See (XI.4.16) for the definition of the operator ∂^-.

potential well corresponding to formation of a solid shell (hence they influence the probability of nucleation, as we see in the following.)

VII.2 Nucleation

In this section we describe nucleation, namely, formation of a new *phase*. [6] This corresponds to decay of a relative minimizer of the free energy functional, as a potential barrier is overcome. As we briefly discuss in Sect. VII.5, this process is essentially stochastic, and the lower the potential barrier, the more likely it is.

Here we discuss *isothermal* nucleation, and postpone to Chap. IX a comparison between isothermal and adiabatic nucleation. We mainly refer to *solid* nucleation, which is more relevant for applications than liquid nucleation.

So far, crystallization and melting have been represented as symmetrical phenomena. In reality, melting is a more stable process than crystallization. This is due to the size of coefficients, and to features that are accounted for neither by the standard Stefan model, nor by the generalization that we propose. These features include differences in the microscopic structure of the phases. For instance, to build a crystalline lattice, molecules must be brought from the liquid to the moving crystal surface, and this requires an *activation energy*. This can have remarkable effects, and indeed is at the basis of glass formation, as we see in the following. In melting there is no analogue of this energy.

Homogeneous and Heterogeneous Nucleation. Usually, one distinguishes between *homogeneous* and *heterogeneous* nucleation of a solid phase. In the first case, the newly formed solid is entirely included in the interior of the liquid; in the second case, the new phase is in contact with an external material, which may be either the container or consist of particles (so-called *impurities)* dispersed in the liquid.

These two modes of heterogeneous nucleation are actually based on the same mechanism: in either case nucleation depends on the (mesoscopic) geometry of the system, and on the contribution given to the free energy by contact with another material. Indeed one might represent dispersed particles as *holes* in the domain Ω, and introduce corresponding boundary contributions into the free energy functional, as we saw previously. In practise, the latter sort of heterogeneous nucleation is hardly distinguishable from homogeneous nucleation.

This model of heterogeneous nucleation requires the use of a mesoscopic (i.e., *small)* scale, as well as an accurate knowledge of the shape of the container and of the location of each impurity. Obviously this approach can be used for direct computation only in simplified settings.

---

[6] Here by *phase* we mean a connected component of either solid or liquid. The same term is also used with the meaning of *state of aggregation* — here either liquid or solid. So, for instance, a solid–liquid system may consist of more than two phases (first meaning).

As a rule, heterogeneous nucleation occurs at smaller undercooling than homogeneous nucleation, and indeed nucleation is often induced by dispersing particles (called *nucleants*) into the liquid. However, some experimental techniques allow the observation of homogeneous nucleation, by inhibiting heterogeneous nucleation.

Heterogeneous Nucleation. Let us compare homogeneous nucleation versus heterogeneous nucleation on a flat substrate; that is, formation of a solid critical ball versus formation of a solid critical cap, respectively. Denoting the contact angle by ω, the ratio between the volume of the cap and that of the ball having the same radius equals $(1+\cos\omega)^2(2-\cos\omega)/4$. A simple calculation (see Exercise 2.3) shows that the formation of a critical cap sitting on a flat surface corresponds to a variation of F_θ proportional to

$$\begin{aligned}\delta\tilde{F}_\theta(R_c,\omega) &:= \delta F_\theta(R_c)\frac{(1+\cos\omega)^2(2-\cos\omega)}{4} \\ &= \frac{4\pi\sigma^3\tau_E^2(1+\cos\omega)^2(2-\cos\omega)}{3L^2\theta^2}.\end{aligned} \quad (2.1)$$

This formula is consistent with (1.3). So here the potential barrier is proportional to the critical volume.

Therefore, for any prescribed undercooling, the free energy barrier for heterogeneous nucleation on the walls of the container is lower than for homogeneous nucleation. Hence heterogeneous nucleation is more likely than homogeneous nucleation, and the larger the contact angle the more likely it is. Moreover, the volume of the critical cap is smaller in a neighbourhood of a convex part of the boundary than near a concave part; this makes heterogeneous nucleation more likely in the first case than in the second. [7]

Obviously this discussion can be extended to the case of a nonuniform temperature, whenever temperature variations are not significant at the mesoscopic length scale.

Limit Cases. So far we assumed that $|\sigma_L - \sigma_S| \leq \sigma$. To examine the limit cases, we move from $|\sigma_L - \sigma_S| < \sigma$ and modify these parameters (obviously this does not correspond to any physical process!).

Let $\sigma_L - \sigma_S \to -\sigma$. Then the contact angle ω vanishes, by (IV.2.4); that is, a solid critical ball *retracts* from Γ until it gets tangent to Γ. In this extreme case heterogeneous and homogeneous nucleation cannot be distinguished. On the other hand, as $\sigma_L - \sigma_S \to \sigma$ the solid phase tends to spread over a flat part of Γ, because of the contact angle condition. In the limit, formation of a thin solid layer on a flat part of Γ causes no variation of the free energy, so that heterogeneous solid nucleation needs no undercooling.

---

[7] This conclusion is consistent with the analysis, e.g., of Chalmers [129; Chap. 3], Flemings [240; Chap. 9], and Woodruff [583; Chap. 2].

If $|\sigma_L - \sigma_S| > \sigma$, then (IV.2.4) becomes meaningless. If $\sigma_S - \sigma_L > \sigma$,[8] any minimizer of the functional F_θ corresponds to a configuration in which the solid phase does not touch the container, independently of the temperature. Hence here no undercooling is required for liquid nucleation to occur. Similarly if $\sigma_L - \sigma_S > \sigma$, solid nucleation occurs without superheating.

Volume Discontinuity. Nucleation is a discontinuous phenomenon, in that it corresponds to a change of the interface topology. In homogeneous nucleation the phase volume is always discontinuous, in heterogeneous nucleation only exceptionally can it be continuous. Let a concave part of the boundary have the critical mean curvature and give no contribution to the free energy, that is, $\sigma_L = \sigma_S$; then nucleation occurs by formation of an *infinitesimal* solid shell. Continuous nucleation does not need any phase fluctuation to occur. However, as this setting is nongeneric, henceforth we regard the phase volume as discontinuous at nucleation.

Critical Temperature of Nucleation. Let us consider the process of slowly and continuously cooling a liquid maintained at a uniform temperature $\theta(t) < 0$. We assume that the material (like a metal and unlike a polymer) is not highly viscous, that the sample is sufficiently large, that pressure is constant, and that heterogeneous nucleation is inhibited.

Under these conditions, by the classical theory of nucleation due to Volmer, Weber, Becker, and Döring,[9] the *nucleation rate* I (namely, the number of nuclei formed in a unit volume in a unit time) is of the form

$$I = C_1 \exp\left(-\frac{C_2}{\theta^2(\tau_E + \theta)}\right), \qquad (2.2)$$

at the beginning of the process. C_1 and C_2 depend on the material, and can be assumed to be positive constants; a typical order of magnitude of C_1 is 10^{33} s^{-1}.[10] This entails that I varies from almost zero to a very large value in a very small range of temperatures; actually, this range is so small that in practise it is often identified with a single value $\theta_c(< 0)$; cf. Fig. 2(a).

The law (2.2) can be compared with (5.4), which we derive in Sect. VII.5.

Growth. Once nuclei have appeared on the mesoscopic length scale, they grow until either they are stopped by impingement on other nuclei, or undercooling is eliminated by release of latent heat.

---

[8] For instance, this happens for gold, see Chalmers [129; p. 84-85].

[9] See, e.g., Chalmers [129; Chap. 3], Christian [135; Chap. 10], Flemings [240; Chap. 9], Turnbull [537], and Woodruff [583; Chap. 2].

[10] C_1 depends on θ, but this has little effect on the order of magnitude of I. Actually, several models of nucleation have been proposed, and they essentially differ in the form of C_1. In solid nucleation, the latter term must include a contribution representing the *activation energy*.

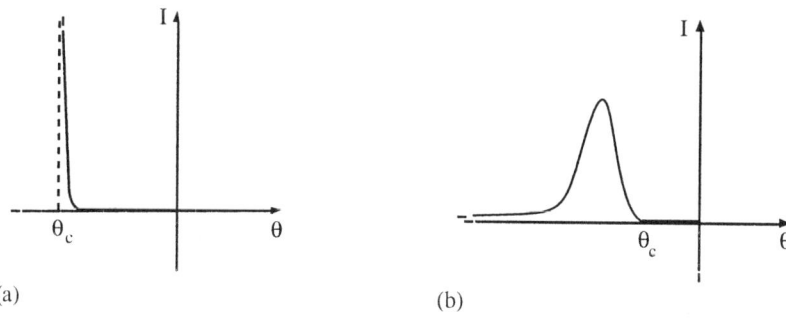

Figure 2. Nucleation rate I versus temperature θ. According to the classical theory of nucleation, cf. (2.2), I grows exponentially as the undercooling (i.e., $-\theta$) increases. In metals I increases very fast in so small a neighbourhood of the critical temperature θ_c, that the corresponding part of the graph can be replaced by a vertical half-line; cf. (**a**).
The qualitative behaviour of glassy and polymeric materials is different, cf. (**b**).

In principle, each nucleus becomes one of the *grains* that constitute the crystal, although the grain structure can be refined during the growth process. Thus the competition between nucleation and growth (see Sect. VII.5) determines the fineness of the grain lattice. In our simplified model, however, we assume that colliding phases coalesce, and keep no trace of the previous structure.

In glassy and polymeric materials, this process is slowed down to a very long time scale, because undercooling causes high viscosity, which reduces the mobility of particles moving to form the crystal lattice; cf. Fig. 2(b). This corresponds to a high *activation energy*. However if, after nucleation, the system is taken to a sufficiently high temperature (below the melting point), then nuclei grow quickly. For these materials (2.2) still holds, but C_1 depends on temperature, is not very large at θ_c, and drops down very fast as θ decreases.

Exercises.

2.1 Consider a material with surface tension coefficients σ, σ_L, σ_S, for liquid–solid, liquid–wall, solid–wall contacts, respectively. Calculate the variation in free energy $\delta \tilde{F}_\theta(R, \omega)$ due to formation of a solid cap on a planar wall of the container; here R is the radius of the cap, ω the contact angle with the substrate; cf. (2.1). Check that the critical radius R_c and the critical angle ω fulfill the Gibbs-Thomson law (IV.2.2) and the contact angle condition (IV.2.4).

2.2 In the setting of the previous exercise, let us define $\delta \tilde{F}_\theta^V(R_c, \omega)$ and $\delta \tilde{F}_\theta^A(R_c, \omega)$ as the volume and area contributions to the critical free energy variation $\delta \tilde{F}_\theta(R_c, \omega)$, respectively. Check that $\delta \tilde{F}_\theta^V(R_c, \omega) = -\frac{2}{3}\delta \tilde{F}_\theta^A(R_c, \omega)$. Show that this relation is due to the fact that $\delta \tilde{F}_\theta^A$ and $\delta \tilde{F}_\theta^V$ are respectively proportional to the square and the cube of the radius.

2.3 On the basis of the previous exercise, check that the formation of a critical solid cap in contact with a flat surface corresponds to a variation of F_θ equal to $\delta \tilde{F}_\theta(R_c, \omega) := \delta F_\theta(R_c)\gamma(\omega)$; here ω is the contact angle, and $\gamma(\omega) := (1 + \cos\omega)^2(2 - \cos\omega)/4$ equals the ratio between the volume of the cap and that of the ball having the same radius. [11]

2.4 For a critical solid cap staying over a nonplanar surface, does the statement of the previous exercise (namely, proportionality between volume of the cap and the corresponding free energy variation) still hold?

VII.3 Stable and Metastable States

In this section we deal with states of either *stable* or *metastable equilibrium,* for a system composed of a homogeneous material capable of attaining two phases, subject to a (possibly nonuniform) temperature field.

Definition. States that can persist indefinitely are said to be **stable.** States capable of persisting just for a limited (either short or long) time, but eventually destined to decay, are said to be **metastable.** [12]

Decay occurs because *fluctuations* allow the system to explore nearby states, see Sect. VII.5. Stability and metastability are closely related to minimization of the free energy functional F_θ, cf. (VI.1.7). We propose a first interpretation of absolute and relative minimizers of this potential, which we amend in the following.

Model 1. *The absolute (relative, resp.) minimizers of F_θ correspond to the states of stable (metastable, resp.) equilibrium.*

Selection of Metastable States. By Proposition VI.4.1, according to Model 1 at any uniform negative temperature the solid phase is stable and the liquid phase is metastable. As we said, the larger the undercooling, the lower the potential barrier that must be overcome to form a supercritical solid ball (or cap or shell), hence the more likely is (either homogeneous or heterogeneous) nucleation to occur. It is then of interest to select the metastable states that can be expected to persist for a prescribed time interval.

Henceforth we assume that we have fixed the time scale T of our observations. To be precise, one should deal with a stochastic formulation, and then consider the states whose *life expectation* is of the order of T. However, here we propose a more naïve approach, and use the term expectation in a more empirical sense.

Temperature Thresholds. Let us consider the process of slowly and continuously cooling a liquid metal that is maintained at a uniform temperature $\theta(t) < 0$, under the conditions that we already specified in the previous section. As we saw, a critical

---

[11] See also Skripov [508; pp. 24, 25].

[12] According to a different convention, the states that we call here metastable are included among the stable ones.

temperature $-a_1^* = -a_1^*(T) \leq 0$ is determined for *homogeneous* solid nucleation. Let $a_2^* = a_2^*(T) \geq 0$ be the critical temperature for homogeneous liquid nucleation.

Let us assume that the temperature thresholds for solid and liquid *heterogeneous* nucleation are $-\hat{a}_1$ and $\hat{a}_2$, respectively. As we saw, heterogeneous nucleation prevails over homogeneous nucleation for a uniform temperature field; that is, $\hat{a}_i < a_i$ for $i = 1, 2$. Anyway the latter can be observed by means of special experimental techniques.

In view of formulating a crude model capable of accounting for both sorts of nucleation, we propose considering two fields $a_i : \Omega \to \mathbf{R}^+$, $i = 1, 2$, which respectively coincide with a_i^* in the bulk, and attain the values $\hat{a}_i$ in a *(thin)* neighbourhood of the boundary. For the sake of simplicity, henceforth we assume that $(a :=)a_1 = a_2$ in the whole Ω, and that

$$a \in L^\infty(\Omega), \quad a > 0 \quad \text{a.e. in } \Omega. \tag{3.1}$$

However our developments can easily be extended to the case of $a_1 \neq a_2$.

(Here we are omitting the dependence on the time scale T, which we regard as prescribed; indeed $a = a(x, T)$.)

Modified Free Energy. In the next section we characterize the states that can be expected to persist on the time scale T, at a prescribed (nonuniform) temperature field u. To that aim, here we modify the free energy potential. Let us fix a function $\psi : \mathbf{R} \times \Omega \to \mathbf{R} \cup \{+\infty\}$ such that

$$\psi(-1, x) = \psi(1, x) = 0, \quad \frac{\partial}{\partial v}\psi(-1, x) = -\frac{\partial}{\partial v}\psi(1, x) = a(x),$$
$$\psi(\cdot, x) \text{ is strictly concave in }]-1, 1[, \quad \psi(v, x) = +\infty \text{ if } |v| > 1, \tag{3.2}$$

for a.a. $x \in \Omega$. For instance, one can set (see Fig. 3)

$$\psi(v, x) := \begin{cases} \frac{a(x)}{2}(1 - v^2) & \text{if } |v| \leq 1, \\ +\infty & \text{if } |v| > 1, \end{cases} \quad \text{for a.a. } x \in \Omega. \tag{3.3}$$

Then we introduce the following *modified free energy* functional:

$$\tilde{F}_\theta(v) := \frac{\sigma}{2} \int_\Omega |\nabla v| + \frac{L}{2\tau_E} \int_\Omega [\psi(v) - \theta v] \, dx + \frac{\sigma_L - \sigma_S}{2} \int_\Gamma v \, d\Gamma \tag{3.4}$$
$$\forall v \in L^\infty(\Omega).$$

Thus $\tilde{F}_\theta(v) < \infty$ if and only if $v \in BV(\Omega)$ and $|v| \leq 1$ a.e. in Ω. This statement should be compared with (VI.1.8).

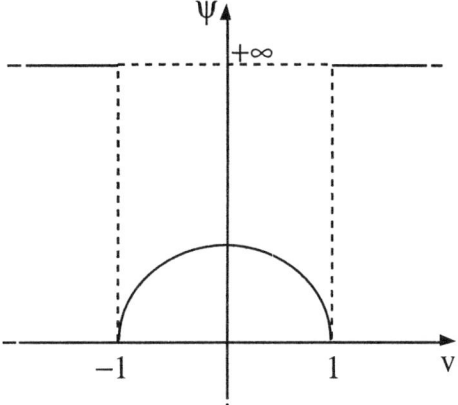

Figure 3. Graph of the function $\psi(\cdot, x) : \mathbf{R} \to \mathbf{R} \cup \{+\infty\}$ defined in (3.3), for a fixed $x \in \Omega$.

Proposition 3.1 *(Existence of Minimizers) Assume that $|\sigma_L - \sigma_S| \leq \sigma$. For any $\theta \in L^1(\Omega)$, there exists an absolute minimizer of $\tilde{F}_\theta$. For appropriate $\theta \in L^1(\Omega)$, there exists a relative minimizer as well.*

Proof. The first statement can be proved by means of the direct method of the calculus of variations, cf. Theorem XI.7.4, thanks to Proposition VI.1.1. (In general this minimizer is not unique since $\tilde{F}_\theta$ is nonconvex.)

To show the second part, let $-a < \theta < 0$ a.e. in Ω. Obviously $\chi = -1$ in Ω (corresponding to a solid) is the *absolute* minimizer. $\chi = 1$ in Ω (corresponding to an undercooled liquid) is a *relative* minimizer in $L^1(\Omega)$, as it is easy to check by an argument similar to that of Proposition VI.4.1. Although a small variation of χ in the L^1-norm may correspond to a small deviation in the L^∞-norm (at variance with what occurs for the unmodified functional F_θ), here the functional does not decrease, because of the condition on θ. □

Notice that relative minimizers of $\tilde{F}_\theta$ no longer occur, if either the total variation term is dropped or ψ is replaced by a convex function.

VII.4 Pure Phases

In this section we propose a model to represent metastable states that are likely to persist on a prescribed time scale.

Unlike F_θ, the functional $\tilde{F}_\theta$ is finite for functions $\chi \in BV(\Omega)$ that attain values in $]-1, 1[$ in some set of positive measure. The physical interpretation of these states in not obvious: apparently they cannot represent a mushy region, as here

VII. Nucleation and Growth

we are dealing with a mesoscopic length scale. However, here we show that this behaviour cannot occur for any either absolute or relative minimizer of $\tilde{F}_\theta$.

Theorem 4.1 *(Absolute Minimizers) Let $\theta \in L^1(\Omega)$. If χ is an absolute minimizer of $\tilde{F}_\theta$, then*

$$|\chi| = 1 \qquad \text{a.e. in } \Omega. \qquad (4.1)$$

Proof. Let us set

$$H_s(\xi) := \begin{cases} 0 & \text{if } \xi < s, \\ 1 & \text{if } \xi \geq s, \end{cases} \qquad \forall s. \xi \in \mathbf{R}. \qquad (4.2)$$

We use the classical *coarea formula* [13]

$$\int_\Omega |\nabla v| = \int_{\mathbf{R}} ds \int_\Omega |\nabla H_s(v)| \; (\leq +\infty) \qquad \forall v \in L^1(\Omega), \qquad (4.3)$$

and the following formula

$$\forall f \in L^1(\Omega), \forall g \in L^\infty(\Omega), \forall c \in \mathbf{R}, \text{ if } f \geq c \text{ a.e. in } \Omega, \text{ then}$$
$$\int_\Omega f(x)g(x)dx = \int_c^{+\infty} ds \int_\Omega H_s(f(x))g(x)dx + c \int_\Omega g(x)dx, \qquad (4.4)$$

which follows from the elementary identity $y = \int_c^{+\infty} H_s(y) ds + c$ for any $y, c \in \mathbf{R}$, $y \geq c$. Obviously, a similar formula holds with Γ in place of Ω.

Let us set

$$B_v(s) := \frac{\sigma}{2} \int_\Omega |\nabla H_s(v)| + \frac{\sigma_L - \sigma_S}{2} \int_\Gamma H_s(v(x)) d\Gamma - \frac{L}{2\tau_E} \int_\Omega \theta(x) H_s(v(x)) dx$$
$$\forall v \in L^1(\Omega), \forall s \in [-1, 1].$$

By (3.4), (4.3) and (4.4) (applied to the integrals over Ω and Γ with $c := -1$), we have

$$\tilde{F}_\theta(\chi) = \int_{-1}^1 B_\chi(s) ds + \frac{L}{2\tau_E} \int_\Omega [\psi(\chi(x), x) + \theta(x)] dx - \frac{\sigma_L - \sigma_S}{2} |\Gamma|. \qquad (4.5)$$

There exists $\hat{s} \in \,]-1, 1[$ such that

$$B_\chi(\hat{s}) \leq \frac{1}{2} \int_{-1}^1 B_\chi(s) ds. \qquad (4.6)$$

---

[13] See Fleming and Rishel [239] and also, e.g., Giusti [268; p. 20], Evans and Gariepy [218; Sect. 5.5], and Ziemer [596; Sect. 2.7].

Setting
$$\hat{\chi}(x) := \begin{cases} 1 & \text{if } \chi(x) \geq \hat{s}, \\ -1 & \text{if } \chi(x) < \hat{s}, \end{cases}$$

for any $s \in [-1, 1]$, a.e. in Ω we have $\hat{\chi} \geq s$ if and only if $\chi \geq \hat{s}$; that is, $H_s(\hat{\chi}) = H_{\hat{s}}(\chi)$. Hence $B_{\hat{\chi}}(s) = B_\chi(\hat{s})$ for any $s \in [-1, 1]$, and (4.6) yields

$$\int_{-1}^{1} \left[B_{\hat{\chi}}(s) - B_\chi(s) \right] ds = 2B_\chi(\hat{s}) - \int_{-1}^{1} B_\chi(s) ds \leq 0. \tag{4.7}$$

As χ is an absolute minimizer of $\tilde{F}_\theta$, by (4.5) and (4.7) we get

$$\int_\Omega \psi(\chi(x), x) dx \leq \int_\Omega \psi(\hat{\chi}(x), x) dx.$$

Moreover, by the definitions of $\hat{\chi}$ and ψ, for a.a. $x \in \Omega$ we have $\psi(\hat{\chi}(x), x) = 0$ and $\psi(\chi(x), x) \geq 0$. Hence $\psi(\hat{\chi}(x), x) = 0$ for a.a. x, and this yields (4.1). □

Remark. A similar result states that *relative* minimizers of $\tilde{F}_\theta$ also fulfill (4.1).[14]

□

Theorem 4.2 *(Relation between F_θ and $\tilde{F}_\theta$)* Let $\theta \in L^1(\Omega)$. Then:

$$\begin{aligned}&\chi \text{ is an absolute minimizer of } \tilde{F}_\theta \text{ iff}\\&\text{it is an absolute minimizer of } F_\theta;\end{aligned} \tag{4.8}$$

$$\begin{aligned}&\text{if } \chi \text{ is a relative minimizer of } \tilde{F}_\theta,\\&\text{then it is a relative minimizer of } F_\theta.\end{aligned} \tag{4.9}$$

The converse of the latter statement does not hold in general.

Proof. Since

$$\begin{aligned}\tilde{F}_\theta(v) &= F_\theta(v) & \forall v \in L^\infty(\Omega), |v| = 1 \text{ a.e. in } \Omega, \\ \tilde{F}_\theta(v) &< F_\theta(v) (= +\infty) & \forall v \in L^\infty(\Omega), |\{x \in \Omega : |v(x)| < 1\}| > 0,\end{aligned} \tag{4.10}$$

the statements (4.8) and (4.9) follow from the previous theorem and its extension to relative minimizers. The final statement can easily be checked if $|\theta| > a$ a.e. in Ω, by the argument of Proposition VI.4.1. □

---

[14] For the proof see V. [560].

We propose an interpretation of this more restricted class of relative minimizers. We recall that the functional $\tilde{F}_\theta$ contains the function a that depends on the time scale T ($a = a(x, T)$).

Model 2. *The absolute minimizers of $\tilde{F}_\theta$ represent the states of stable equilibrium. Given a time scale T, the relative minimizers of $\tilde{F}_\theta$ represent the metastable states that can be expected to persist on that time scale.*

This is consistent with *Model 1* and with experimental evidence for a uniform temperature field θ. According to this model solid nucleation starts when the undercooled liquid fails to be a (global) relative minimizer of $\tilde{F}_\theta$.

If a is uniform in Ω, then it can easily be identified by cooling the system at a uniform temperature; in fact a coincides with the nucleation temperature, as $(\partial/\partial v)[\psi(v, \cdot) - \theta v] = 0$ iff $|\theta| = a$, by (3.2).

VII.5 From Nucleation to Growth

Nucleation is a *microscopic stochastic* process. A precise *deterministic* description of phase nucleation at a mesoscopic length scale does not seem feasible, nevertheless here we propose a simplified approach, in order to get a *qualitative* understanding of the phenomenon. This is only meant as a naïve surrogate of a more precise model, which should involve stochastic differential equations.

A Criterion for Deterministic Evolution. In Sect. V.6 we derived the following result, for thermodynamical systems in which the state variables other than the temperature do not flow, cf. (V.6.13)$_1$:

$$\text{under the action of a prescribed temperature field,} \atop \text{a thermodynamical system tends to reduce its free energy;} \quad (5.1)$$

this potential was defined in (VI.1.7). More suggestively, one can say that the tendency to decrease the free energy *drives* the evolution. This law applies to a broad class of processes. In the next section we represent the evolution of solid–liquid systems by a *gradient flow* with respect to the functional $F_\theta(\chi)$.

As we discussed in Sect. VI.4, the nonconvex functional F_θ may admit relative minimizers, which correspond to the *wells* of the

$$\text{graph of } F_\theta := \{(\chi, F_\theta(\chi)) : \chi \in L^1(\Omega)\} \subset L^1(\Omega) \times (\mathbf{R} \cup \{+\infty\}).$$

These states are stationary with respect to the gradient dynamics, and indeed may persist for some time. However, *fluctuations* (i.e. random variations of the state variables) occur and eventually allow the system to explore states outside the potential well, from which descent along the graph of the potential F_θ continues.

We are especially concerned with evolution of an undercooled liquid. We assume that the temperature field is prescribed, and neglect the *activation energy* needed for solid nucleation.

The Radial Setting. Homogeneous (heterogeneous, resp.) nucleation corresponds to decay of a local (global, resp.) relative minimizer of the free energy functional.

Let us consider *homogeneous* nucleation. As we saw, in an isotropic material, newly born nuclei can be assumed to be radially symmetric. In this radial setting, smooth descent along the graph of δF_θ (defined in (1.1), cf. Fig. 1) can be described by the O.D.E.

$$c\frac{dR}{dt} = -\frac{\delta F'_\theta(R)}{8\pi R^2} = -\frac{\sigma}{R} - \frac{L\theta}{2\tau_E} \quad \text{where } R > 0, \tag{5.2}$$

where c is a positive coefficient. This is a special example of *(mean) curvature flow* (with forcing term), and is discussed in the next section.

As it is easy to see, at any temperature this equation cannot account for the exit from the potential well $R = 0$. Actually, nucleation is due to fluctuations, which can be represented by the simple *stochastic* O.D.E. [15]

$$cdR = -\left(\frac{\sigma}{R} + \frac{L\theta}{2\tau_E}\right) dt + \varepsilon dW; \tag{5.3}$$

here ε is a small parameter, and W represents the *Wiener process*. Note that in the deterministic limit we retrieve (5.2). [16]

Instead of entering into the analysis of the equation (5.3), we just draw some qualitative conclusions. It is known that the probability of getting to the rim of the potential well in a small time interval δt is proportional to $\delta t \exp\left[-\delta F_\theta(R_c(\theta))/k\tau\right]$, cf. (1.3). [17] At that point, descent outside the well is still governed by fluctuations for a short range of values of R, as $\delta F'_\theta(R_c(\theta)) = 0$. Anyway, the evolution from $R = R_c$ to, say, $R = R_1 := 3\sigma\tau_E/L|\theta|$ (notice that $\delta F_\theta(R_1) = \delta F_\theta(0) = 0$) is more likely (actually, much more likely) than the initial *uphill* fluctuation from $R = 0$ to $R = R_c$. If $\delta F'_\theta(R_1)$ is not too small, then the gradient term prevails over fluctuations, and the nucleus grows by mean curvature flow.

Fluctuations and Singularities in Phase Transitions. Although statistical mechanics [18] provides the appropriate framework to deal with fluctuations, let us consider the following simplified criterion, which is consistent with the previous analysis of the radial setting: [19]

---

[15] See, e.g., Gihman and Skorohod [265].

[16] Qualitatively similar results would be obtained if the stochastic term εdW were added to the equation (6.5).

[17] See, e.g., Freidlin and Wentzell [243].

[18] See, e.g., Landau and Lifshitz [340] and Huang [302].

[19] This issue may be compared with the discussion of Christian [135; Chap. 13].

"For any $\chi_1, \chi_2 \in \text{Dom}(F_\theta)$, if $F_\theta(\chi_2) < F_\theta(\chi_1)$ and there exists a path from χ_1 to χ_2 along which F_θ is nondecreasing, then in a small time interval δt the density of probability of a phase fluctuation $\chi_1 \to \chi_2$ is proportional to $\delta t \exp\{[F_\theta(\chi_1) - F_\theta(\chi_2)]/k\tau\}$." Here k is the *Boltzmann constant*.

A precise formulation of the preceding statement would require the use of a probability measure in $L^1(\Omega)$, and would go far beyond the limits of this essay.

Time Scale of Nucleation. Let us assume that in the initial stage nucleation is a *Poisson process*. By (1.2), the expected time for homogeneous nucleation in a prescribed region is then proportional to

$$\exp\left\{-\frac{\delta F_\theta(R_c(\theta))}{k\tau}\right\} = \exp\left\{-\frac{16\pi\sigma^3\tau_E^2}{3kL^2\theta^2(\tau_E + \theta)}\right\}, \tag{5.4}$$

consistently with the rate law (2.2). For a flat substrate characterized by a contact angle ω, by (2.1) the expected time for heterogeneous nucleation is proportional to

$$\exp\left\{-\frac{\delta \tilde{F}_\theta(R_c(\theta),\omega)}{k\tau}\right\} = \exp\left\{-\frac{4\pi\sigma^3\tau_E^2(1+\cos\omega)^2(2-\cos\omega)}{3kL^2\theta^2(\tau_E + \theta)}\right\}. \tag{5.5}$$

Either quantity can be regarded as the time scale of the respective nucleation mode.

In this derivation we have neglected the *activation energy*, which is required to take particles to the surface of the growing crystal and to fit them into the lattice. This can be represented by means of a temperature-dependent pre-exponential factor in the preceding expressions of the nucleation rate. At fast cooling rates, when a high undercooling is reached, for certain materials the mobility of the particles is much reduced by viscosity. The activation energy can then become so large that solidification is slowed down to a long time scale, and a glass be formed; see also Sects. V.1 and VII.2. However, here we deal with slow cooling rates.

Solid nucleation occurs with exchange of latent heat of phase transition, and in principle this modifies the *mesoscopic* temperature field (so-called *recalescence*). However, at the initial stage of the process the released heat is quite small, as it is proportional to the nucleated volume, and does not sensibly affect the *macroscopic* temperature. At a latter stage of growth, the released heat becomes relevant, and phase transition must be coupled with the energy balance equation. The question then arises of matching the two phases of the process: (i) isothermal nucleation, and (ii) front motion coupled with heat diffusion.

Conclusions. The preceding discussion yields the following picture of phase transition. Extrapolating the radially symmetrical case, under the action of a slowly changing temperature field, one can distinguish the following modes:

(i) *Discontinuous Phase Transition* (e.g., nucleation). This is due to fluctuations, which remove the system from relative minimizers of the free energy. At the microscopic length scale this process is stochastic, but its time scale can be determined.

(ii) *Continuous Phase Transition* (i.e. front motion). This corresponds to descent along the graph of the free energy functional; at the mesoscopic length scale, this can be represented by a deterministic law.

Since the temperature field is affected by the latent heat released by phase transition, the growth law must be coupled with the energy equation. Such a model is studied in the next chapter.

VII.6 Mean Curvature Flow

In this section we assume that continuous phase transition is governed by a *gradient flow* with respect to the potential F_θ in $L^1(\Omega)$, and derive the law of mean curvature flow with forcing term.

Gradient Flow. At first let us consider a general framework. Let Ψ be a functional defined in a Banach space B with dual B'. By *gradient flow* with respect to Ψ we mean a law of the form

$$a(u)\frac{du}{dt} + \nabla\Psi(u) \ni 0 \qquad \text{in } B'. \tag{6.1}$$

Here $a(u)$ is a positive coefficient, and ∇ represents some sort of (possibly multivalued) differential operator; for instance, the *Gâteaux differential*, the usual subdifferential of convex analysis, or the operator ∂^-; see Sect. XI.4. Loosely speaking, this equation describes descent of the representative point $(u, \Psi(u)) \in B \times \mathbf{R}$ along the graph of the functional Ψ in the steepest direction of B'. This represents relaxation towards equilibrium. [20]

In concrete cases, there is some freedom in the choice of the function space B, of the state variable u, of the functional Ψ, of the differential operator ∇, and of the relaxation coefficient $a(u)$. Examples include the *Allen-Cahn* (or *Landau-Ginzburg*) *equation* [21] (VI.5.2) and the analogous inclusion (VI.5.3). These correspond to $B := L^2(\Omega)$ and $\Psi := J_{i\varepsilon}$ for $i = 1, 2$, respectively, see (VI.5.1).

By taking $B := H^{-1}(\Omega)$ and $\Psi := J_{1\varepsilon}$, one gets

$$-c\Delta^{-1}\frac{\partial\chi}{\partial t} - \varepsilon a \Delta\chi + \frac{4}{\varepsilon}\chi(\chi^2 - 1) = \frac{L\theta}{2\tau_E} \qquad \text{in } H_0^1(\Omega), \tag{6.2}$$

where $-\Delta^{-1}$ is defined as in (II.5.9).

---

[20] One might account for oscillations by including a term bd^2u/dt^2 ($b > 0$) in the left side of (6.1). In space-distributed systems, typically this would yield wave propagation.

[21] See, e.g., Allen and Cahn [7].

As we saw in Sect. V.6, cf. (V.6.24), equations like (6.2) can be used whenever the function χ represents a conserved quantity. Indeed, by applying the operator $-\Delta$ to both sides of (6.2) and defining the flux $\vec{j}$ in an obvious way, (6.2) can be written as a conservation law, $c\partial\chi/\partial t + \nabla \cdot \vec{j} = 0$ in Q, without any source term. A typical example is offered by *phase separation* in an undercooled *eutectic* binary system. [22] If the system does not exchange matter with the exterior, the difference of the concentrations of two components is a conserved quantity. This process can be described by a classical equation, due to Cahn and Hilliard, similar to (6.2). [23]

The gradient flow for the modified free energy $\tilde{F}_\theta$, cf. (3.4), with $B := L^2(\Omega)$ is equivalent to the following variational inequality [24]

$$\begin{cases} |\chi| \leq 1 \text{ a.e. in } \Omega; \forall v \in BV(\Omega), \text{ such that } |v| \leq 1 \text{ a.e. in } \Omega, \\ c\int_\Omega \frac{\partial \chi}{\partial t}(\chi - v)dx + \frac{\sigma}{2}\int_\Omega (|\nabla\chi| - |\nabla v|) + \frac{\sigma_L - \sigma_S}{2}\int_\Gamma (\chi - v)d\Gamma \\ \leq \frac{L}{2\tau_E}\int_\Omega (\theta + a\chi)(\chi - v)dx \qquad \text{a.e. in }]0, T[. \end{cases} \qquad (6.3)$$

On the other hand, it is not sensible to consider the gradient flow for the free energy F_θ, cf. (VI.1.7), with $B := L^p(\Omega)$, for any $p \in [1, +\infty[$. Indeed $\partial\chi/\partial t \in L^p(\Omega)'$ and $|\chi| = 1$ a.e. in Q only if $\partial\chi/\partial t = 0$ identically in Q. [25]

Therefore we look for a different gradient dynamics for the free energy F_θ.

Gradient Flow for the Free Energy: First Alternative. Here we consider two alternative dynamics of gradient flow for the free energy of solid–liquid systems.

If the temperature θ (< 0) is uniform and the solid phase occupies a ball at $t = 0$, we can expect that this shape is maintained at least for some time, and only the radius $R = R(t)$ changes. This simple setting is especially convenient for developing a heuristic approach. So we start from the radial case, and then extrapolate our conclusions to the general setting.

Let us define $\delta F_\theta(R)$ as in (1.1). As a gradient flow for δF_θ, at first let us consider the equation

$$aR'(t) = -\delta F'_\theta(R(t)) = -8\pi\sigma R(t) - \frac{4\pi L\theta}{\tau_E}R(t)^2, \qquad (6.4)$$

where a is a positive constant. Since $1/R(t)$ equals the mean curvature $\kappa(t)$, this also reads

$$\frac{a}{8\pi}\kappa'(t) - \sigma\kappa(t) = \frac{L\theta}{2\tau_E}. \qquad (6.5)$$

---

[22] See, e.g., Astarita [29; Sect. 9.3], Lupis [369; Sect. 3.4], and Kittel and Krömer [325; Chap. 11].

[23] See, e.g., Cahn-Hilliard [117] and Cahn [116].

[24] This variational inequality is coupled with the energy balance equation (IV.4.5) in Sect. IX.2; see also V. [559].

[25] Nevertheless, one might consider the gradient flow for F_θ with $B = H^{-1}(\Omega)$, if χ were a conserved quantity.

As θ is constant, the solution of this simple O.D.E. is

$$\kappa(t) = \left(\kappa(0) + \frac{L\theta}{2\sigma\tau_E}\right)\exp\frac{8\pi\sigma t}{a} - \frac{L\theta}{2\sigma\tau_E}.$$

So

$$\kappa(0) > -\frac{L\theta}{2\sigma\tau_E} \quad\Rightarrow\quad \kappa(t) \to +\infty \quad\text{as}\quad t \to +\infty,$$

$$\kappa(0) < -\frac{L\theta}{2\sigma\tau_E} \quad\Rightarrow\quad \kappa(t) \to 0 \quad\text{in finite time.}$$

By (1.2), this reads

$$\begin{aligned}R(0) < R_c &\Rightarrow R(t) \to 0 \text{ in infinite time,}\\ R(0) > R_c &\Rightarrow R(t) \to +\infty \text{ in finite time.}\end{aligned} \quad (6.6)$$

This is not the physical behaviour.

Gradient Flow for the Free Energy: Second Alternative. Let us denote by A the area of the sphere of radius R, and express the free energy as a function of A:

$$\delta F_\theta(R) = \sigma A + \frac{L\theta}{6\tau_E\sqrt{\pi}}A^{3/2} =: \delta\hat{F}_\theta(A).$$

The gradient flow for $\delta\hat{F}_\theta(A)$ reads

$$cA'(t) = -\delta\hat{F}'_\theta(A(t)) = \sigma + \frac{L\theta}{4\tau_E\sqrt{\pi}}\sqrt{A}, \qquad (6.7)$$

where c is a positive constant, and is equivalent to

$$8\pi c R'(t) = -\frac{\sigma}{R(t)} - \frac{L\theta}{2\tau_E}. \qquad (6.8)$$

Here $R = R_c$ is still an unstable equilibrium point, but the qualitative behaviour of the solution is at variance with (6.6):

$$\begin{aligned}R(0) < R_c &\Rightarrow R(t) \to 0 \text{ in finite time,}\\ R(0) > R_c &\Rightarrow R(t) \to +\infty \text{ in infinite time.}\end{aligned} \quad (6.9)$$

This kinetic law is usually regarded as physically adequate. Note that the *Lyapunov function* of this dynamics is $L(R) = \sigma\log R + L\theta R/2\tau_E$ for any $R > 0$, which has a singularity in the origin. In this setting nucleation can be hardly represented as a continuous process.

Notice that (6.4) and (6.8) are equivalent, for a suitable choice either of $a = a(R)$ or of $c = c(A)$; so they just differ in the linearization.

Mean Curvature Flow. If θ is nonconstant but radially symmetrical with respect to the center of the initial ball, then the growing solid phase preserves the spherical shape. Then θ must be replaced by $\theta(R(t), t)$.

For a space- and time-dependent temperature, in place of (6.8) it seems appropriate to consider the following more general law of *mean curvature flow* with forcing term

$$cv_S = \sigma\kappa + \frac{L\theta}{2\tau_E} \qquad \text{on } S. \tag{6.10}$$

Here v_S is the normal component of the velocity of the moving solid–liquid interface, which is assumed positive when the liquid advances through the solid; for example, $v_S = -R'(t)$ for a solid ball. This kinetic law seems appropriate for describing phase transition.

Phase transition causes exchange of latent heat, hence (6.10) must be coupled with the energy balance equation, cf. (IV.4.5).

VII.7 Nonlinear Mean Curvature Flow

The mean curvature flow equation (6.10) accounts neither for phase nucleation nor for other singularities in phase transition. Therefore here we propose the following *nonlinear mean curvature flow* equation with forcing term

$$\alpha(v_S) \ni \sigma\kappa + \frac{L\theta}{2\tau_E} \qquad \text{on } S, \tag{7.1}$$

where v_S is defined as previously and [26]

$$\alpha : \mathbf{R} \to 2^{\mathbf{R}} \text{ is a nonconstant, bounded, maximal monotone graph.} \tag{7.2}$$

For instance, one can consider

$$\tilde{\alpha}(\xi) := \begin{cases} -M & \text{if } \xi < -\frac{M}{c}, \\ c\xi & \text{if } -\frac{M}{c} \leq \xi \leq \frac{M}{c}, \\ M & \text{if } \xi > \frac{M}{c}, \end{cases} \qquad \forall \xi \in \mathbf{R}, \tag{7.3}$$

where M and c are positive constants.

The boundedness of α has a regularizing effect in space, since by (7.1) the mean curvature is uniformly bounded whenever the same holds for θ. On the other hand,

---

[26] See Sect. XI.5.

this implies a loss of regularity for the front velocity v_S, and in fact evolution can be discontinuous. One can conclude that here space singularities do not appear on the moving surface, since (loosely speaking) they can be excluded by means of jumps of the phase.

Evolution Modes. The equation (7.1) accounts for the following modes of phase transition:

(i) *Mean Curvature Flow,* corresponding to $|\sigma\kappa + L\theta/2\tau_E| = |cv_S| \leq M$. This is smooth in time and occurs at almost any instant.

(ii) *Singular Evolution.* For instance, nucleation occurs in the bulk of an either undercooled or superheated phase, as $|\theta|$ overcomes the threshold $2\tau_E M/L$. *Formally,* this corresponds to $|v_S| = +\infty$.

Singularities in evolution may appear in several forms: either as phase nucleation, as phase annihilation, as merging of two separate phases, as splitting of a single phase, or as other changes in the phase topology; see Fig. 4. [27]

In (7.2) the maximal monotone graph α is allowed to be multi-valued. Let us assume that $\alpha(0) = [a, b]$, with $a < 0 < b$; for instance, $\alpha := \tilde{\alpha} + \text{sign}$, where $\tilde{\alpha}$ is as in (7.3) and "sign" is the sign graph, cf. (XI.5.3). The sign-term represents a sort of *dry friction* which opposes the interface motion. This can represent *hysteresis* in front evolution, as we see in the next section, and is reminiscent of the pinning effect that impurities exert on domain walls in ferromagnetism.

This model can be extended to nonhomogeneous materials, by allowing α to depend on x explicitly.

Topological Changes and Volume Discontinuities. According to this model, the phase volume does not need to be continuous in time. For instance, let us consider nucleation of a solid phase in an undercooled liquid. By (7.1), the mean curvature is bounded, hence nucleation can only occur by instantaneous formation of a solid phase in a set of strictly positive (mesoscopic) volume, like a ball of radius larger than the critical value.

More generally, we conjecture that (7.1) implies that [28]

$$\text{phase volume is discontinuous,} \qquad (7.4)$$
$$\text{as any change occurs in the phase topology.}$$

Of course, here we are dealing with the mesoscopic length scale, because of the size of the surface tension coefficient σ. On the macroscopic scale, usually phase nucleation occurs without any volume change.

---

[27] On the basis of the theory of maximal monotone operators, see, e.g., Brézis [90], one is induced to exclude the occurrence of the intermediate regime corresponding to $M = |\sigma\kappa + L\theta/2\tau_E| < |cv_S| < +\infty$.

[28] This is further discussed in Sects. VIII.1 and VIII.2. See also V. [568, 569].

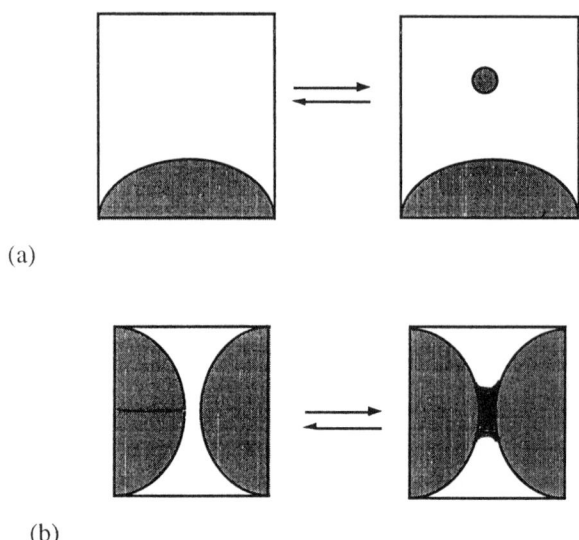

(a)

(b)

Figure 4. Examples of singular evolution. (**a**) represents nucleation (from left to right) and the opposite phenomenon of annihilation (from right to left).

In (**b**) we see *bridging* (from left to right): a solid bridge is instantaneously formed between two colliding domains. Conversely (see (**b**) from right to left) a domain can instantaneously *split* into two connected components. This might also represent a change in the connection multiplicity (i.e., appearing or breaking of a *handle*).

In any of these cases, according to this model, phase transition occurs instantaneously in a set of *positive* volume.

VII.8 Hysteresis in Front Motion

In this section we outline a model that includes hysteresis in the motion of smooth phase interfaces.

Cahn's Model of Front Motion. At small undercooling an atomically smooth interface can only advance by forming a new layer. This occurs by *two-dimensional nucleation:* a solid island is formed on the interface, and then expands by *lateral growth*. [29]

Cahn proposed a phenomenological model for motion of a (flat) interface, which advances or recedes with constant area A. He represented the interface free energy

---

[29] See Cahn [115], Chalmers [129; p. 44], Christian [135; Sect. 53], Flemings [240; p. 304], and Woodruff [583; Sect. 8.4].

$\hat{F}_\theta$ as a function $\hat{F}_0$ of the normal advancement coordinate x of the interface (oriented towards the liquid phase), plus a term proportional to the relative temperature θ and to x, plus an unessential term independent of x:

$$\hat{F}_\theta(x) := \hat{F}_0(x) + \frac{LA}{2\tau_E}\theta x \; (+ \text{ Constant}) \qquad \forall x \in \mathbf{R}. \tag{8.1}$$

If $\hat{F}_0(x) := \sigma(x)A$, this is consistent with the extension of (VI.1.7) to a non-homogeneous material. On the contrary, here the function $\hat{F}_0$ is assumed to be periodic, with period equal to the thickness of an interface layer; for instance, $\hat{F}_0(x) := \sigma A + b\sin(x/\lambda)$, where b is a positive constant, cf. Fig. 5(a).

It is assumed that the interface advances through local minimizers of $\hat{F}_\theta$. If $\theta < \theta_c := -2\tau_E \max \hat{F}_0'/LA \, (< 0)$, then $\hat{F}_\theta$ is a decreasing function of x, and the interface moves continuously; cf. Fig. 5(b). On the other hand, if $\theta_c < \theta < 0$, then the front can only advance by repeated two-dimensional solid nucleation and lateral growth, and x must equal an integral number of molecular spacings. Melting is similarly described for $\theta > 0$.

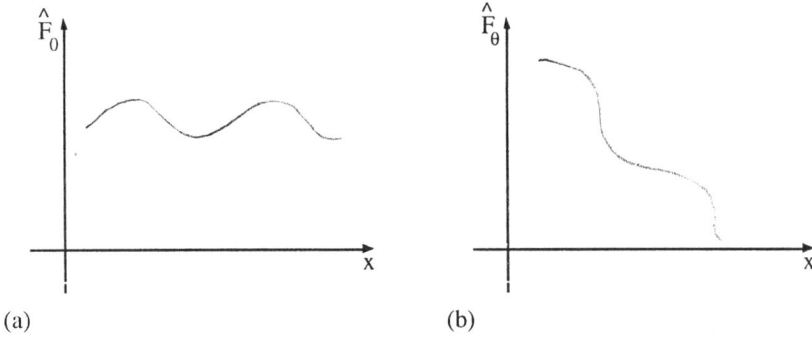

(a) (b)

Figure 5. Illustration of Cahn's model of interface motion. (a) represents the dependence of the free energy versus the normal advancement coordinate x of the interface, in case of no undercooling. In presence of a sufficiently large undercooling, the profile is antimonotone, cf. (b).

Front Motion with Hysteresis. If the critical temperature for two-dimensional nucleation is smaller than θ_c, then the solid front can only advance if $\theta < \theta_c$. If we have a similar behaviour for melting, with a threshold $\hat{\theta}_c > 0$, then front motion exhibits a *hysteresis* effect:

$$\begin{array}{l}\text{the liquid phase advances if } \theta > \hat{\theta}_c, \text{ and recedes if } \theta < \theta_c; \\ \text{if } \theta_c < \theta < \hat{\theta}_c \text{ the front is in equilibrium.}\end{array} \tag{8.2}$$

More precisely, in the latter case the equilibrium is stable if $\theta = 0$, and metastable if $\theta \neq 0$. Usually the undercooling $-\theta_c(> 0)$ needed to displace a flat interface is much smaller than the threshold for homogeneous nucleation.

Henceforth we assume that $\hat{\theta}_c = -\theta_c$, for the sake of simplicity.

Cahn's model deals with the *atomic* length scale: two relative minimizers of $\hat{F}_0$ are just a distance δ of atomic order apart. Dealing with a mesoscopic length scale, we must pass to the limit as $\delta \to 0$, keeping $(\max \hat{F}_0 - \min \hat{F}_0)/\delta$ constant; in fact this ratio is proportional to $\max \hat{F}_0'(x)$, which in turn is proportional to θ_c. Thus on the mesoscopic length scale the dependence of the interface free energy on the normal advancement coordinate is lost, but the model retains the threshold $|\theta_c|$.

On the basis of this picture, we propose the following condition, which accounts for hysteresis in front motion and generalizes the Gibbs-Thomson law:

$$\frac{L}{2\tau_E}\theta + \sigma\kappa \in c\,\text{sign}(v_S) \qquad \text{on } \mathcal{S}. \tag{8.3}$$

Here v_S denotes the normal velocity of the melting front (assumed negative for solidification), and $c := \max \hat{F}_0'/A$. Hence the maximal deviation from the temperature prescribed by the Gibbs-Thomson law is $2\tau_E c/L = 2\tau_E \max \hat{F}_0'/LA = |\theta_c|$.

(8.3) is equivalent to the variational inequality

$$\left(\frac{L}{2\tau_E}\theta + \sigma\kappa\right)(v_S - \xi) \geq c|v_S| - c|\xi| \qquad \forall \xi \in \mathbf{R}, \text{ on } \mathcal{S}. \tag{8.4}$$

It is easy to see that for a planar front we retrieve the behaviour described by Cahn, whereas for $c = 0$ we get the standard Gibbs-Thomson law (VI.4.3).

This behaviour is similar to that prescribed by *dry friction*. This evolution mode is *rate independent*: the size of the front velocity does not depend on θ, at variance with kinetic undercooling and phase relaxation, cf. Sect. V.1. It has a regularizing effect, since it damps small oscillations of θ near the value 0.

More general laws can be obtained by combining relaxation with hysteresis. For instance, one can include a *curvature flow* term, and get a law of the form (7.1), with $\alpha := \tilde{\alpha} + c\,\text{sign}$, where $\tilde{\alpha}$ is as in (7.3). Indeed, both the mean curvature flow (6.10) and the hysteresis motion (8.3) can be regarded as special cases of (7.1).

VII.9 Comments

In this chapter we dealt with metastable states and nucleation phenomena. Here are the main issues:

(i) the role of surface tension and nonconvexity of the free energy in nucleation;

(ii) the modelling of stable and metastable states in solid–liquid systems;

(iii) Theorem 4.1, based on the coarea formula, for minimization of a nonconvex functional including the total variation;

(iv) the interplay between discontinuous stochastic nucleation and continuous deterministic phase growth;

(v) the derivation of the time scale of nucleation;

(vi) the derivation of the law of mean curvature flow as a gradient flow for the free energy, and its generalization to include nucleation;

(vii) Cahn's model of hysteresis in front motion.

Phase Transition Modes. The above discussion yields the following picture of phase nucleation and growth.

Systems consisting of materials capable of attaining two phases can be in metastable equilibrium; an undercooled liquid is a typical example. At prescribed temperature, those states correspond to relative minimizers of the free energy, which can be represented as a functional of the phase variable χ. The occurrence of such states is due to the nonconvexity of that functional, and the compactness provided by the perimeter term. Without the latter, minimizing sequences would not necessarily preserve the nonconvex constraint "$|\chi| = 1$ a.e. in Ω" in the limit.

The naïve approach of Sect. VII.5 surrogates a more accurate analysis, which should involve statistical mechanics and stochastic differential equations. A wide class of processes governed by descent along a *steep* nonconvex potential may be expected to exhibit continuous deterministic evolution via a gradient flow, and discontinuous stochastic evolution via a Poisson process. These two modes of evolution should be represented as extreme cases of a more general law, which should also express their *competition* in an intermediate range. In the radial case we considered the simple stochastic O.D.E. (5.3); but in a general geometry this setting should be represented by a stochastic P.D.E.. Such an equation should then be coupled with the heat equation. This all is an open problem.

Mean Curvature Flow. Studies on this topic have flourished in the last ten years, after the pioneering contributions of Brakke [84], Gage [259], Grayson [273], Huisken [303], Barles [58], Sethian [500], Osher and Sethian [437], Evans and Spruck [220], Chen, Giga and Goto [131], and others. For instance, see the reviews of Evans, Soner and Souganidis in [219], Ilmanen [307], and the proceeding volumes [105, 172].

In most of the works that deal with this topic, the evolving surface is represented either as a level set of a function of space and time, or via a time dependent *approximate characteristic function* (or *phase field*). The latter is a continuous function that is equal to ± 1, with the exception of a *thin* transition zone which is interpreted as a neighbourhood of the evolving surface. The concept of *viscosity solution* has been used. An alternative approach using (exact) *characteristic functions* has been studied by Almgren, Taylor and Wang [9], Almgren and Wang [10], Luckhaus and Sturzenhecker [366], and V. [567, 568]. Phase transition coupled with mean curvature flow has been studied in the radial case by V. [554].

A discussion about gradient flow for interface free energies and various generalizations of the mean curvature flow can be found in Cahn and Taylor [118, 530].

Nucleation. This phenomenon is especially important for applications, and has been extensively studied by physicists and material scientists, see the *Book Selection*.

In Sect. VII.2 we considered nonpolymeric materials, such as metals; interesting phenomena also occur in polymers; see, e.g., Andreucci et al. [20, 21], Ziabicki [595], and Fasano [222] for a survey.

The equation (8.3) has been studied by V. [562] for a prescribed evolution of θ. The modified law of mean curvature flow (7.1) has been dealt with by V. [568] for a given forcing term, and has been coupled with heat diffusion in V. [569]. Results in the direction of Theorem 4.1, based on a generalization of the classical coarea formula, can be found in V. [558, 560, 561].

Chapter VIII. The Stefan-Gibbs-Thomson Problem with Nucleation

Outline

Phase transitions are studied at a *mesoscopic* length scale, distinguishing front motion from nucleation and other discontinuous transitions. A generalization of the Stefan model, which accounts for both modes of phase transition, is formulated in the framework of Sobolev spaces. Surface tension effects are represented by the Gibbs-Thomson law.

Existence of a solution is proved via approximation by time-discretization, derivation of uniform estimates, and passage to the limit.

This approach is also applied to the *Mullins-Sekerka problem*.

Prerequisites. Knowledge of basic function spaces is assumed (cf., e.g., Sects. XI.1 and XI.2). Acquaintance with methods of analysis of nonlinear P.D.E.s in Sobolev spaces is needed to read Sects. VIII.3 and VIII.4.

Reference is made to material of Chaps. IV, VI, and VII.

VIII.1 Modes of Phase Transition

In this chapter we deal with the Stefan problem with surface tension, which we introduced in Sect. IV.2. In this section we discuss front motion and singular phase transition.

Let $\Omega \subset \mathbf{R}^3$ be a bounded domain, fix any $T > 0$, and set $Q := \Omega \times]0, T[$. We use the same notation as in Chap. IV, and assume that the (positive) coefficients $C_V, L, k, \sigma, \tau_E$ are constant, and that the first three of them are phase-independent. As we saw, the energy balance and the Fourier conduction law yield

$$C_V \frac{\partial \theta}{\partial t} - k \Delta \theta = f \qquad \text{a.e. in } Q \setminus \mathcal{S}. \tag{1.1}$$

Moreover, the temperature is continuous across solid–liquid interfaces.

For a.a. $t \in]0, T[$, let us denote by $[\cdot]_{\mathcal{S}_t}$ the jump across $\mathcal{S}_t$, that is, the difference between the trace from the liquid phase Ω_t^+ and that from the solid phase Ω_t^-. Let

$\vec{v}$ be the velocity of the interface, and $\vec{\nu} \in \mathbf{R}^3$ a unit vector normal to $\mathcal{S}_t$. The Stefan condition and the Gibbs-Thomson law, respectively, read

$$k \left[\frac{\partial \theta}{\partial \nu} \right]_{\mathcal{S}_t} = -L\vec{v} \cdot \vec{\nu} \qquad \text{on } \mathcal{S}_t, \text{ a.e. in }]0, T[, \tag{1.2}$$

$$\kappa = -\frac{L}{2\tau_E \sigma} \theta \qquad \text{on } \mathcal{S}_t, \text{ a.e. in }]0, T[. \tag{1.3}$$

By Proposition IV.4.1, setting $u := C_V \theta + (L/2)\chi$, the equations (1.1) and (1.2) are formally equivalent to the energy balance equation

$$\frac{\partial u}{\partial t} - k\Delta\theta = f \qquad \text{in } \mathcal{D}'(Q). \tag{1.4}$$

As we see in the next sections, the following regularity properties look natural as surface tension effects are inserted into the model:

$$\begin{aligned} &\theta \in L^2\left(0, T; H^1(\Omega)\right), \quad \chi \in L^\infty(Q) \cap L^1(0, T; BV(\Omega)), \\ &u \in L^2(Q) \cap H^1\left(0, T; H^{-1}(\Omega)\right) \quad \left(\text{if } f \in L^2\left(0, T; H^{-1}(\Omega)\right)\right). \end{aligned} \tag{1.5}$$

Consistently with our discussion of Sect. IV.3, we deal with a *mesoscopic* model. As we did in the previous chapter, we distinguish two modes of phase transition.

Continuous Phase Transition (i.e., *Front Motion*). In this case phase transition is *isothermal;* this entails that the density of internal energy u is discontinuous across $\mathcal{S}_t$, cf. Fig. 1(a). At almost any instant, energy conservation and local equilibrium (i.e., free energy minimization), respectively, yield the Stefan condition (1.2) and the Gibbs-Thomson law (1.3).

Discontinuous Phase Transition (e.g., *Phase Nucleation*). Nucleation, namely, onset of an either liquid or solid connected component, is a discontinuous phenomenon, for it entails a change of the interface topology.

Discontinuities in phase transition can also occur in other ways, and can a priori be excluded only in special cases. For instance, a connected component may annihilate; it may also split into two components, or conversely the latter may merge into a single one; a simply connected component may form a ring, or conversely, and so on; cf. Fig. VII.4. Although we often refer to nucleation, most of our discussion carries over to more general singularities.

Adiabatic Nucleation. As we saw in Sect. VII.2, as a rule homogeneous phase nucleation cannot be continuous in time, but must occur by instantaneous transition in a set, A say, with positive volume. [1]

---

[1] We saw that this statement fails in special geometries, for instance in presence of holes having the shape of a critical ball. However, it holds if Ω fulfils the following *internal ball condition*
$$\forall x \in \Omega, \exists y \in \Omega : x \in B_{R^*}(y) \subset \Omega, \text{ where } R^* := 3\sigma\tau_E/L|\theta|.$$

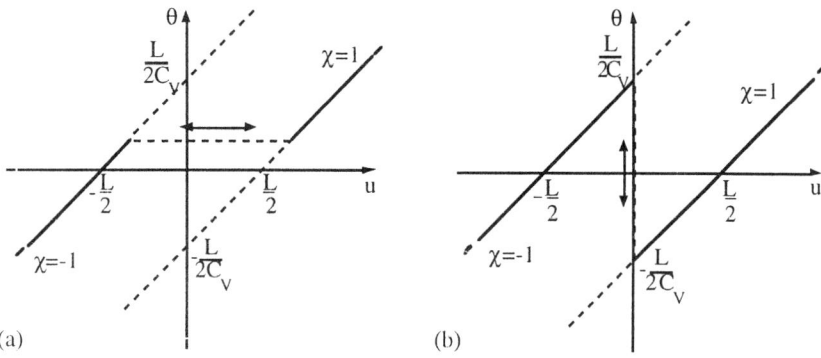

Figure 1. Temperature versus density of internal energy for basic modes of phase transition, according to the first model of this chapter.

For *front motion,* the temperature θ is continuous in space and time at the interface S between the phases. Hence the internal energy density u has a jump in space and time across S, cf. (**a**). By the Gibbs-Thomson law (1.3), the mean curvature of the interface is proportional to $-\theta$.

On the other hand, any instant of *phase nucleation* corresponds to a jump in time of the temperature in a subset of Ω of strictly positive measure. Here the internal energy density is continuous in space and time, cf. (**b**).

So, according to this model, front motion is isothermal and continuous, whereas nucleation is adiabatic and discontinuous. Moreover, here nucleation is reversible, and transforms an undercooled liquid into a superheated solid.

If a phase discontinuity occurred at constant temperature, u would also jump in A. But this is excluded by the energy balance (1.4), and by the regularity (1.5), since a finite amount of heat cannot be diffused instantaneously. Hence, in this framework, discontinuous phase transition can only occur at constant internal energy, that is, *adiabatically.* [2]

We want to describe this setting quantitatively. At first, let us set

$$\Phi(\chi) := \sigma P(\chi) + \frac{\sigma_L - \sigma_S}{2} \int_\Gamma \chi d\Gamma \qquad \forall \chi \in \mathrm{Dom}(P), \qquad (1.6)$$

where Γ is the boundary of Ω; see (VI.1.5) for the definition of the functional P. We assume that $|\sigma_L - \sigma_S| \leq \sigma$, so that the functional Φ is lower semicontinuous with respect to the strong topology of $L^1(\Omega)$, cf. Proposition VI.1.1.

To account for adiabatic nucleation, here we propose to minimize the functional

$$K_u(\chi) := \Phi(\chi) - \frac{L}{2T_E C_V} \int_\Omega u\chi dx, \qquad (1.7)$$

at constant u. This is equivalent to minimizing the free energy potential $F_\theta(\chi)$ at constant u, since $-\theta\chi = -(1/C_V)\left(u\chi - (L/2)\chi^2\right) = -u\chi/C_V + L/2C_V$, as $|\chi| = 1$.

---

[2] Nonadiabatic nucleation is treated in Sect. IX.1 by means of a different model.

This variational principle yields (VI.4.14), but not the Gibbs-Thomson law; however, the minimization is consistent with that law; see Proposition VI.4.4. Therefore in our model we impose the minimization of $K_u(\chi)$, and also require the Gibbs-Thomson law separately.

Discontinuous Nucleation.

Proposition 1.1 *(Discontinuous Nucleation) Let (θ, χ) be radially symmetric. Assume that $\chi(r,t) = -1$ iff $r < R(t)$, and $R : [0,T] \to \mathbf{R}^+$ is nondecreasing, $R(t) = 0$ if $t < t_0$, $R(t) > 0$ if $t > t_0$.*
If θ^0 and f are uniformly bounded, then the function R is discontinuous at t_0.

Proof. At first, let us consider the auxiliary function $\hat{\theta}$ such that

$$\begin{cases} C_V \dfrac{\partial \hat{\theta}}{\partial t} - k\Delta\hat{\theta} = f & \text{in } Q, \\ \hat{\theta} = \theta^0 & \text{in } \Omega \times \{0\}, \\ \hat{\theta} = 0 & \text{on } \Gamma \times]0, T[. \end{cases} \quad (1.8)$$

Note that $\hat{\theta}$ is uniformly bounded in Q. As R is nondecreasing, we have $\partial\chi/\partial t \leq 0$ in $\mathcal{D}'(Q)$. Then (1.4) and (1.8)$_1$ yield

$$C_V \frac{\partial}{\partial t}(\theta - \hat{\theta}) - k\Delta(\theta - \hat{\theta}) = -\frac{L}{2}\frac{\partial \chi}{\partial t} \geq 0 \quad \text{in } \mathcal{D}'(Q);$$

hence $\theta \geq \hat{\theta} \geq -\|\hat{\theta}\|_{L^\infty(Q)}$ a.e. in Q, by the maximum principle.

As $\theta(R(t),t) < 0$ for $t > t_0$, by the Gibbs-Thomson law (1.3), we have

$$R(t) = -\frac{2\sigma T_E}{L\theta(R(t),t)} \geq \frac{2\sigma T_E}{L\|\hat{\theta}\|_{L^\infty(Q)}} \quad \forall t > t_0. \quad (1.9)\square$$

By a simple comparison procedure based on the maximum principle, one can exclude continuous nucleation for more general geometries as well. By a similar argument, one can show that annihilation can only occur discontinuously.

Other Phase Discontinuities. In the next section, cf. Proposition 2.1, under natural conditions on the data, we show that if two solid balls grow until they eventually coalesce, then at some time they must be instantaneously joined by a short *neck*, similar to that of Fig. VII.4(b). Similarly, if a connected solid phase splits into two connected components, as soon as a sufficiently thin neck is formed, it instantaneously disappears. [3]

---

[3] One might expect that the necks that are formed are *short*, and that those that disappear are *thin*. This would require a hysteresis mechanism that is not present in this model.

Therefore nucleation, annihilation, phase merging and splitting are accompanied by discontinuities in the phase volume. [4] By extrapolation, we then conjecture that under smoothness assumptions

> changes in the phase topology only occur by instantaneous phase transition in a mesoscopic region having positive volume. (1.10)

Nucleation Thresholds. Let us consider a single-phase system at a uniform temperature θ. If $\sigma_L = \sigma_S$, then $\Phi(\chi) = 0$, and the minimization of K_u (cf. (1.7)) yields

$$\text{either } \chi = -1 \text{ in } \Omega \text{ and } u \leq 0 \text{ (whence } \theta \leq \tfrac{L}{2C_V}),$$
$$\text{or } \chi = 1 \text{ in } \Omega \text{ and } u \geq 0 \text{ (whence } \theta \geq -\tfrac{L}{2C_V}). \quad (1.11)$$

This is consistent with the model of Sect. VI.2, which prescribes adiabatic nucleation at $u = 0$. When and where this occurs, $\chi = -1$ and $\theta = L/2C_V$ are instantaneously changed to $\chi = 1$ and $\theta = -L/2C_V$, or conversely. Therefore $L/2C_V$ is the maximum undercooling/superheating attainable prior to nucleation, and the phase versus temperature dependence exhibits a *hysteresis* effect, whereas u uniquely determines the phase (for $u \neq 0$). Here the two temperature thresholds are symmetric just because we assumed that the two phases have the same *heat capacity*. However, our model can easily be extended if this restriction is dropped.

As we saw, the temperature is discontinuous at the boundary of the nucleated phase. This induces high temperature gradients, and by this takes the system far from equilibrium, at least in a small set and for a short time. Space discontinuities of this sort are consistent with the regularity $\theta \in L^2\left(0,T; H^1(\Omega)\right)$ only if they occur in a subset of $]0,T[$ of vanishing measure. After nucleation the temperature discontinuity is immediately smoothed down, and the internal energy density becomes discontinuous at phase interfaces.

Drawbacks. The preceding model can be objected to on the following ground:
  (i) it prescribes an undercooling threshold for solid nucleation which is much larger than is experimentally observed for water and many other materials;
  (ii) it requires the nucleated solid to be highly superheated;
  (iii) it allows for instantaneus nucleation of macroscopic regions, even of the whole sample if it is cooled uniformly.

Actually, this model accounts for *adiabatic* nucleation, whereas *nonadiabatic* nucleation is usually observed. Modifications are introduced in Sect. IX.1, in order to remove these drawbacks. Anyway, this model appears to be a reasonable starting point, and in Sects. VIII.3 and VIII.4 we see that its analysis is already challenging.

---

[4] We still assume that Ω fulfills the *internal ball condition*, which we recalled in one of the previous footnotes.

VIII.2 Formulation of the Problem

The Basic Problem. In this section we formulate an analytical problem that accounts for:
 (i) heat diffusion in the interior of each phase,
 (ii) exchange of latent heat and the *Gibbs-Thomson law* at phase interfaces,
 (iii) adiabatic nucleation as outlined in the previous section.

Let us assume that Ω is a bounded domain of Lipschitz class, and

$$u^0 \in L^2(\Omega), \qquad f \in L^2\left(0, T; H^{-1}(\Omega)\right). \tag{2.1}$$

For the sake of simplicity, we consider the homogeneous Dirichlet condition for θ; however, our developments can be extended to different boundary conditions. We denote the duality pairing between $H^{-1}(\Omega)$ and $H_0^1(\Omega)$ by ${}_{H^{-1}(\Omega)}\langle \cdot, \cdot \rangle_{H_0^1(\Omega)}$.

Problem 2.1 *(Stefan-Gibbs-Thomson Problem with Adiabatic Nucleation) To find* $\theta \in L^2\left(0, T; H_0^1(\Omega)\right)$ *and* $\chi \in L^1(0, T; BV(\Omega))$ *such that, setting* $u := C_V \theta + (L/2)\chi$ *a.e. in* Q,

$$\iint_Q \left(-u \frac{\partial \xi}{\partial t} + k \nabla \theta \cdot \nabla \xi \right) dx\, dt$$
$$= \int_0^T {}_{H^{-1}(\Omega)}\langle f, \xi \rangle_{H_0^1(\Omega)}\, dt + \int_\Omega u^0(x) \xi(x, 0)\, dx \tag{2.2}$$
$$\forall \xi \in H^1(Q), \xi = 0 \text{ on } (\Gamma \times]0, T[) \cup (\Omega \times \{T\}),$$

$$\Phi(\chi) - \Phi(v) \le \frac{L}{2\tau_E C_V} \int_\Omega u(\chi - v)\, dx \qquad \forall v \in \text{Dom}(P), \text{ a.e. in }]0, T[. \tag{2.3}$$

Hence for almost any fixed $t \in]0, T[$, *setting* $\Omega_t^+ := \{x \in \Omega : \chi(x, t) = 1\}$ *and denoting by* $\mathcal{S}_t$ *the boundary of* Ω_t^+ *in* Ω, *the following holds: for any* $x \in \mathcal{S}_t$, *there exist* $a, b \in \mathbf{R}$, *an open set* $G \subset \mathbf{R}^2$, *and a function* $\psi : G \to]a, b[$ *of class* C^1, *such that, possibly after a coordinate rotation,*

$$\Omega_t^+ \cap (G \times]a, b[) = \{(y_1, y_2, z) \in G \times]a, b[: z < \psi(y_1, y_2)\}.$$

Setting $\tilde{\nabla} := (\partial/\partial y_1, \partial/\partial y_2)$ *and* $\theta \circ \psi : y \mapsto \theta(y, \psi(y), t)$, *it is then required that*

$$\tilde{\nabla} \cdot \frac{\tilde{\nabla} \psi}{\sqrt{1 + |\tilde{\nabla} \psi|^2}} = -\frac{L}{\sigma \tau_E} \theta \circ \psi \qquad \text{in } H^{-1}(G). \tag{2.4}$$

Remark. Previously we stated that $\mathcal{S}_t$ is a surface of class C^1 for almost any $t \in]0, T[$. Actually, (2.3) and the regularity of θ yield more regularity for $\mathcal{S}_t$.

Note that $H^1(\Omega) \subset L^6(\Omega)$ by Sobolev inclusion. [5] Hence

$$\theta(\cdot, t) \in L^6(\Omega) \qquad \text{for a.a. } t \in]0, T[, \tag{2.5}$$

whence also $u(\cdot, t) \in L^6(\Omega)$. By Proposition XI.8.3, (2.3) then implies that

$$S_t \text{ is a surface of class } C^{1,1/4}, \text{ for a.a. } t \in]0, T[. \tag{2.6}$$

If instead Ω were a bidimensional set, we would have $H^1(\Omega) \subset L^p(\Omega)$ for any $p < \infty$, and S_t would be a surface of class $C^{1,\gamma}$ for any $\gamma < \frac{1}{2}$ and almost any $t \in]0, T[$. [6] □

Interpretation. The equation (2.2) yields the energy balance equation

$$\frac{\partial u}{\partial t} - k\Delta\theta = f \qquad \text{in } \mathcal{D}'(Q). \tag{2.7}$$

Comparing the terms of this equation, we get that $u \in H^1(0, T; H^{-1}(\Omega))$. The weak equation (2.2) and the differential equation (2.7) yield the initial condition

$$u|_{t=0} = u^0 \qquad \text{in } \Omega \text{ (in the sense of the traces).} \tag{2.8}$$

The variational inequality (2.3) is equivalent to the conditions

$$|\chi| = 1 \qquad \text{a.e. in } Q, \tag{2.9}$$

$$\begin{cases} \forall v \in BV(\Omega) \text{ such that } |v| = 1 \text{ a.e. in } \Omega, \\ \dfrac{\sigma}{2} \int_\Omega (|\nabla\chi| - |\nabla v|) + \dfrac{\sigma_L - \sigma_S}{2} \int_\Gamma (\chi - v) d\Gamma \\ \leq \dfrac{L}{2\tau_E C_V} \int_\Omega u(\chi - v) dx \qquad \text{a.e. in }]0, T[. \end{cases} \tag{2.10}$$

The onset of any region in which $|\chi| \neq 1$ is excluded, consistently with the fact that this model is set at a mesoscopic length scale; see Sect. IV.3.

As far as the moving interface is concerned, by Proposition VI.4.4, (2.3) only implies that

$$\left|\frac{2\sigma\tau_E}{L}\kappa + \theta\right| \leq \frac{L}{2C_V} \qquad \text{on } S_t, \text{ a.e. in }]0, T[. \tag{2.11}$$

A more precise condition is provided by (2.4), which expresses a weak form of the *Gibbs-Thomson law*

$$\kappa = -\frac{L}{2\sigma\tau_E}\theta \qquad \text{on } S. \tag{2.12}$$

---

[5] See Theorem XI.2.1.
[6] All the results of this chapter hold in two dimensions as well.

We see in the following that (2.3) also implies other conditions, which concern nucleation and other discontinuities in phase transition.

Let us denote by $\omega(x,t)$ the angle formed by the normal to $\mathcal{S}_t$, oriented towards Ω_t^+, and the outward normal to Γ. By Theorem VI.4.4, (2.3) also yields the *contact angle condition*

$$\cos\omega = \frac{\sigma_S - \sigma_L}{\sigma} \qquad \text{on } \mathcal{S}_t \cap \Gamma, \text{ a.e. in }]0,T[. \tag{2.13}$$

Finally, the condition $\theta \in L^2\left(0,T; H_0^1(\Omega)\right)$ obviously implies that

$$\theta = 0 \qquad \text{on } \Gamma \text{ (in the sense of the traces), a.e. in }]0,T[. \tag{2.14}$$

Discontinuous Coalescence. Now we draw two results from Theorem XI.8.5. At first, we prove a statement that we already mentioned in the previous section. We denote the initial temperature field by θ^0.

Proposition 2.1 *(Discontinuous Coalescence) Let θ^0 and f be uniformly bounded from below. Assume that the solution of Problem 2.1 represents two solid balls, which grow until they eventually coalesce.*

Then the distance $\delta(t)$ between the two solid regions is a discontinuous function of time. More precisely, there exists $c > 0$ such that $\delta(t) \notin]0,c[$ for any t.

This means that, as the balls get sufficiently close, they are instantaneously joined by a short neck.

Outline of the Proof. For any $p \geq 1$ and any constant $M > 0$ let us multiply the energy balance equation by $-\left[(\theta - M)^-\right]^p$, and consider the term

$$X_t := \int_0^t {}_{H^{-1}(\Omega)}\langle \frac{\partial \chi}{\partial t}, -\left[(\theta - M)^-\right]^p \rangle_{H_0^1(\Omega)} d\tau$$

(for the moment let us assume that this integral is well-defined).

Let us denote by $\hat{t}$ the supremum of the instants in which the two balls are disjoint and set $\hat{Q} := \Omega \times]0,\hat{t}[$. By the Gibbs-Thomson law, we have $\theta < 0$ on the boundary of the solid balls; moreover $\partial \chi/\partial t \leq 0$ in $\mathcal{D}'(\hat{Q})$. Hence $X_{\hat{t}} \geq 0$, and a uniform estimate for $\int_\Omega \left[(\theta - M)^-\right]^{p+1} dx$ can easily be derived in $]0,\hat{t}[$, via the energy balance equation. Passing to the limit as $p \to +\infty$, we then have ess $\inf_{\hat{Q}} \theta > -\infty$, so that (XI.8.2) is fulfilled with C independent of t. Hence by Theorem XI.8.1 we conclude that $\mathcal{S}_t$ is a γ-almost minimal boundary for any $\gamma < \frac{1}{2}$, for almost any $t \in]0,T[$.

By contradiction, let us now assume that the balls grow continuously until they meet at the instant $\hat{t}$. Then for any $\varepsilon > 0$ there exist an $r > 0$, an interval $]t_0, \hat{t}[$, and for almost any $t \in]t_0, \hat{t}[$ a point x_t on the solid–liquid interface, such that the volume fraction of the ball $B_r(x_t)$ occupied by the liquid phase is less than ε. Now this contradicts Theorem XI.8.5.

This argument can be made precise by using a suitable approximation of (θ, χ), such that X_t is meaningful. $\square$

A similar procedure can be used to prove a statement analogous to Proposition 2.1 about phase splitting: if a connected solid phase splits into two connected components, as soon as a sufficiently thin neck is formed, it instantaneously disappears. This conclusion can easily be extended to other geometries, since the two merging regions only need to be smooth in a neighbourhood of the contact zone.

The next result shows that, even if either θ^0 or f is not uniformly bounded, we can estimate the ratio between phase volumes.

Proposition 2.2 *Let* (θ, χ) *be a solution of Problem 2.1, and set* $K := \frac{L}{\sigma \tau_E} \left(\frac{4\pi}{3}\right)^{5/6}$,

$$R(t) := \left(\frac{\pi}{4K \|\theta(\cdot, t)\|_{L^6(\Omega)}}\right)^2 \quad \text{for a.a. } t \in]0, T[. \tag{2.15}$$

Then $\frac{1}{R} \in L^1(0, T)$, *and*

$$\frac{\pi r^3}{6} \leq |\Omega^+ \cap B_r(x)| \leq \frac{7\pi r^3}{6} \quad \forall x \in \Omega, \forall r \leq R(t), \text{ for a.a. } t \in]0, T[. \tag{2.16}$$

Proof. By (2.5), (XI.8.5) yields (XI.8.2) with $\gamma = 1/4$ and $C = K \|\theta(\cdot, t)\|_{L^6(\Omega)}$. (2.16) then follows from Theorem XI.8.5. $\square$

A Stronger Formulation. In Problem 2.1 only pure phases are admitted, cf. (2.9), and at almost any instant the phases are separated by a differentiable surface. It then seems natural to compare this problem with a stronger formulation, which we derive under the hypothesis that the solution fulfills further regularity conditions. However, it is not obvious that the latter conditions can be fulfilled.

Let us assume that phase discontinuities just occur in a *finite* set J of instants, and that the space-time interface S is of class C^1 with respect to time, in $]0, T[\setminus J$. If $f \in L^2(Q)$, the equation (2.7) yields the heat equation (1.1) in the interior of each phase, and the Stefan condition (1.2) at solid–liquid interfaces; see Proposition IV.4.1.

If we assume that $\chi^0 = -1$ a.e. in $\{x \in \Omega : u^0(x) < 0\}$ and $\chi^0 = 1$ a.e. in $\{x \in \Omega : u^0(x) > 0\}$, then (2.8) is *formally* equivalent to

$$\theta|_{t=0} = \theta^0, \quad \chi|_{t=0} = \chi^0 \quad \text{a.e. in } \Omega. \tag{2.17}$$

The condition $\theta \in L^2\left(0, T; H^1_0(\Omega)\right)$ obviously implies

$$[\theta]_{S_t} = 0 \quad \text{on } S_t, \text{ a.e. in }]0, T[. \tag{2.18}$$

We can now introduce a different formulation of our problem. Let

$$f \in L^2(Q), \quad \theta^0 \in L^2(\Omega), \quad \chi^0 \in BV(\Omega),$$
$$|\chi^0| = 1 \text{ a.e. in } \Omega, \quad \theta^0 \chi^0 \geq 0 \text{ a.e. in } \Omega, \tag{2.19}$$

and set $u^0 := C_V \theta^0 + (L/2)\chi^0$ a.e. in Ω.

Problem 2.2 *(Stronger Formulation of the Stefan-Gibbs-Thomson Problem with Adiabatic Nucleation) To find a measurable field $\theta : Q \to \mathbf{R}$ and two disjoint open sets $Q_1, Q_2 \subset Q$ such that the following conditions are fulfilled:*
(i) $S_t := \{x \in \Omega : (x,t) \in \partial Q_1 \cap \partial Q_2\}$ is a surface of class C^1, for almost any $t \in]0, T[$, and (1.3) holds;
(ii) $\partial \theta / \partial t$, $\Delta \theta \in L^2(Q \setminus S)$, $\nabla \theta \in L^2(Q; \mathbf{R}^3)$, $\vec{v} \cdot \vec{\nu}$ (normal interface velocity) $\in L^2(S_t)$ for a.a. $t \in]0, T[$, and (1.1) and (1.2) are fulfilled;
(iii) setting $\chi = 1$ in Q_1, $\chi = -1$ in Q_2, $u := C_V \theta + (L/2)\chi$ in Q, (2.3) holds;
(iv) (2.13), (2.14) and (2.17) are fulfilled.

The sets of conditions (i), (ii) and (iv) respectively correspond to the Gibbs-Thomson law, energy conservation, and boundary conditions; (2.3) accounts for nucleation and other singularities in phase evolution.

A More General Model. For any $\mu \in [0, 1]$, we consider the following problem.

Problem 2.1$_\mu$ *To find θ and χ fulfilling the conditions of Problem 2.1, with (2.3) replaced by the following variational inequality:*

$$\Phi(\chi) - \Phi(v) \leq \frac{L}{2\tau_E C_V} \int_\Omega \left(u + \frac{\mu L}{2}\chi\right)(\chi - v)dx \quad \forall v \subset \text{Dom}(P), \tag{2.20}$$
$$\text{a.e. in }]0, T[.$$

Obviously for $\mu = 0$ we retrieve Problem 2.1. The interpretation of Problem 2.1$_\mu$ is similar to that of Problem 2.1; these problems only differ in the nucleation thresholds and in the temperature of the new born phase. It is easy to see that, according to Problem 2.1$_\mu$, a liquid at a uniform temperature can be undercooled down to $\theta = -(1+\mu)L/2C_V$, and symmetrically a solid can be superheated up to $\theta = (1+\mu)L/2C_V$, before (adiabatic) nucleation occurs; see Fig. 2. Here μ represents a material parameter.

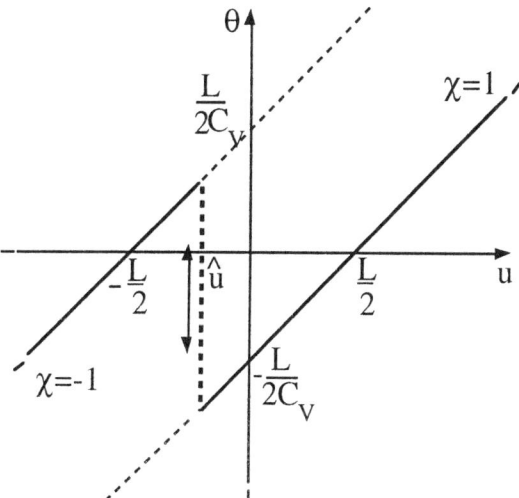

Figure 2. Temperature versus internal energy density at nucleation according to Problem 2.1$_\mu$. Here $\hat{u} := -\mu L/2$. For $\mu = 1$ solid nucleation occurs at $\theta = 0$ *(hypercooling)*, and the newborn solid is not superheated. Similarly the nucleated liquid is not undercooled.

For $\mu = 1$ these nucleation thresholds are twice those of the original formulation of Problem 2.1. In the case of undercooling, this behaviour has been observed for some materials under special experimental conditions, and is called *hypercooling*. Here solid nucleation occurs without formation of any metastable phase, and the new phase is at the uniform temperature $\theta = 0$.

VIII.3 Some Auxiliary Results

In this section we present two results that are used in the next section to prove the existence of a solution of Problem 2.1.

Lemma 3.1 [7] *There exists a constant $C > 0$ such that, for any $v \in H^1(\mathbf{R}^3) \cap L^1(\mathbf{R}^3)$ and $\varphi \in L^1(\mathbf{R}^3)$, if $\varphi(x) \in \{-2, 0, 2\}$ for a.e. $x \in \mathbf{R}^3$, then*

$$\int_{\mathbf{R}^3} |\varphi| dx \leq 4 \int_{\mathbf{R}^3} |v + \varphi| dx + C \left(\int_{\mathbf{R}^3} |\nabla v|^2 dx \cdot \int_{\mathbf{R}^3} |v + \varphi| dx \right)^{3/4}. \quad (3.1)$$

Proof. We denote by $|A|$ the ordinary three-dimensional Lebesgue measure of a (measurable) set A, and by C_1, C_2 suitable positive constants. We also use notations like $\{\alpha < v < \beta\} := \{x \in \mathbf{R}^3 : \alpha < v(x) < \beta\}$.

---

[7] See Luckhaus [361].

By the *Sobolev inequality* (XI.2.3), there exists a constant $C_1 > 0$ such that, setting $g(x) := \min\left\{[v(x) - \tfrac{1}{2}]^+, 1\right\}$ for any $x \in \mathbf{R}^3$,

$$\left(\int_{\mathbf{R}^3} g(x)^{3/2} dx\right)^{2/3} \leq C_1 \int_{\mathbf{R}^3} |\nabla g|\, dx = C_1 \int_{\{1/2 < v < 3/2\}} |\nabla v|\, dx$$

$$\leq C_1 \left(|\{\tfrac{1}{2} < v < \tfrac{3}{2}\}|\int_{\mathbf{R}^3} |\nabla v|^2 dx\right)^{1/2};$$

hence, as $g = 1$ in $\{v > 3/2\}$, we have

$$|\{\tfrac{1}{2} < v < \tfrac{3}{2}\}|^{2/3} \leq \left(\int_{\mathbf{R}^3} g(x)^{3/2} dx\right)^{2/3}$$

$$\leq C_1 \left(|\{\tfrac{1}{2} < v < \tfrac{3}{2}\}|\int_{\mathbf{R}^3} |\nabla v|^2 dx\right)^{1/2}.$$

After writing the same formula for $-v$, we get

$$|\{|v| > \tfrac{3}{2}\}|^{2/3} \leq C_1 \left(|\{\tfrac{1}{2} < v < \tfrac{3}{2}\}|\int_{\mathbf{R}^3} |\nabla v|^2 dx\right)^{1/2}.$$

Notice that $|v + \varphi| \geq \tfrac{1}{2}$ whenever either $\tfrac{1}{2} \leq |v| \leq \tfrac{3}{2}$, or $|v| \leq \tfrac{3}{2}$ and $\varphi \neq 0$. Then we get

$$\frac{1}{2}\int_{\mathbf{R}^3} |\varphi|\, dx \leq |\{|v| > \tfrac{3}{2}\} \cap \{|\varphi| \neq 0\}| + |\{|v| \leq \tfrac{3}{2}\} \cap \{|\varphi| \neq 0\}|$$

$$\leq C_1^{3/2}\left(|\{\tfrac{1}{2} < v < \tfrac{3}{2}\}|\int_{\mathbf{R}^3} |\nabla v|^2 dx\right)^{3/4} + 2\int_{\mathbf{R}^3} |v + \varphi|\, dx$$

$$\leq C_1^{3/2}\left(2\int_{\mathbf{R}^3} |v + \varphi|\, dx \int_{\mathbf{R}^3} |\nabla v|^2 dx\right)^{3/4} + 2\int_{\mathbf{R}^3} |v + \varphi|\, dx,$$

that is, (3.1). □

The latter result is applied through the following lemma.

Lemma 3.2 [8] *Let the sequences $\{\theta_m\} \subset L^2\left(0, T; H_0^1(\Omega)\right)$, $\{\chi_m\} \subset L^1(Q)$ be such that $|\chi_m| = 1$ a.e. in Q for any $m \in \mathbf{N}$, and*

$$\theta_m \to \theta \qquad \text{weakly in } L^2\left(0, T; H_0^1(\Omega)\right), \qquad (3.2)$$

---

[8] See Luckhaus [361].

VIII.3 Some Auxiliary Results

$$\chi_m \to \chi \quad \text{weakly star in } L^\infty(Q), \tag{3.3}$$

$$C_V \theta_m + \frac{L}{2}\chi_m \to C_V \theta + \frac{L}{2}\chi \quad \text{strongly in } L^3\left(0, T; L^1(\Omega)\right). \tag{3.4}$$

Then

$$\chi_m \to \chi \quad \text{strongly in } L^q(Q), \forall q < +\infty. \tag{3.5}$$

Proof. For the sake of simplicity, here we replace C_V and $L/2$ by 1. However, the argument can easily be extended to the general setting.

$$\tilde{\theta}_m := \theta_m, \quad \tilde{\chi}_m := \chi_m \quad \text{in } \Omega \times]0, T[;$$
$$\tilde{\theta}_m := 0, \quad \tilde{\chi}_m := 0 \quad \text{in } (\mathbf{R}^3 \setminus \Omega) \times]0, T[.$$

Note that $\tilde{\theta}_m \in L^2\left(0, T; H^1\left(\mathbf{R}^3\right)\right)$. Let us fix any $\sigma \in \,]0, T/2[$ and any $\delta = (\delta_x, \delta_t) \in \mathbf{R}^4$ such that $|\delta_t| < \sigma$. For almost any $(x,t) \in Q^\sigma_\infty := \mathbf{R}^3 \times]\sigma, T - \sigma[$, we set

$$\hat{\chi}^\delta_m(x,t) := \tilde{\chi}_m(x + \delta_x, t + \delta_t) - \tilde{\chi}_m(x, t),$$
$$\hat{\theta}^\delta_m(x,t) := \tilde{\theta}(x + \delta_x, t + \delta_t) - \tilde{\theta}(x, t).$$

Note that $\hat{\chi}^\delta_m \in \{-2, 0, 2\}$ a.e. in Q^σ_∞. We can now apply Lemma 3.1 for $\varphi = \hat{\chi}^\delta_m$ and $v = \hat{\theta}^\delta_m$. By (3.1) and the Schwarz-Hölder inequality, we have

$$\iint_{Q^\sigma_\infty} |\hat{\chi}^\delta_m| dx dt \leq 4 \iint_{Q^\sigma_\infty} |\hat{\theta}^\delta_m + \hat{\chi}^\delta_m| dx dt$$
$$+ C \int_\sigma^{T-\sigma} \left(\int_{\mathbf{R}^3} |\nabla \hat{\theta}^\delta_m|^2 dx \cdot \int_{\mathbf{R}^3} |\hat{\theta}^\delta_m + \hat{\chi}^\delta_m| dx\right)^{3/4} dt$$
$$\leq 4 \iint_{Q^\sigma_\infty} |\hat{\theta}^\delta_m + \hat{\chi}^\delta_m| dx dt \tag{3.6}$$
$$+ C \left\| \|\nabla \hat{\theta}^\delta_m\|^{3/2}_{L^2(\mathbf{R}^3)} \right\|_{L^{4/3}(\sigma, T-\sigma)} \left\| \|\hat{\theta}^\delta_m + \hat{\chi}^\delta_m\|^{3/4}_{L^1(\mathbf{R}^3)} \right\|_{L^4(\sigma, T-\sigma)}$$
$$\leq 4 \iint_{Q^\sigma_\infty} |\hat{\theta}^\delta_m + \hat{\chi}^\delta_m| dx dt$$
$$+ C \|\nabla \hat{\theta}^\delta_m\|^{3/2}_{L^2(Q^\sigma_\infty; \mathbf{R}^3)} \|\hat{\theta}^\delta_m + \hat{\chi}^\delta_m\|^{3/4}_{L^3(\sigma, T-\sigma; L^1(\mathbf{R}^3))}.$$

Note that

$$\|\nabla \hat{\theta}^\delta_m\|_{L^2(Q^\sigma_\infty; \mathbf{R}^3)} \leq 2\|\nabla \tilde{\theta}_m\|_{L^2(Q; \mathbf{R}^3)} \leq \text{Constant}.$$

Moreover by the Fréchet-Riesz-Kolmogorov criterion (see Theorem XI.3.3), (3.4) implies

$$\iint_{Q^\sigma_\infty} |\hat{\theta}^\delta_m + \hat{\chi}^\delta_m| dx dt \to 0, \quad \|\hat{\theta}^\delta_m + \hat{\chi}^\delta_m\|_{L^3(\sigma, T-\sigma; L^1(\mathbf{R}^3))} \to 0$$

as $\delta \to 0$, uniformly for $m \in \mathbf{N}, \forall \sigma > 0$.

Hence, by (3.6), we get

$$\iint_{Q_\infty^\sigma} |\hat{\chi}_m^\delta| dx dt \to 0 \qquad \text{as } \delta \to 0, \text{ uniformly for } m \in \mathbf{N}, \forall \sigma > 0.$$

As the χ_ms are uniformly bounded in $L^\infty(Q)$, by the afore-mentioned compactness criterion we get that $\chi_m \to \chi$ strongly in $L^1(Q)$. This yields (3.5). □

VIII.4 Existence Result

In this section we prove that Problem 2.1 has at least one solution, and point out the simple modifications that allow us to extend the result to Problem 2.1_μ.

Theorem 4.1 *(Existence) Assume that*

$$f \in L^2\left(0, T; H^{-1}(\Omega)\right), \quad u^0 = C_V \theta^0 + \tfrac{L}{2}\chi^0 \quad \text{a.e. in } \Omega, \qquad (4.1)$$
$$\theta^0 \in L^2(\Omega), \quad \chi^0 \in \text{Dom}(P).$$

Then Problem 2.1 has at least one solution such that moreover

$$\theta \in L^\infty\left(0, T; L^2(\Omega)\right), \qquad \chi \in L^\infty(0, T; BV(\Omega)) \qquad (4.2)$$

Hence S_t is a manifold of class $C^{1,1/4}$ for almost any $t \in]0, T[$, as we saw.

Proof. [9] This argument is rather technical, mainly because of the necessity of deriving the *nucleation condition* (2.3) and the nonconvex constraint $|\chi| = 1$ a.e. in Q (which is implicit in the definition of Φ). The latter is obtained by means of the results of the previous section.

For the reader's convenience, we split this proof into several steps.

(i) Approximation. In view of the extension we consider in Sect. IX.1, we introduce the following notation

$$(w, v) := \int_\Omega wv\, dx \qquad \forall w, v \in L^2(\Omega). \qquad (4.3)$$

We point out the steps in our argument where we use this specific choice of the continuous bilinear form $(\cdot, \cdot)$.

---

[9] This proof is partly based on that of Luckhaus [361] and on the Luckhaus results of Sect. VIII.3.

VIII.4 Existence Result

Let us fix any $m \in \mathbf{N}$, set $h := T/m$ and

$$_{H^{-1}(\Omega)}\langle Aw, v\rangle_{H_0^1(\Omega)} := k\int_\Omega \nabla w \cdot \nabla v\, dx \qquad \forall w, v \in H_0^1(\Omega).$$

For any $v \in L^2(\Omega)$, there exists a (unique) $w \in H_0^1(\Omega) \cap H^2(\Omega)$ such that $C_V w + (h/2)Aw = v$ in $H^{-1}(\Omega)$. This defines a linear and positive continuous operator

$$J_h := \left(C_V I + \frac{h}{2}A\right)^{-1} : L^2(\Omega) \to H_0^1(\Omega) \cap H^2(\Omega) : v \mapsto w$$

($I :=$ identity). Note that by (4.3) $(J_h v, v) \geq 0$ for any $v \in L^2(\Omega)$. For any $m \in \mathbf{N}$ we also set

$$f_m^n(x) := \frac{1}{h}\int_{(n-1)h}^{nh} f(x, \tau)d\tau \ (\in H^{-1}(\Omega)) \qquad \text{for } n = 1, \ldots, m.$$

Possibly replacing θ^0 by a suitable approximate sequence, we can assume that $\theta^0 \in H_0^1(\Omega)$.

We can now introduce a time-discretized problem. Its unusual form reveals the care that is needed to handle the nonconvex constraint $|\chi| = 1$ a.e. in Q. [10]

Problem 2.1$_m$ *To find $\chi_m^n \in L^\infty(\Omega) \cap BV(\Omega)$ for $n = 1, \ldots, m$, such that the following holds. Setting $\theta_m^0 := \theta^0$, $\chi_m^0 := \chi^0$, and recursively for $n = 1, \ldots, m$,*

$$g_m^n := C_V \theta_m^{n-1} + \frac{L}{2}\chi_m^{n-1} - \frac{h}{2}A\theta_m^{n-1} + hf_m^n \qquad \text{in } H^{-1}(\Omega), \tag{4.4}$$

$$\theta_m^n := J_h\left(g_m^n - \frac{L}{2}\chi_m^n\right) \qquad \text{a.e. in } \Omega, \tag{4.5}$$

$$Z_m^n(v) := \tau_E \Phi(v) + \frac{L}{4}\left(J_h\left(\frac{L}{2}v - g_m^n\right) - \theta_m^{n-1}, v - \chi_m^{n-1}\right) \tag{4.6}$$
$$\forall v \in \text{Dom}(P),$$

it is required that

$$Z_m^n(\chi_m^n) = \inf Z_m^n \qquad \text{for } n = 1, \ldots, m. \tag{4.7}$$

Applying Theorem XI.7.4 with $B = BV(\Omega)$ and $B_0 = L^1(\Omega)$, it is not difficult to see that this problem has at least one solution. In general this is not unique, as the functional Z_m^n is nonconvex.

---

[10] For instance, setting $\tilde{\Phi}(v) := \Phi(v) + \int_\Omega v^2 dx$ for any $v \in \text{Dom}(P)$, notice that $\tilde{\Phi}(v) = \inf \tilde{\Phi} \not\Rightarrow \partial\Phi(v) + 2v \ni 0$, because of the nonconvexity.

(ii) A Priori Estimates. The equation (4.5) is equivalent to the following *Crank-Nicolson scheme*

$$C_V \frac{\theta_m^n - \theta_m^{n-1}}{h} + \frac{L}{2}\frac{\chi_m^n - \chi_m^{n-1}}{h} + A\frac{\theta_m^n + \theta_m^{n-1}}{2} = f_m^n \quad \text{in } H^{-1}(\Omega), \tag{4.8}$$

$$\text{for } n = 1, \ldots, m.$$

Multiplying this equation by $h(\theta_m^n + \theta_m^{n-1})/2$, we get

$$\frac{C_V}{2}\left(\|\theta_m^n\|_{L^2(\Omega)}^2 - \|\theta_m^{n-1}\|_{L^2(\Omega)}^2\right) + \frac{L}{4}\left(\chi_m^n - \chi_m^{n-1}, \theta_m^n + \theta_m^{n-1}\right)$$
$$+ \frac{hk}{4}\|\nabla(\theta_m^n + \theta_m^{n-1})\|_{L^2(\Omega;\mathbf{R}^3)}^2 \leq \frac{h}{2} {}_{H^{-1}(\Omega)}\langle f_m^n, \theta_m^n + \theta_m^{n-1}\rangle_{H_0^1(\Omega)}$$

$$\text{for } n = 1, \ldots, m.$$

By (4.7) we have $Z_m^n(\chi_m^n) \leq Z_m^n(\chi_m^{n-1})$; that is, by (4.5),

$$\tau_E\left[\Phi(\chi_m^n) - \Phi(\chi_m^{n-1})\right] \leq \frac{L}{4}\left(\theta_m^n + \theta_m^{n-1}, \chi_m^n - \chi_m^{n-1}\right) \tag{4.9}$$

$$\text{for } n = 1, \ldots, m.$$

The two latter inequalities yield

$$\frac{C_V}{2}\left(\|\theta_m^n\|_{L^2(\Omega)}^2 - \|\theta_m^{n-1}\|_{L^2(\Omega)}^2\right) + \tau_E\left[\Phi(\chi_m^n) - \Phi(\chi_m^{n-1})\right]$$
$$+ \frac{hk}{4}\|\nabla(\theta_m^n + \theta_m^{n-1})\|_{L^2(\Omega;\mathbf{R}^3)}^2 \leq \frac{h}{2} {}_{H^{-1}(\Omega)}\langle f_m^n, \theta_m^n + \theta_m^{n-1}\rangle_{H_0^1(\Omega)} \tag{4.10}$$

$$\text{for } n = 1, \ldots, m;$$

then by a simple calculation we get

$$\max_{n=1,\ldots,m}\|\theta_m^n\|_{L^2(\Omega)}, \; h\sum_{n=1}^m \|\nabla\theta_m^n\|_{L^2(\Omega;\mathbf{R}^3)}^2, \; \max_{n=1,\ldots,m}\Phi(\chi_m^n) \tag{4.11}$$

$$\leq \text{Constant (independent of } m).$$

Setting $u_m^n := C_V\theta_m^n + (L/2)\chi_m^n$ a.e. in Ω for $n = 0, \ldots, m$, by (4.11) and by comparison in (4.8), we have

$$h\sum_{n=1}^m \|u_m^n\|_{BV(\Omega)}^2, \; h\sum_{n=1}^m \left\|\frac{u_m^n - u_m^{n-1}}{h}\right\|_{H^{-1}(\Omega)}^2 \leq \text{Constant}. \tag{4.12}$$

(iii) Further Notation. Let us set

$$\hat{\theta}_m := \text{linear interpolate in } [0, T] \text{ of } \{\theta_m^n(\cdot, nh) := \theta_m^n\}_{n=0,\ldots,m}, \text{ a.e. in } \Omega, \tag{4.13}$$

$$\theta_m(\cdot, t) := \theta_m^n \quad \text{for } (n-1)h < t \leq nh, n = 1, \ldots, m, \text{ a.e. in } \Omega; \quad (4.14)$$

define similarly $\hat{\chi}_m, \chi_m, \hat{u}_m, u_m, \hat{f}_m, f_m$, and so on. Set also $\tau_{-h}v(t) := v(t-h)$ for any $v : \mathbf{R} \to \mathbf{R}$ and any $h, t \in \mathbf{R}$. Then (4.8) can be written in the form

$$\frac{\partial \hat{u}_m}{\partial t} + A\frac{\theta_m + \tau_{-h}\theta_m}{2} = f_m \quad \text{in } H^{-1}(\Omega), \text{ a.e. in }]0, T[; \quad (4.15)$$

moreover, (4.11) and (4.12) yield

$$\|\hat{\theta}_m\|_{L^\infty(0,T;L^2(\Omega))\cap L^2(0,T;H^1(\Omega))}, \|\hat{\chi}_m\|_{L^\infty(Q)\cap L^\infty(0,T;BV(\Omega))},$$
$$\|\hat{u}_m\|_{L^\infty(0,T;L^2(\Omega))\cap L^2(0,T;BV(\Omega))\cap H^1(0,T;H^{-1}(\Omega))} \leq \text{Constant.} \quad (4.16)$$

Clearly these estimates for $\hat{\theta}_m$ and $\hat{\chi}_m$ hold also for θ_m and χ_m, respectively.

(iv) Some Convergence Properties. By (4.16), there exist θ, χ, u such that, possibly taking $m \to \infty$ along a subsequence,

$$\hat{\theta}_m, \theta_m \to \theta \quad \text{weakly star in } L^\infty\left(0, T; L^2(\Omega)\right) \cap L^2\left(0, T; H^1(\Omega)\right), \quad (4.17)$$

$$\hat{\chi}_m, \chi_m \to \chi \quad \text{weakly star in } L^\infty(Q) \cap L^\infty(0, T; BV(\Omega)), \quad (4.18)$$

$$\hat{u}_m \to u \quad \text{weakly star in } L^\infty\left(0, T; L^2(\Omega)\right) \cap L^2(0, T; BV(\Omega)) \quad (4.19)$$
$$\cap H^1\left(0, T; H^{-1}(\Omega)\right).$$

By taking $m \to \infty$ in (4.15) we get (2.7), hence (2.2).

By Lions-Aubin's Theorem XI.3.5 (ii), we have

$$\begin{cases} L^2(0, T; BV(\Omega)) \cap H^1\left(0, T; H^{-1}(\Omega)\right) \subset L^2\left(0, T; L^1(\Omega)\right) \\ \text{with compact injection.} \end{cases} \quad (4.20)$$

Therefore by (4.19), possibly extracting a further subsequence,

$$u_m \to u \quad \text{a.e. in } Q. \quad (4.21)$$

Hence, since the sequence $\{u_m\}$ is uniformly bounded in $L^\infty\left(0, T; L^2(\Omega)\right)$, by Proposition XI.3.11 we have

$$\hat{u}_m := C_V \hat{\theta}_m + \frac{L}{2}\hat{\chi}_m \to u \quad \text{strongly in } L^q\left(0, T; L^r(\Omega)\right), \quad (4.22)$$
$$\forall q < +\infty, \forall r < 2,$$

and the same holds for u_m.

The equation (4.7) implies $|\chi_m^n| = 1$ a.e. in Ω for $n = 1, \ldots, m$; hence $|\chi_m| = 1$ a.e. in Q. We can then apply Lemma 3.2 (with nonnormalized coefficients), getting

$$\chi_m \to \chi \quad \text{strongly in } L^q(Q), \forall q < +\infty. \tag{4.23}$$

Hence, possibly taking $m \to \infty$ along a subsequence, we have $\chi_m \to \chi$ and $\theta_m \to \theta$ a.e. in Q.

By (4.21), we also have $\theta_m \to \theta$ a.e. in Q.

By Sobolev inclusion $H^1(\Omega) \subset L^6(\Omega)$. [11] By space interpolation, [12] (4.17) then implies

$$\hat{\theta}_m, \theta_m \to \theta \quad \text{weakly star in } L^p\left(0, T; L^{2+\varepsilon}(\Omega)\right) \cap L^{2+\varepsilon}\left(0, T; L^r(\Omega)\right), \tag{4.24}$$
$$\text{with } \varepsilon = \varepsilon(p) > 0, \varepsilon = \varepsilon(r) > 0, \forall p < +\infty, \forall r < 6.$$

Hence, by Proposition XI.3.11,

$$\theta_m \to \theta \quad \text{strongly in } L^q\left(0, T; L^2(\Omega)\right) \cap L^2\left(0, T; L^r(\Omega)\right), \tag{4.25}$$
$$\forall p < +\infty, \forall r < 6.$$

(v) Towards the Proof of (2.3). By (4.7), through some tedious calculations, we derive the following variational inequality

$$\begin{aligned}
0 \geq Z_m^n(\chi_m^n) - Z_m^n(v) &= \tau_E \left[\Phi(\chi_m^n) - \Phi(v)\right] \\
&\quad - \frac{L}{4}\left(\theta_m^n + \theta_m^{n-1}, \chi_m^n - \chi_m^{n-1}\right) \\
&\quad + \frac{L}{4}\left(\theta_m^n + \theta_m^{n-1} - \frac{L}{2}J_h(v - \chi_m^n), v - \chi_m^{n-1}\right) \\
&= \tau_E \left[\Phi(\chi_m^n) - \Phi(v)\right] - \frac{L}{4}\left(\theta_m^n + \theta_m^{n-1}, \chi_m^n - v\right) \\
&\quad - \frac{L^2}{8}\left(J_h(v - \chi_m^n), \chi_m^n - \chi_m^{n-1}\right) - \frac{L^2}{8}\left(J_h(v - \chi_m^n), v - \chi_m^n\right) \\
&= \tau_E \left[\Phi(\chi_m^n) - \Phi(v)\right] \\
&\quad - \frac{L}{4}\left(\theta_m^n + \theta_m^{n-1} - \frac{L}{2}J_h(\chi_m^n - \chi_m^{n-1}), \chi_m^n - v\right) \\
&\quad - \frac{L^2}{8}\left(J_h(v - \chi_m^n), v - \chi_m^n\right) \quad \forall v \in \text{Dom}(P), \text{ for } n = 1, \ldots, m.
\end{aligned} \tag{4.26}$$

Note that (4.15) yields

$$\left(C_V I + \frac{h}{2}A\right)(\theta_m^n + \theta_m^{n-1}) + \frac{L}{2}(\chi_m^n - \chi_m^{n-1}) = hf_m^n + 2C_V\theta_m^{n-1}$$
$$\text{in } H^{-1}(\Omega), \text{ for } n = 1, \ldots, m,$$

---

[11] See Theorem XI.2.1.

[12] See, e.g., Bergh and Löfström [72] and Triebel [536].

that is,
$$-\frac{L}{2} J_h(\chi_m^n - \chi_m^{n-1}) = \theta_m^n + \theta_m^{n-1} - h J_h f_m^n - 2C_V J_h \theta_m^{n-1}$$
$$\text{a.e. in } \Omega, \text{ for } n = 1, \ldots, m.$$

So, setting
$$\tilde{\theta}_m^n := \theta_m^n + \theta_m^{n-1} - \frac{h}{2} J_h f_m^n - C_V J_h \theta_m^{n-1} \qquad \text{a.e. in } \Omega, \text{ for } n = 1, \ldots, m,$$

by (4.26) we have
$$\tau_E \left[\Phi(\chi_m^n) - \Phi(v) \right] \le \frac{L}{2} \left(\tilde{\theta}_m^n, \chi_m^n - v \right) + \frac{L^2}{8} \left(J_h(v - \chi_m^n), v - \chi_m^n \right)$$
$$\forall v \in \text{Dom}(P), \text{ for } n = 1, \ldots, m;$$

that is, setting $\tilde{\theta}_m(\cdot, t) := \tilde{\theta}_m^n$ a.e. in Ω for $(n-1)h < t \le nh$, $n = 1, \ldots, m$,
$$\tau_E \left[\Phi(\chi_m) - \Phi(v) \right] \le \frac{L}{2} \left(\tilde{\theta}_m, \chi_m - v \right) + \frac{L^2}{8} \left(J_h(v - \chi_m), v - \chi_m \right) \quad (4.27)$$
$$\forall v \in \text{Dom}(P), \text{ a.e. in }]0, T[.$$

(vi) Proof of (2.3). Note that
$$h J_h f_m \to 0 \qquad \text{strongly in } L^2 \left(0, T; H_0^1(\Omega) \right) \subset L^2 \left(0, T; L^6(\Omega) \right).$$

Hence by (4.25) we have
$$\tilde{\theta}_m \to \theta \qquad \text{strongly in } L^q \left(0, T; L^2(\Omega) \right) \cap L^2 \left(0, T; L^r(\Omega) \right), \qquad (4.28)$$
$$\forall p < +\infty, \forall r < 6.$$

Taking $m \to \infty$ in (4.27), we then get
$$\tau_E \left[\Phi(\chi) - \Phi(v) \right] \le \frac{L}{2} (\theta, \chi - v) + \frac{L^2}{8 C_V} (v - \chi, v - \chi) \qquad (4.29)$$
$$\forall v \in \text{Dom}(P), \text{ a.e. in }]0, T[.$$

Henceforth we use the specific form (4.3) of the bilinear form $(\cdot, \cdot)$. Note that for any $\chi, v \in \text{Dom}(P)$, as $|\chi| = |v| = 1$ a.e. in Q, we have
$$(v - \chi)^2 = 2\chi(\chi - v) + v^2 - \chi^2 = 2\chi(\chi - v) \qquad \text{a.e. in } Q$$

(this point is crucial). Hence

$$\int_\Omega (v - \chi)^2 dx = 2 \int_\Omega \chi(\chi - v) dx \qquad \text{a.e. in } \Omega, \forall \chi, v \in \text{Dom}(P), \qquad (4.30)$$

and so (4.29) can be written in the form

$$\begin{aligned}
\tau_E \left[\Phi(\chi) - \Phi(v) \right] &\leq \frac{L}{2} \int_\Omega \left(\theta + \frac{L}{2C_V} \chi \right) (\chi - v) dx \\
&= \frac{L}{2C_V} \int_\Omega u(\chi - v) dx \qquad \forall v \in \text{Dom}(P), \text{ a.e. in }]0, T[.
\end{aligned} \qquad (4.31)$$

(vii) Approximate Gibbs-Thomson Law. Let us fix any $m \in \mathbf{N}$. Note that, for almost any t, if a sequence $\{v_j\}$ is uniformly bounded in $L^\infty(\Omega)$ and $v_j \to \chi_m$ strongly in $L^1(\Omega)$, then $J_h(v_j - \chi_m) \to 0$ strongly in $L^\infty(\Omega)$. Hence

$$\frac{|(J_h(v_j - \chi_m), v_j - \chi_m)|}{\|v_j - \chi_m\|_{L^1(\Omega)}} \leq \|J_h(v_j - \chi_m)\|_{L^\infty(\Omega)} \to 0 \qquad (4.32)$$

$$\text{a.e. in }]0, T[.$$

Therefore, recalling the definition (VI.1.7) of F_θ, (4.26) yields [13]

$$\liminf \frac{F_{\tilde\theta_m}(v) - F_{\tilde\theta_m}(\chi_m)}{\|v - \chi_m\|_{L^1(\Omega)}} \geq 0 \qquad \text{as } v \to \chi_m \text{ strongly in } L^1(\Omega). \qquad (4.33)$$

For any $m \in \mathbf{N}$ and almost any $t \in]0, T[$, let us denote the boundary of the set $\Omega_m^+(t) := \{x \in \Omega : \chi_m(x, t) = 1\}$ in Ω by $\mathcal{S}_m(t)$ and its mean curvature by κ_m.

By Proposition XI.8.3, $\mathcal{S}_m(t)$ is of class $C^{1,1/4}$ for almost any $t \in]0, T[$. By Theorem VI.4.2 and (4.33), the approximate solution fulfills the following approximate Gibbs-Thomson law

$$\kappa_m = -\frac{L}{2\sigma\tau_E} \tilde\theta_m \qquad \text{a.e. on } \mathcal{S}_m. \qquad (4.34)$$

Note that (4.32) is not fulfilled with J_h replaced by $\lim_{h \to 0} J_h = C_V^{-1} I$, cf. (4.30). Hence the previous procedure cannot be used to derive the Gibbs-Thomson law from the limit variational inequality (4.31).

(viii) Parametric Representation of the Interface. In order to derive the Gibbs-Thomson law (2.4), we locally represent $\mathcal{S}_m(t)$ as a Cartesian surface, and then apply the classical regularity theory of *almost minimal boundaries,* cf. Sect. XI.8.

---

[13] In terms of the subdifferential operator ∂^-, cf. (XI.4.16), (4.33) also reads $\partial^- F_{\tilde\theta_m}(\chi_m) \ni 0$ in $L^\infty(\Omega)$.

By (4.23) and (4.29), there exists $\tilde{\Lambda} \subset\,]0,T[$ such that $|\,]0,T[\,\setminus\tilde{\Lambda}\,| = 0$, and, possibly extracting a further subsequence, as $m \to \infty$

$$\chi_m(\cdot, t) \to \chi(\cdot, t) \qquad \text{strongly in } L^1(\Omega), \forall t \in \tilde{\Lambda}, \tag{4.35}$$

$$\|\theta_m(\cdot, t)\|_{L^6(\Omega)} \leq \text{Constant (independent of } m), \forall t \in \tilde{\Lambda}. \tag{4.36}$$

Let us fix any $\tilde{t} \in \tilde{\Lambda}$. As we saw, $\mathcal{S}_{\tilde{t}}$ can be locally represented as the Cartesian graph of a function of class $C^{1,1/4}$. More precisely, for any $x \in \mathcal{S}_{\tilde{t}}$, there exists an open neighbourhood $K_x^{\tilde{t}}$ of x in $\mathbf{R}^3$ such that $\mathcal{S}_{\tilde{t}} \cap K_x^{\tilde{t}}$ is the graph of a function $\varphi_x^{\tilde{t}} : G_x^{\tilde{t}}\,(\subset \mathbf{R}^2) \to \mathbf{R}$ of class $C^{1,1/4}$; that is, possibly after an appropriate rotation of the axes, $\mathcal{S}_{\tilde{t}} \cap K_x^{\tilde{t}} = \{(y, \varphi_x^{\tilde{t}}(y)) : y \in G_x^{\tilde{t}}\}$. As $\mathcal{S}_{\tilde{t}}$ is compact, we can extract a finite covering $\{K_j := K_{x_j}^{\tilde{t}}\}_{j \in J}$ from $\{K_x^{\tilde{t}} : x \in \mathcal{S}_{\tilde{t}}\}$.

By (4.35) and (4.36), we can apply Theorem XI.8.4. So, provided that m is sufficiently large ($m \geq M$, say), we have:

(1) $\mathcal{S}_m(\tilde{t}) \subset \cup_{j \in J} K_j$;
(2) for any $j \in J$, $\mathcal{S}_m(\tilde{t}) \cap K_j$ is the Cartesian graph of a function $\varphi_{j,m} : G_j \to \mathbf{R}$;
(3) for any $j \in J$, as $m \to \infty$

$$\varphi_{j,m} \to \varphi_j \qquad \text{strongly in } C^{1,1/4}(\bar{G}_j); \tag{4.37}$$

(4) for any $j \in J$, $\mathcal{S}_t \cap K_j$ is the Cartesian graph of φ_j.

(ix) Proof of (2.4). Let us set $\tilde{\nabla} := (\partial/\partial y_1, \partial/\partial y_2)$, and $A_{j,m} := \left(\Omega_m^+(\tilde{t}) \triangle \Omega^+(\tilde{t})\right) \cap (G_j \times \mathbf{R})$ (here $\triangle$ represents the symmetric difference).

Possibly excluding another set of instants of vanishing measure, by (4.34), respectively, by (4.23) and (4.28), we can assume that at $t = \tilde{t}$

$$\sigma \iint_{G_j} \frac{\tilde{\nabla}\varphi_{j,m} \cdot \tilde{\nabla}\xi}{\sqrt{1 + |\tilde{\nabla}\varphi_{j,m}|^2}} dy_1 dy_2 = \frac{L}{\tau_E} \iint_{G_j} (\theta_m \circ \varphi_{j,m}) \xi \, dy_1 dy_2 \tag{4.38}$$

$$\forall \xi \in H_0^1(G_j),$$

$$\iint_{G_j} |\theta_m \circ \varphi_{j,m} - \theta \circ \varphi_j| dy_1 dy_2 \leq \iint_{A_{j,m}} \left|\frac{\partial \theta_m}{\partial y_3}\right| dy_1 dy_2$$

$$\leq \|\theta_m\|_{H_0^1(\Omega)} |A_{j,m}|^{1/2} \to 0 \qquad \forall j \in J, \text{ as } m \to \infty.$$

Moreover, by (4.37) we have

$$\frac{\tilde{\nabla}\varphi_{j,m}}{\sqrt{1 + |\tilde{\nabla}\varphi_{j,m}|^2}} \to \frac{\tilde{\nabla}\varphi_j}{\sqrt{1 + |\tilde{\nabla}\varphi_j|^2}} \qquad \text{uniformly in } G_j, \forall j \in J.$$

We can then pass to the limit in (4.38). This yields the Gibbs-Thomson law (2.4) for any $\tilde{t} \in \tilde{\Lambda}$. □

Remark. It is easy to see that the perimeters converge for almost any $t \in]0, T[$, more precisely,

$$\Phi(\chi_m) \to \Phi(\chi) \quad \text{and} \quad P(\chi_m) \to P(\chi) \qquad \text{a.e. in }]0, T[. \tag{4.39}$$

By taking $v = \chi$ in (4.25), we get

$$\limsup_{m \to \infty} \Phi(\chi_m) \leq \Phi(\chi) \qquad \text{a.e. in }]0, T[.$$

On the other hand, as $|\chi_m| = |\chi| = 1$ a.e. in Q, we have

$$\liminf_{m \to \infty} P(\chi_m) = \frac{1}{2} \liminf_{m \to \infty} \mathrm{Var}(\chi_m) \geq \frac{1}{2} \mathrm{Var}(\chi) = P(\chi) \qquad \text{a.e. in }]0, T[.$$

Moreover $\int_\Gamma \chi_m d\Gamma \to \int_\Gamma \chi d\Gamma$ a.e. in $]0, T[$. Hence the two latter formulae yield (4.39). □

The preceding existence result can be extended in several ways. Here are two simple examples.

Temperature- and Phase-Dependent Conductivity. Let us assume that $k = \hat{k}(\theta, \chi)$, with $\hat{k} \in C^0(\mathbf{R} \times [0, 1])$ bounded by two positive constants. In order to approximate the corresponding Problem 2.1, it suffices to set recursively

$$\tilde{A}_m^n(\theta, v) := \int_\Omega \hat{k}\left(\theta_m^{n-1}, \chi_m^{n-1}\right) \nabla \theta \cdot \nabla v \, dx \qquad \forall \theta, v \in H_0^1(\Omega), \tag{4.40}$$
$$\text{for } n = 1, \ldots, m,$$

and to replace J_h by $\tilde{J}_{m,h}^n := \left(C_V I + (h/2)\tilde{A}_m^n\right)^{-1}$. The previous argument holds almost unchanged, and the strong convergence of θ_m and χ_m allows us to pass to the limit in the $\hat{k}$-term.

Problem with Hypercooling. The proof of Theorem 4.1 can also be extended to show existence of a solution of Problem 2.1_μ.

Theorem 4.2 *For any $\mu \in [0, 1]$, if (4.1) is satisfied, then Problem 2.1_μ has at least one solution that fulfills (4.2).*

Proof. For any $m \in \mathbf{N}$, we consider an approximate problem similar to Problem 2.1_m, but with g_m^n and Z_m^n replaced by

$$g_m^n := C_V \theta_m^{n-1} + \frac{L}{2}\chi_m^{n-1} - (1 - \mu)\frac{h}{2}A\theta_m^{n-1} + hf_m^n \qquad \text{a.e. in } \Omega, \tag{4.41}$$

$$Z_m^n(v) := \tau_E \Phi(v)$$
$$+ \frac{L}{4} \left((1+\mu) J_h \left(\frac{L}{2} v - g_m^n \right) - (1-\mu) \theta_m^{n-1}, v - \chi_m^{n-1} \right) \quad (4.42)$$
$$\forall v \in \text{Dom}(P).$$

Here (4.5) yields the following discretized equation, in place of (4.8):
$$C_V \frac{\theta_m^n - \theta_m^{n-1}}{h} + \frac{L}{2} \frac{\chi_m^n - \chi_m^{n-1}}{h} + A \frac{(1+\mu)\theta_m^n + (1-\mu)\theta_m^{n-1}}{2} = f_m^n \quad (4.43)$$
$$\text{in } H^{-1}(\Omega), \text{ for } n = 1, \ldots, m.$$

The one multiplies this equation by $h[(1+\mu)\theta_m^n + (1-\mu)\theta_m^{n-1}]/2$, and goes on as previously, with θ_m^{n-1} replaced by $\mu\theta_m^n + (1-\mu)\theta_m^{n-1}$. So, in place of (4.15), one gets
$$\frac{\partial \hat{u}_m}{\partial t} + A \frac{(1+\mu)\theta_m + (1-\mu)\tau_{-h}\theta_m}{2} = f_m \quad \text{in } H^{-1}(\Omega), \quad (4.44)$$
$$\text{a.e. in }]0,T[,$$

and, instead of (4.26),
$$\tau_E [\Phi(\chi_m) - \Phi(v)]$$
$$\leq \frac{L}{2} \left(\tilde{\theta}_m, \chi_m - v \right) + (1+\mu) \frac{L^2}{8} (J_h(v - \chi_m), v - \chi_m) \quad (4.45)$$
$$\forall v \in \text{Dom}(P), \text{ a.e. in }]0,T[.$$

Taking $m \to \infty$, in place of (4.29) one then gets
$$\tau_E [\Phi(\chi) - \Phi(v)] \leq \frac{L}{2} (\theta, \chi - v) + (1+\mu) \frac{L^2}{8C_V} (v - \chi, v - \chi) \quad (4.46)$$
$$\forall v \in \text{Dom}(P), \text{ a.e. in }]0,T[,$$

which yields (2.19). □

VIII.5 The Mullins-Sekerka Problem

In this section we deal with the Hele-Shaw problem (see Sect. IV.7), including the Gibbs-Thomson law (2.12) and the contact angle condition (2.13). This is known as the *Hele-Shaw problem with surface tension*, or also as the *Mullins-Sekerka problem*, cf. Sect. IV.7.

This setting exhibits several analogies with the Stefan problem. However, here some time regularity of the temperature is lost; on the other hand, some time regularity is gained for the phase function, and this suffices to exclude nucleation and the onset of other singularities in phase transition.

Let us assume that $|\sigma_L - \sigma_S| \leq \sigma$ and
$$f \in L^2\left(0,T; H^{-1}(\Omega)\right), \quad \chi^0 \in \text{Dom}(P). \quad (5.1)$$

Problem 5.1 *(Mullins-Sekerka Problem) To find $\theta \in L^2\left(0, T; H_0^1(\Omega)\right)$ and $\chi \in L^1(0, T; BV(\Omega))$ such that $|\chi| = 1$ a.e. in Q, and*

$$\iint_Q \left(-\frac{L}{2}\chi\frac{\partial \xi}{\partial t} + k\nabla\theta \cdot \nabla\xi\right) dxdt$$
$$= \int_0^T {}_{H^{-1}(\Omega)}\langle f, \xi\rangle_{H_0^1(\Omega)} dt + \int_\Omega \frac{L}{2}\chi^0(x)\xi(x, 0) dx \quad (5.2)$$
$$\forall \xi \in H^1(Q), \xi = 0 \text{ on } (\Gamma \times]0, T[) \cup (\Omega \times \{T\}),$$

Moreover, it is required that for almost any fixed $t \in]0, T[$, setting $\Omega_t^+ := \{x \in \Omega : \chi(x, t) = 1\}$ and denoting by S_t the boundary of Ω_t^+ in Ω, the following holds: for any $x \in S_t$, there exist $a, b \in \mathbf{R}$, an open set $G \subset \mathbf{R}^2$, and a function $\psi: G \to]a, b[$ of class C^1, such that, possibly after a coordinate rotation,

$$\Omega_t^+ \cap (G \times]a, b[) = \{(y_1, y_2, z) \in G \times]a, b[: z < \psi(y_1, y_2)\}.$$

Setting $\tilde{\nabla} := (\partial/\partial y_1, \partial/\partial y_2)$ and $\theta \circ \psi : y \mapsto \theta(y, \psi(y), t)$, it is then required that

$$\tilde{\nabla} \cdot \frac{\tilde{\nabla}\psi}{\sqrt{1 + |\tilde{\nabla}\psi|^2}} = -\frac{L}{\sigma \tau_E} \theta \circ \psi \quad \text{in } H^{-1}(G). \quad (5.3)$$

Remark. The interpretation of the latter problem is similar to that of Problem 2.1. Notice that in place of (2.7) and (2.8) we have

$$\frac{L}{2}\frac{\partial \chi}{\partial t} - k\Delta\theta = f \quad \text{in } \mathcal{D}'(Q), \quad (5.4)$$

$$\chi|_{t=0} = \chi^0 \quad \text{in } \Omega \text{ (in the sense of the traces),} \quad (5.5)$$

and this yields $\chi \in H^1\left(0, T; H^{-1}(\Omega)\right)$. This time regularity prevents χ from being discontinuous in any positive volume set. Therefore the phase volume is continuous as nucleation and other singularities (e.g., merging or splitting of connected components) occur in the phase evolution.

Here the condition (2.3) is replaced by the pointwise constraint on $|\chi|$ (which was implicit in (2.3), and here is required explicitly). Indeed, in the limit as $C_V \to 0$, the nucleation temperature $\theta_c = L/2C_V$ diverges, and (2.3) is reduced to the condition $\Phi(\chi) < +\infty$ a.e. in $]0, T[$. □

Theorem 5.1 *(Existence) Assume that (5.1) is satisfied. Then Problem 5.1 has at least one solution such that moreover*

$$\chi \in L^\infty(0, T; BV(\Omega)). \quad (5.6)$$

Hence S_t is a manifold of class $C^{1,1/4}$ for almost any $t \in \,]0,T[$.

Outline of the Proof. We indicate two possible arguments.

The first one simply consists of passing to the limit as $C_V \to 0$ in Problem 2.1. By the argument of Theorem 4.1, it appears that the estimates on $C_V \theta + (L/2)\chi \in H^1\left(0,T;H^{-1}(\Omega)\right), \theta \in L^2\left(0,T;H_0^1(\Omega)\right), \chi \in L^\infty(0,T;BV(\Omega))$ are uniform with respect to C_V. Hence, possibly extracting a subsequence, $C_V \theta + (L/2)\chi$ converges strongly in $L^2(Q)$, as $C_V \to 0$. Since $C_V \theta \to 0$, χ converges strongly in $L^2(Q)$; hence the limit function fulfills the nonconvex constraint $|\chi| = 1$ a.e. in Q. (5.2) is easily derived, and the argument based on the local Cartesian representation of S can also be used to get (5.3).

In the alternative, one can directly prove existence of a solution by an argument similar to that of Theorem 4.1, setting $J_h := 2A^{-1}/h$. Actually, here the proof is simpler, since steps (v) and (vi) of Sect. VIII.4 are not needed. Moreover, a comparison in the approximate equation yields a uniform estimate for χ_m in $H^1\left(0,T;H^{-1}(\Omega)\right)$, which coupled with the uniform estimate for χ_m in $L^\infty(0,T;BV(\Omega))$ yields strong convergence in $L^2(Q)$, without the need of applying the results of Sect. VIII.3. □

VIII.6 Comments

In this chapter we illustrated an approach which is under current investigation; see V. [566 – 569]. One of the main issues of this formulation is the representation of the phase configuration by means of a characteristic function. A similar direction of analysis has also been pursued by Almgren, Taylor and Wang [9], Almgren and Wang [10], Luckhaus [361, 362, 363], and Luckhaus and Sturzenhecker [366].

Here we dealt with the (stationary) Gibbs-Thomson law. The corresponding dynamical equation, namely, the mean curvature flow with forcing term, cf. Sects. VII.6 and VII.7, has been coupled with the heat equation by Soner [512] and V. [568, 569].

Micro-Scale Models. In the last years, a considerable amount of research has been devoted to another class of models, in which the phase distribution is represented by a function that can also attain intermediate values between -1 and 1, in transition layers. [14] Models of this sort necessarily deal with a finer length scale than that which we referred to here; for the sake of classification, we labelled these scales as microscopic and mesoscopic, respectively.

The latter models can be grouped into two classes, on the basis of a distinction outlined in Sects. V.6 and VII.6; see Hohenberg and Halperin [294].

---

[14] These sets cannot be interpreted as mushy regions, because they occur at a much finer length scale.

Phase transitions in which the phase is characterized by a conserved variable include phase separation, for which a classical equation was proposed by Cahn and Hilliard. See, for example, Cahn-Hilliard [117], Cahn [116], Novick-Cohen and Segel [431], Von Wahl [572], Elliott [208, 210], Blowey and Elliott [75], Bates and Fife [61], Kenmochi and Niezgódka [319], and Alt and Pawlow [13, 14, 15, 16].

Phenomena in which the phase variable is not conserved include solid–liquid transitions, which can be described by the so-called *phase-field* model, and related approaches. See, for example, Fix [237, 238], Caginalp [110, 111, 112], Elliott and Zheng [213], Zheng [594], Penrose and Fife [451, 452], Sprekels and Zheng [513], Caginalp and Xie [114], Kenmochi and Niezgódka [318], Laurençot [342], Colli and Sprekels [141, 142], and Kenmochi [317].

The analysis of these models has recently been reviewed by Brokate and Sprekels [98].

Chapter IX. Two-Scale Models of Phase Transitions

Outline

The model of phase transitions including nucleation and surface tension effects introduced in Chap. VIII is here amended. A *mean field* approach is introduced, via convolution with a *Gaussian kernel,* to account for *nonadiabatic* nucleation at small undercooling. This can be interpreted as a transformation from mesoscopic to macroscopic length scale, and yields a *two-scale Stefan problem.*

A model of phase-field type including the perimeter functional is also briefly studied, as well as another model of mean field type.

A microscopic model of ferromagnetism due to Landau and Lifshitz is then outlined. These models of solid–liquid and ferromagnetic systems are briefly compared.

Prerequisites. Knowledge of basic function spaces is assumed.
Reference is made to material of Chaps. IV and VI through VIII.

IX.1 Two-Scale Stefan Problem and Nonadiabatic Nucleation

In this chapter we continue our analysis of the Stefan problem with surface tension. In this section we continue our discussion about Problem VIII.2.1.

Nucleation Thresholds. As we saw, cf. (VIII.1.11), in a one-phase system at uniform temperature, the condition (VIII.2.3) corresponds to prescribing that nucleation of the solid (liquid, resp.) phase only occurs when the undercooling (superheating, resp.) reaches the threshold $|\theta| = L/2C_V$. This is not satisfactory, since so large an undercooling is exceptional: it can only be attained if the material is absolutely pure, the walls are perfectly polished, and other extreme conditions are fulfilled.

Experimentally it is hard to meet these requirements, and indeed for most materials the limit undercooling is never reached. For instance, for water at room pressure $L \sim 80$ cal cm^{-3}, $C_V = 1$ cal K^{-1} cm^{-3}; hence $L/2C_V \sim 40$ K, [1] and it is difficult to attain this undercooling. So the question arises of devising a more realistic model, capable of accounting for nucleation at smaller undercooling.

---

[1] Here [cal] stands for calorie, [cm] for centimeter, and [K] for Kelvin degree, as usual.

Fast Heat Diffusion. For a moment, let us discuss the physical process neglecting surface tension. Let us consider a completely liquid system that is slowly cooled while keeping the temperature uniform in space. We want to understand why the model of Sect. VIII.1 cannot represent nucleation at *small* undercooling. As we saw, (VIII.2.3) accounts for *adiabatic* nucleation, which means that the energy density u is unchanged at the nucleation instant, cf. Fig. VIII.1(b). Hence a jump of χ from 1 to -1 implies a jump of u from θ to $\theta + L/C_V$ (so called *recalescence* effect). If solid nucleation occurred at some temperature $\theta \in]-L/2C_V, 0[$, it would yield a highly superheated solid, and the potential K_u, cf. (VIII.1.7), would not be minimized.

It might be argued that usually *nuclei* (namely, newborn solid phases) are very small (remember that the model which we are discussing deals with a mesoscopic length scale), hence the released latent heat is also very small. Therefore the latter should be quickly removed by diffusion, because of the high temperature gradient occurring at the boundary of the nucleus R. So actually the temperature would increase by an amount smaller than L/C_V, in a region larger than R. The smaller and the more fingered and irregular is the region R, the faster is heat diffusion and decay of the temperature jump. Clearly this process is nonadiabatic. This behaviour cannot be accounted for in our Stefan-type model, since within that framework nucleation is instantaneous, whereas diffusion is not.

In order to overcome this difficulty, here we propose to use *convolution* in space with a *Gaussian* kernel, to represent fast diffusion of latent heat occurring at solid nucleation. This *mean-field-type model* accounts for a temperature increase in a larger region than that of nucleation. Then we include surface tension effects. In this way, a sort of competition arises between surface tension and fast heat diffusion: surface tension prevents the region R from being too small and too irregular, whereas the need of diffusing latent heat makes R as small and irregular as possible. This also applies to liquid nucleation; in that case, latent heat must be *absorbed* quickly.

Modified Problem. Let us fix a (small) positive *interaction distance* β, and introduce the *Gaussian* kernel

$$\eta(x)(=\eta_\beta(x)) := (\beta\sqrt{\pi})^{-3} \exp\left(-\frac{|x|^2}{\beta^2}\right) \qquad \forall x \in \mathbf{R}^3. \tag{1.1}$$

We denote the convolution operation by $*$:

$$(v * w)(x) := \int_{\mathbf{R}^3} v(x-y) w(y) dy \qquad \forall x \in \Omega, \forall v \in L^1(\Omega),$$
$$\forall w \in L^\infty(\mathbf{R}^3); \tag{1.2}$$

here v has been extended with value 0 outside Ω. As $\int_{\mathbf{R}^3} \eta(x) dx = 1$, the convolution with η represents a weighted space average, which is *essentially* confined to a ball with radius of order β. Note that $v * \eta_\beta * \eta_\beta = v * \eta_{\sqrt{2}\beta}$ for any $v \in L^1(\Omega)$.

If we set $\tau := \beta^2 C_V/4k$, then for any smooth function $w^0 : \mathbf{R}^3 \to \mathbf{R}$ the function $w := w^0 * \eta$ is such that

$$\begin{cases} C_V \dfrac{\partial w}{\partial \tau} - k\Delta w = 0 & \forall x \in \mathbf{R}^3, \forall \tau > 0, \\ w(x, 0) = w^0(x) & \forall x \in \mathbf{R}^3. \end{cases}$$

The quantity $\beta^2 C_V/4k$ defines a *fast* time scale, which we regard as characteristic of nucleation. Within the model we are developing, it is more important to know that scale than to identify the exact kinetics of nucleation.

For any constant $\lambda \in [0, 1]$, let us define the operator $\mathcal{T}_\lambda : L^2(\Omega) \to L^2(\Omega)$ and the scalar product $\langle \cdot, \cdot \rangle_\eta$ as follows:

$$\mathcal{T}_\lambda u := \lambda u + (1-\lambda) u * \eta_{\sqrt{2}\beta} \quad \text{in } \Omega, \forall u \in L^2(\Omega), \tag{1.3}$$

$$\langle u, v \rangle_\eta := \int_\Omega (u * \eta)(v * \eta) dx \left(= \int_\Omega u(v * \eta_{\sqrt{2}\beta}) dx \right) \tag{1.4}$$
$$\forall u, v \in L^2(\Omega).$$

In $L^2(\Omega)$ this product determines a norm equivalent to the standard one, provided that $\lambda > 0$.

Let us assume that (VIII.2.1) is satisfied. We can now amend Problem VIII.2.1.

Problem 1.1$_\lambda$ *To find $\theta \in L^2\left(0, T; H_0^1(\Omega)\right)$ and $\chi \in L^1(0, T; BV(\Omega))$ such that (VIII.2.2) is fulfilled, and (in place of (VIII.2.3) and (VIII.2.4), respectively),*

$$\begin{aligned} \Phi(\chi) - \Phi(v) &+ \frac{(1-\lambda)L^2}{8\tau_E C_V} \left(\langle \chi, \chi \rangle_\eta - \langle v, v \rangle_\eta \right) \\ &\leq \frac{L}{2\tau_E C_V} \int_\Omega (\mathcal{T}_\lambda u)(\chi - v) dx \quad \forall v \in \text{Dom}(P), \text{ a.e. in }]0, T[, \end{aligned} \tag{1.5}$$

$$\tilde{\nabla} \cdot \frac{\tilde{\nabla}\psi}{\sqrt{1+|\tilde{\nabla}\psi|^2}} = -\frac{L}{\sigma\tau_E}(\mathcal{T}_\lambda\theta) \circ \psi \quad \text{in } H^{-1}(G). \tag{1.6}$$

Interpretation. For $\lambda = 1$ we retrieve Problem VIII.2.1, so we just discuss the case of $\lambda \in [0, 1[$. The equation (1.6) is a weak formulation of the following *modified Gibbs-Thomson law*

$$\kappa = -\frac{L}{2\sigma\tau_E}\mathcal{T}_\lambda\theta \quad \text{on } \mathcal{S}. \tag{1.7}$$

The interpretation of (1.5) is the main issue here. [2]

---

[2] A simpler setting is obtained for $\lambda = 0$, hence the reader might find it convenient at first to interpret some formulae for $\lambda = 0$.

Competing Functionals. Let us set

$$K_u^\lambda(\chi) := \Phi(\chi) + \frac{(1-\lambda)L^2}{8\tau_E C_V}\langle \chi, \chi \rangle_\eta - \frac{L}{2\tau_E C_V}\int_\Omega (\mathcal{T}_\lambda u)\chi\,dx \qquad (1.8)$$
$$\forall v \in \text{Dom}(P).$$

This potential can be compared with $K_u = K_u^1$, cf. (VIII.1.7). The equation (1.5) reads

$$K_u^\lambda(\chi) = \inf K_u^\lambda \qquad \text{a.e. in }]0, T[. \qquad (1.9)$$

The first two terms occurring in the definition of K_u^λ are in competition. The first one tends to separate the phases by a *smooth* boundary, whereas the second one tends to mix them and to form a rough interface. The latter tendency is consistent with the need of quickly diffusing the latent heat released at solid nucleation, which we discussed before. It then appears that the solid nucleation temperature is a decreasing function of λ. This also applies to liquid nucleation, where latent heat must be absorbed quickly.

As β vanishes η converges to a *Dirac measure*, hence $\mathcal{T}_\lambda v \to v$ and $\langle v, v \rangle_\eta \to \int_\Omega v^2 dx = |\Omega|$ for any $v \in \text{Dom}(P)$. So in the limit we retrieve the potential K_u but for an unessential additive constant.

An Equivalent Formulation ... Notice that for any $\chi, v \in L^2(\Omega)$

$$\langle v, v \rangle_\eta - \langle \chi, \chi \rangle_\eta = \langle \chi - v, \chi - v \rangle_\eta - 2\langle \chi, \chi - v \rangle_\eta, \qquad (1.10)$$

whence

$$\frac{L}{2\tau_E C_V}\int_\Omega (\mathcal{T}_\lambda u)(\chi - v)dx + \frac{(1-\lambda)L^2}{8\tau_E C_V}\left(\langle v, v\rangle_\eta - \langle \chi, \chi\rangle_\eta\right)$$
$$= \frac{L}{2\tau_E C_V}\left\{\int_\Omega \left(C_V T_\lambda \theta + \frac{\lambda L}{2}\chi\right)(\chi - v)dx + \frac{(1-\lambda)L}{2}\langle \chi, \chi - v\rangle_\eta\right\}$$
$$+ \frac{(1-\lambda)L^2}{8\tau_E C_V}\left(\langle \chi - v, \chi - v\rangle_\eta - 2\langle \chi, \chi - v\rangle_\eta\right)$$
$$= \frac{L}{2\tau_E C_V}\int_\Omega \left(C_V T_\lambda \theta + \frac{\lambda L}{2}\chi\right)(\chi - v)dx + \frac{(1-\lambda)L^2}{8\tau_E C_V}\langle \chi - v, \chi - v\rangle_\eta.$$

Therefore (1.5) is equivalent to

$$\Phi(\chi) - \Phi(v) \leq \frac{L}{2\tau_E C_V}\int_\Omega \left(C_V T_\lambda \theta + \frac{\lambda L}{2}\chi\right)(\chi - v)dx$$
$$+ \frac{(1-\lambda)L^2}{8\tau_E C_V}\langle \chi - v, \chi - v\rangle_\eta \qquad \forall v \in \text{Dom}(P), \text{ a.e. in }]0, T[. \qquad (1.11)$$

... and Its Interpretation. We have

$$\|(w - v) * \eta\|_{L^\infty(\Omega)} \to 0 \qquad \text{as } v \to w \text{ strongly in } L^1(\Omega), w, v \in \text{Dom}(P);$$

hence

$$\langle w - v, w - v \rangle_\eta \le \|(w - v) * \eta\|_{L^\infty(\Omega)} \|(w - v) * \eta\|_{L^1(\Omega)}$$
$$= o\left(\|w - v\|_{L^1(\Omega)}\right) \qquad \forall w, v \in \text{Dom}(P). \tag{1.12}$$

On the other hand, we have

$$\|w - v\|_{L^2(\Omega)}^2 \ne o\left(\|w - v\|_{L^1(\Omega)}\right) \qquad \forall w, v \in \text{Dom}(P).$$

Indeed for any $w, v \in \text{Dom}(P)$, $|w| = |v| = 1$ a.e. in Ω, whence $\|w - v\|_{L^2(\Omega)}^2 = 2\|w - v\|_{L^1(\Omega)}$.

The inequality (1.12) is the main analytical difference due to the use of convolution, as we can see in the argument of the existence result in the following.

By (1.12), (1.11) yields [3]

$$\sigma \partial^- \Phi(\chi) \ni \frac{L}{2\tau_E C_V}\left(C_V T_\lambda \theta + \frac{\lambda L}{2}\chi\right) \qquad \text{in } L^\infty(\Omega). \tag{1.13}$$

This condition is consistent with the modified Gibbs-Thomson law (1.6), cf. Proposition VI.4.4, and even implies the latter if $\lambda = 0$.

In conclusion, $\lambda \ne 1$ corresponds to the presence of a free energy term competing against the perimeter, and implies a correction of the interface temperature.

An Ideal Experiment. As before, let us consider the cooling of a liquid system that is maintained at a uniform temperature $\theta = \theta(t)$. The latter field is *almost equal* to $T_\lambda \theta(t)$, but in a thin neighbourhood of the boundary Γ. For a moment, let us neglect the perimeter term. As $T_\lambda \theta(t)$ descends below the value $-\lambda L/2C_V$, the field $\chi = 1$ no longer fulfills (1.13), which is equivalent to (1.5). Let us see how nucleation of the solid phase can then occur in a *small* region R at $T_\lambda \theta(t) = -\lambda L/2C_V$, cf. Fig. 1. Notice that here the nucleation threshold is smaller by a factor λ than that which we derived in the previous section.

As we saw, as nucleation occurs in a set R, there χ jumps to -1 and θ to $-\lambda L/2C_V + L/C_V$. Let us assume that the region R is much smaller than the *interaction distance* β. Accordingly, denoting by D_R the diameter of R, we consider the limit behaviour as $D_R/\beta \to 0$. We have

$$\text{as } \frac{D_R}{a} \to 0, \quad \theta * \eta \to -\frac{\lambda L}{2C_V} \quad \text{in } R;$$

hence, after nucleation,

$$T_\lambda \theta \to \lambda\left(-\frac{\lambda L}{2C_V} + \frac{L}{C_V}\right) + (1-\lambda)\left(-\frac{\lambda L}{2C_V}\right) = \frac{\lambda L}{2C_V} \qquad \text{in } R.$$

Hence, as $D_R/a \to 0$, $C_V T_\lambda \theta + (\lambda L/2)\chi \to 0$ in R after nucleation, since $\chi = -1$ in R. We conclude that (1.13) (i.e., (1.5)) is fulfilled if the perimeter term is neglected.

---

[3] See (XI.4.16) for the definition of the subdifferential operator ∂^-.

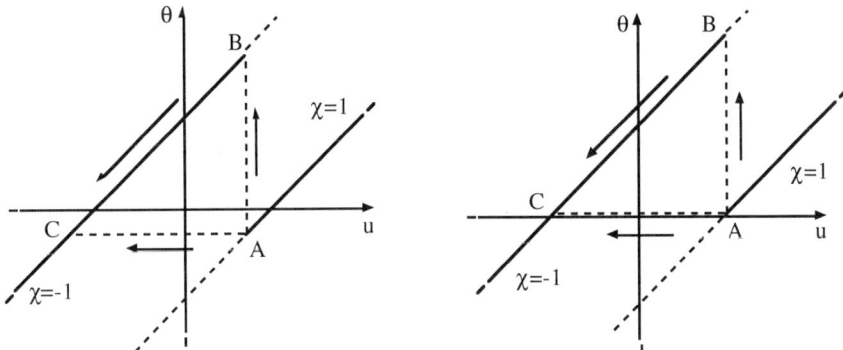

Figure 1. Temperature versus internal energy density for solid nucleation in a uniformly undercooled liquid, according to Problem 1.1_λ in (a).

The nucleated phase moves from $\theta = -\lambda L/2C_V$, $\chi = 1$ (state A) to $\theta = (L/C_V)(1 - \lambda/2)$, $\chi = -1$ (state B). Fast heat diffusion follows, and the state $\theta = -\lambda L/2C_V$, $\chi = -1$ (state C) is quickly reached. This is represented by the convolution with the Gaussian kernel η, cf. (1.1); β^2 is the time scale of nucleation.

This picture describes macroscopic quantities, and here the small undercooling due to the Gibbs-Thomson law is neglected. On the corresponding time scale the transition from A to C appears as instantaneous.

The limit case $\lambda = 0$ is outlined in (b).

The case in which the perimeter term is included in the free energy functional can be similarly discussed, provided that D_R is so large that in (1.5) the volume term prevails over the perimeter. This seems to be a reasonable assumption, since the surface tension coefficient σ is very small. Here we get a smaller value of the nucleation temperature: surface tension acts against nucleation.

If the temperature is uniform in a subregion $\Omega' \subset \Omega$, then nucleation simultaneously occurs at several sites in Ω'. Macroscopically, this may appear as the formation of a *mushy region*. The distance among these nuclei increases as the interaction distance β increases, since β determines the size of the region affected by recalescence.

Theorem 1.1 *(Existence) Let $\lambda \in]0, 1]$. Under the hypothesis (VIII.4.1), Problem 1.1_λ has a solution that fulfills the regularity conditions (VIII.4.2).*

Outline of the Proof. This is similar to that of Theorem VIII.4.1. Here we just point out the main differences.

We fix any $m \in \mathbf{N}$, set $h := T/m$, and introduce a discretized problem that is formally identical to Problem VIII.2.1_m, but where the scalar product $(\cdot, \cdot)$ is replaced by

$$(u, v)_\lambda := \lambda \int_\Omega uv\,dx + (1 - \lambda)\langle u, v\rangle_\eta \left(= \int_\Omega (\mathcal{T}_\lambda u)v\,dx \right) \quad (1.14)$$
$$\forall u, v \in L^2(\Omega)$$

in (VIII.4.6). J_h is positive definite also with this substitution, hence the discretized problem has a solution.

To derive a priori estimates, we then multiply the discretized equation (VIII.4.8) by $(h/2)\mathcal{T}_\lambda(\theta_m^n + \theta_m^{n-1})$ and integrate over Ω. Thus we get an inequality of the form (VIII.4.10), with $\mathcal{T}_\lambda \theta_m^n$ in place of θ_m^n. As $\lambda > 0$, this yields the estimate (VIII.4.11).

The integral equation (VIII.2.2) follows as before. By the procedure we used in the proof of Theorem VIII.4.1, we get (VIII.4.29) with $(\cdot, \cdot)_\lambda$ in place of $(\cdot, \cdot)$. By (VIII.4.30), this yields

$$\Phi(\chi) - \Phi(v) \leq \frac{L}{2\tau_E}(\theta, \chi - v)_\lambda + \frac{\lambda L^2}{4\tau_E C_V} \int_\Omega \chi(\chi - v) dx$$
$$+ \frac{(1-\lambda)L^2}{8\tau_E C_V} \langle \chi - v, \chi - v \rangle_\eta \qquad \forall v \in \mathrm{Dom}(P), \text{ a.e. in }]0, T[; \qquad (1.15)$$

this inequality coincides with (1.11), which in turn is equivalent to (1.5). Finally, the procedure that we used in part (vi) of the proof of Theorem VIII.4.1 yields (VIII.2.4), with θ replaced by $\mathcal{T}_\lambda \theta$. □

Scale Transformation. On the macroscopic length scale, the surface tension coefficient σ is very small, hence the interface mean curvature can be very large and the interface can look rather irregular. Indeed, as we saw, the Gibbs-Thomson law holds at a *mesoscopic* scale, which is essentially determined by the size of σ, and is intermediate between the macroscopic scale of laboratory measurements and the microscopic scale of molecular phenomena. So, in the integral $\int_\Omega u\chi dx$ occurring in the potential $K_u(\chi)$ (cf. (1.7)), u and χ are mesoscopic variables; accordingly, the model of Sects. VIII.1 and VIII.2 can be regarded as mesoscopic.

Let us now consider the model of this section. The convolution with a Gaussian function $\eta = \eta_\beta$ can be regarded as a *scale transformation* from the finer to the coarser length scale, in this case from mesoscopic to macroscopic variables. For $\lambda = 0$ the potential $K_u^\lambda(\chi)$, cf. (1.8), contains the term $\int_\Omega \mathcal{T}_\lambda u\chi dx = \int_\Omega (u*\eta)(\chi*\eta)dx$. Here $u*\eta$ and $\chi*\eta$ are macroscopic variables; in $\int_\Omega \mathcal{T}_\lambda u\chi dx$ values of λ between 0 and 1 correspond to a mixture of mesoscopic and macroscopic contributions. As we saw, this allows us to account for a range of undercooling thresholds, although it does not seem very natural from the scaling viewpoint.

Therefore the Problem$_\lambda$ can be regarded as a two-scale model of phase transitions. The *interaction constant* β determines the macroscopic length scale; as we assumed in our previous discussion, β is assumed to be much larger than the capillary length σ.

Case of $\lambda = 0$. On account of the interpretation in terms of macroscopic and mesoscopic scales, it looks natural to exclude any mesoscopic part in $\mathcal{T}_\lambda \theta$ by taking $\lambda = 0$. In this case by the discussion of the previous section it appears that, for a sufficiently large β, nucleation is represented as an almost isothermal process:

the larger is β, the nearer is the phenomenon to being isothermal. Moreover, here nucleation occurs without undercooling or superheating (but for the small amount prescribed by the Gibbs-Thomson law on the boundary of the nucleated phase); cf. Fig. 1(b).

For $\lambda = 0$, difficulties arise in proving existence of a solution of Problem 1.1_λ along the lines of Theorem VIII.1.1, because in this case the uniform estimate for $\hat{u}_m$ in $H^1\left(0, T; H^{-1}(\Omega)\right)$ is lost, and consequently one cannot apply Lemma 3.2 to derive strong convergence of χ_m. Therefore we modify the Gibbs-Thomson law, by introducing a term capable of providing some *time compactness* for the sequence $\{\chi_m\}$.

The most natural approach consists of using a law like the *mean curvature flow*, cf. (VII.6.8), since the front velocity can provide the desired compactness. However, a rather different formulation is needed to represent that law in terms of the characteristic function χ. [4]

Hysteresis in Front Motion. In the alternative, one can deal with a hysteresis law like (VII.8.3), which can be regarded as a simplified model of *two-dimensional nucleation,* as we briefly discussed in Sect. VII.8. On the basis of the developments of the previous section, one can also include the scale transformation $v \mapsto v * \eta$ in (VII.8.3). Let us denote by "sign" the sign graph, cf. (XI.5.3), and by $v_\mathcal{S}$ the (normal) speed of the melting front $\mathcal{S}_t$ for any $t \in [0, T]$; here melting corresponds to $v_\mathcal{S} > 0$, and solidification to $v_\mathcal{S} < 0$. We fix a Gaussian kernel $\eta = \eta_\beta$ as in (1.1), a positive constant c, and consider the inclusion

$$\frac{L}{2\tau_E} \theta * \eta_{\sqrt{2}\beta} + \sigma\kappa \in c\,\text{sign}(v_\mathcal{S}) \qquad \text{on } \mathcal{S}, \qquad (1.16)$$

which is equivalent to the following variational inequality

$$\left(\frac{L}{2\tau_E} \theta * \eta_{\sqrt{2}\beta} + \sigma\kappa\right)(v_\mathcal{S} - \xi) \geq c|v_\mathcal{S}| - c|\xi| \qquad \text{on } \mathcal{S}, \forall \xi \in \mathbf{R}. \qquad (1.17)$$

This corresponds to taking $\lambda = 0$ in (1.6), and adding the sign-term.

This condition can be expressed in a weak form, and then inserted into Problem VIII.2.1, in place of (VIII.2.3) and (VIII.2.4).

This problem has a solution such that $\chi \in BV\left(0, T; L^1(\Omega)\right) \cap L^\infty(0, T; BV(\Omega))$ ($=: X$). This can be proved via approximation by time-discretization, derivation of a priori estimates, and passage to the limit. We do not pursue this issue further, and just notice that in this case the results of Sect. VIII.3 are not needed, since (VIII.4.23) follows from a uniform estimate in X.

---

[4] See Luckhaus and Sturzenhecker [366] and V. [568, 569].

IX.2 Another Model with Surface Tension

In this section we briefly discuss a model obtained by coupling the energy balance law (VIII.1.4) with the phase dynamics (VII.6.3); the latter is a gradient flow for the free energy functional $\tilde{F}_\theta$, cf. (VII.3.4).

Let us assume that

$$\begin{aligned} \chi^0 \in BV(\Omega), & \quad |\chi^0| \leq 1 \quad \text{a.e. in } \Omega, \\ u^0 \in L^2(\Omega), & \quad f \in L^2\left(0, T; H^{-1}(\Omega)\right). \end{aligned} \tag{2.1}$$

For any $\varepsilon > 0$ we introduce the following weak formulation.

Problem 2.1$_\varepsilon$ *To find* $\theta_\varepsilon \in L^2\left(0, T; H_0^1(\Omega)\right)$ *and* $\chi_\varepsilon \in H^1\left(0, T; L^2(\Omega)\right) \cap L^1(0, T; BV(\Omega))$ *such that, setting* $u_\varepsilon := C_V \theta_\varepsilon + (L/2)\chi_\varepsilon$ *a.e. in* Q,

$$\iint_Q \left(-u_\varepsilon \frac{\partial \xi}{\partial t} + k \nabla \theta_\varepsilon \cdot \nabla \xi\right) dx\, dt$$
$$= \int_0^T {}_{H^{-1}(\Omega)}\langle f, \xi \rangle_{H_0^1(\Omega)} dt + \int_\Omega u^0(x) \xi(x, 0) dx \tag{2.2}$$
$$\forall \xi \in H^1(Q), \xi = 0 \text{ on } (\Gamma \times]0, T[) \cup (\Omega \times \{T\}),$$

$$|\chi_\varepsilon| \leq 1 \quad \text{a.e. in } Q, \tag{2.3}$$

$$\varepsilon \int_\Omega \frac{\partial \chi_\varepsilon}{\partial t}(\chi_\varepsilon - v) dx + \frac{\sigma}{2} \int_\Omega \left(|\nabla \chi_\varepsilon| - |\nabla v|\right)$$
$$+ \frac{\sigma_L - \sigma_S}{2} \int_\Gamma (\chi_\varepsilon - v) d\Gamma \leq \frac{L}{2\tau_E} \int_\Omega (\theta_\varepsilon + a\chi_\varepsilon)(\chi_\varepsilon - v) dx \tag{2.4}$$
$$\forall v \in H^1(\Omega), |v| \leq 1 \text{ a.e. in } \Omega, \text{ a.e. in }]0, T[,$$

$$\chi_\varepsilon(\cdot, 0) = \chi^0 \quad \text{a.e. in } \Omega. \tag{2.5}$$

Interpretation. As we saw, (2.2) is equivalent to the energy balance equation (VIII.2.7) coupled with the initial condition (VIII.2.8).

By setting $\mathcal{V}(v) := \int_\Omega |\nabla v|$ for any $v \in BV(\Omega)$, and denoting by $I_{[-1,1]}$ the indicator function of the interval $[-1, 1]$, [5] the system (2.3), (2.4) is equivalent to the differential inclusion

$$\varepsilon \frac{\partial \chi_\varepsilon}{\partial t} + \partial \left(\frac{\sigma}{2}\mathcal{V} + I_{[-1,1]}\right)(\chi_\varepsilon) \ni \frac{L}{2\tau_E}(\theta_\varepsilon + a\chi_\varepsilon) \quad \text{in } BV(\Omega)', \tag{2.6}$$
$$\text{a.e. in }]0, T[.$$

---

[5] In the sense of Sect. XI.4.

Notice that this represents a gradient flow for the functional $\tilde{F}_\theta$, rather than F_θ; in fact here we are not assuming that χ_ε is a characteristic function. However, in the stationary setting these functionals have the same minimizers; see Theorem VII.4.2. [6]

Let us denote by Problem 2.1 the reduced problem, obtained by setting $\varepsilon = 0$ in Problem 2.1$_\varepsilon$ and dropping the requirement that $\chi_\varepsilon \in H^1\left(0, T; L^2(\Omega)\right)$ as well as the initial condition (2.5). Obviously, taking the limit as $\varepsilon \to 0$ corresponds to regarding the kinetics of phase transition as instantaneous.

These problems deal with a mesoscopic length scale, since they contain the total variation functional multiplied by the surface tension coefficient. On the other hand, the phase field model is set at a microscopic length-scale, since interfaces are represented by transition layers. In the stationary setting, the passage from the microscopic to the mesoscopic scale is represented by the Γ-limit as $\varepsilon \to 0$, see Theorem VI.5.1.

Theorem 2.1 *(Existence, Uniqueness and Limit Behaviour)* [7]
(i) If (2.1) holds, then Problem 2.1$_\varepsilon$ has one and only one solution for any $\varepsilon > 0$.
(ii) There exist θ, χ such that, as $\varepsilon \to 0$ along a suitable sequence,

$$\theta_\varepsilon \to \theta \quad \text{weakly star in } L^\infty\left(0, T; L^2(\Omega)\right) \cap L^2\left(0, T; H^1(\Omega)\right), \quad (2.7)$$

$$\chi_\varepsilon \to \chi \quad \text{weakly star in } L^\infty(Q) \cap L^\infty(0, T; BV(\Omega)). \quad (2.8)$$

(iii) If $a \leq L/2C_V$, then (θ, χ) is a solution of Problem 2.1.

Outline of the Proof. The existence and uniqueness of the solution are easily proved by means of the procedure outlined in Sect. V.1. In particular, a priori estimates corresponding to (2.7) and (2.8) can be derived by multiplying the energy balance equation (VIII.2.7) by θ_ε and (2.6) by $\partial \chi_\varepsilon / \partial t$, adding the resulting identities, and integrating in time.

In order to pass to the limit as $\varepsilon \to 0$, notice that u_ε is uniformly bounded in $L^2(0, T; BV(\Omega)) \cap H^1\left(0, T; H^{-1}(\Omega)\right)$, by (2.7), (2.8) and comparison in (2.2); moreover, the latter space is continuously and compactly included in $L^1(Q)$. Hence along a suitable subsequence we have

$$\iint_Q \left(C_V \theta_\varepsilon + \frac{L}{2} \chi_\varepsilon\right) \chi_\varepsilon \, dx dt \to \iint_Q \left(C_V u + \frac{L}{2} \chi\right) \chi \, dx dt. \quad (2.9)$$

---

[6] Because of the regularity $\chi_\varepsilon \in H^1\left(0, T; L^2(\Omega)\right)$, χ_ε is a characteristic function only if it is stationary. However, one might expect that in typical situations $|\chi| \simeq 1$ at the exception of thin transition layers.

[7] See V. [559].

Therefore, since we assumed that $a \leq L/2C_V$, we have

$$\limsup_{\varepsilon \to 0} \iint_Q (\theta_\varepsilon + a\chi_\varepsilon)(\chi_\varepsilon - v)dxdt$$
$$= \frac{1}{C_V} \lim_{\varepsilon \to 0} \iint_Q \left(C_V \theta_\varepsilon + \frac{L}{2}\chi_\varepsilon\right)(\chi_\varepsilon - v)dxdt$$
$$- \left(\frac{L}{2C_V} - a\right) \liminf_{\varepsilon \to 0} \iint_Q \left(|\chi_\varepsilon|^2 - \chi_\varepsilon v\right) dxdt \qquad (2.11)$$
$$\leq \frac{1}{C_V} \iint_Q \left(C_V \theta + \frac{L}{2}\chi\right)(\chi - v)dxdt$$
$$- \left(\frac{L}{2C_V} - a\right) \iint_Q \left(|\chi|^2 - \chi v\right) dxdt = \iint_Q (\theta + a\chi)(\chi - v)dxdt.$$

This allows to pass to the limit in (2.4). □

A similar argument can be used for other problems of phase field type.

IX.3 A Mean Field Model

In this section we outline a model of phase transitions without surface tension.

Let us define the kernel $\eta = \eta_\beta$ as in (1.1), drop the total variation term, and replace $a\chi_\varepsilon$ by the *mean field* $a\chi_\varepsilon * \eta * \eta (= a\chi_\varepsilon * \eta_{\sqrt{2}\beta})$ in (2.4). This yields the inclusion

$$\varepsilon \frac{\partial \chi_\varepsilon}{\partial t} + \partial I_{[-1,1]}(\chi_\varepsilon) \ni \frac{L}{2\tau_E}(\theta_\varepsilon + a\chi_\varepsilon * \eta * \eta) \qquad \text{a.e. in } Q. \qquad (3.1)$$

Because of the convolution, here nucleation occurs only if the critical undercooling $\theta_\varepsilon = -a$ is attained in a sufficiently large region.

Let us assume that

$$\chi^0 \in L^\infty(\Omega), \quad |\chi^0| \leq 1 \quad \text{a.e. in } \Omega, \quad u^0 \in L^2(\Omega), \quad f \in L^2(Q). \qquad (3.2)$$

For any $\varepsilon > 0$ we introduce the following weak formulation.

Problem 3.1$_\varepsilon$ *To find $\theta_\varepsilon \in L^2\left(0, T; H_0^1(\Omega)\right)$ and $\chi_\varepsilon \in H^1\left(0, T; L^2(\Omega)\right)$ such that, setting $u_\varepsilon := C_V \theta_\varepsilon + (L/2)\chi_\varepsilon$ a.e. in Q, (2.2), (2.3), (2.5) hold, and*

$$\varepsilon \int_\Omega \frac{\partial \chi_\varepsilon}{\partial t}(\chi_\varepsilon - v)dx \leq \frac{L}{2\tau_E} \int_\Omega (\theta_\varepsilon + a\chi_\varepsilon * \eta * \eta)(\chi_\varepsilon - v)dx \qquad (3.3)$$
$$\forall v \in L^\infty(\Omega), |v| \leq 1 \text{ a.e. in } \Omega, \text{ a.e. in }]0, T[.$$

Let us denote by Problem 3.1 the reduced problem, obtained by setting $\varepsilon = 0$ in Problem 3.1$_\varepsilon$, dropping the requirement that $\chi_\varepsilon \in H^1\left(0,T;L^2(\Omega)\right)$ and the initial condition (2.5).

Theorem 3.1 *(Existence, Uniqueness and Limit Behaviour)*
(i) If (3.2) holds, then Problem 3.1$_\varepsilon$ has one and only one solution for any $\varepsilon > 0$.
(ii) If moreover

$$a < \frac{L}{2C_V}, \qquad \theta^0 := \frac{1}{C_V}\left(u^0 - \frac{L}{2}\chi^0\right) \in H^1_0(\Omega), \tag{3.4}$$

then there exist θ, χ such that, as $\varepsilon \to 0$ along a suitable sequence,

$$\theta_\varepsilon \to \theta \qquad \text{weakly star in } H^1\left(0,T;L^2(\Omega)\right) \cap L^\infty\left(0,T;H^1_0(\Omega)\right), \tag{3.5}$$

$$\chi_\varepsilon \to \chi \qquad \text{weakly star in } L^\infty(Q), \tag{3.6}$$

$$\chi_\varepsilon \ast \eta \ast \eta \to \chi \ast \eta \ast \eta \qquad \text{weakly in } H^1\left(0,T;H^r(\Omega)\right), \forall r > 0. \tag{3.7}$$

Moreover, (θ, χ) is a solution of Problem 3.1.

Outline of the Proof. Part (i) is proved by a standard procedure. In particular, estimates for θ_ε in $L^\infty\left(0,T;L^2(\Omega)\right) \cap L^2\left(0,T;H^1(\Omega)\right)$ and for χ_ε in $H^1\left(0,T;L^2(\Omega)\right)$ can *formally* be derived by multiplying the energy balance equation (VIII.2.7) by θ_ε and (3.1) by $\partial\chi_\varepsilon/\partial t$, adding the resulting identities, and integrating in time. [8] The estimate on θ_ε is also uniform with respect to ε.

Let us consider the proof of part (ii).

In order to derive a further a priori estimate, *formally* let us multiply the energy balance equation (VIII.2.7) by $\partial\theta_\varepsilon/\partial t$, differentiate (3.1) in time, multiply it by $(L/2)\partial\chi_\varepsilon/\partial t$, sum these formulae, and integrate in time. This yields

$$C_V \iint_Q \left(\frac{\partial\theta_\varepsilon}{\partial t}\right)^2 dx dt + \frac{k}{2}\int_\Omega |\nabla\theta_\varepsilon|^2 dx - \frac{k}{2}\int_\Omega |\nabla\theta^0|^2 dx$$
$$+ \frac{\varepsilon L}{4}\int_\Omega \left(\frac{\partial\chi_\varepsilon}{\partial t}(x,T)\right)^2 dx - \frac{\varepsilon L}{4}\int_\Omega \left(\frac{\partial\chi_\varepsilon}{\partial t}(x,0)\right)^2 dx \tag{3.8}$$
$$\leq \iint_Q f \frac{\partial\theta_\varepsilon}{\partial t} dx dt + \frac{La}{2}\iint_Q \left(\frac{\partial\chi_\varepsilon}{\partial t} \ast \eta\right)^2 dx dt.$$

---

[8] This procedure, as well as that outlined in the following, can be justified via time discretization, by a technique similar to that illustrated in Chap. II.

By extending v to $\mathbf{R}^3$ with value 0 and using the properties $\int_{\mathbf{R}^3} \eta(y)dy = 1, \eta \geq 0$ and the convexity of the square function, we have

$$\int_\Omega v^2 dx = \int_{\mathbf{R}^3} \eta(y)dy \int_{\mathbf{R}^3} v(\xi)^2 d\xi = \int_{\mathbf{R}^3} dy\, \eta(y) \int_{\mathbf{R}^3} v(x-y)^2 dx$$
$$= \int_{\mathbf{R}^3} dx \int_{\mathbf{R}^3} v(x-y)^2 \eta(y) dy \geq \int_{\mathbf{R}^3} dx \left(\int_{\mathbf{R}^3} v(x-y)\eta(y) dy \right)^2 \quad (3.9)$$
$$= \int_\Omega (v * \eta)^2 dx \qquad \forall v \in L^2(\Omega).$$

(In passing we notice that the equality holds only if the extended function v is constant in $\mathbf{R}^3$, that is, only if it vanishes in Ω.) Hence

$$\iint_Q \left(\frac{\partial \theta_\varepsilon}{\partial t} * \eta \right)^2 dxdt \leq \iint_Q \left(\frac{\partial \theta_\varepsilon}{\partial t} \right)^2 dxdt. \quad (3.10)$$

Note that

$$\frac{La}{2} \iint_Q \left(\frac{\partial \chi_\varepsilon}{\partial t} * \eta \right)^2 dxdt = \frac{2a}{L} \iint_Q \left(\frac{\partial u_\varepsilon}{\partial t} * \eta - C_V \frac{\partial \theta_\varepsilon}{\partial t} * \eta \right)^2 dxdt. \quad (3.11)$$

Moreover, by a simple comparison in the energy balance equation (VIII.2.7) and by the first set of a priori estimates, we see that $\partial u_\varepsilon / \partial t$ is uniformly bounded in $L^2\left(0,T; H^{-1}(\Omega)\right)$. Hence $(\partial u_\varepsilon/\partial t) * \eta$ is uniformly bounded in $L^2(Q)$.

The latter statement and (3.8), (3.10), (3.11) yield an inequality of the form

$$C_V \left(1 - \frac{2aC_V}{L} \right) \iint_Q \left(\frac{\partial \theta_\varepsilon}{\partial t} \right)^2 dxdt + \frac{k}{2} \int_\Omega |\nabla \theta_\varepsilon|^2 dx$$
$$\leq C_1 \left[\iint_Q \left(\frac{\partial \theta_\varepsilon}{\partial t} \right)^2 dxdt \right]^{1/2} + C_2, \quad (3.12)$$

with C_1 and C_2 independent of ε. By the assumption (3.4) on a, this yields a uniform estimate on θ_ε corresponding to the convergence (3.5). This estimate and the uniform boundedness of u_ε in $H^1\left(0,T; H^{-1}(\Omega)\right)$ yield an estimate for χ_ε in the latter space. Therefore $\chi_\varepsilon * \eta$ is uniformly bounded in $H^1(0,T; H^r(\Omega))$ for any $r > 0$, and (3.7) holds.

By (3.5) through (3.7) it is easy to pass to the limit in Problem 3.1$_\varepsilon$ as $\varepsilon \to 0$ along a suitable sequence. $\square$

One can show that the solution of Problem 3.1 is unique. The argument (which will not be presented here) is based on a technique which has some similarities with that used for the latter estimate.

IX.4 Micromagnetics

In this section we outline a classical model of ferromagnetism, known as *micromagnetics*, [8] and compare it with the mesoscopic model of solid–liquid systems that we have been studying in the last two chapters.

In Sect. IV.8, we briefly considered the evolution of a ferromagnetic material at a macroscopic length scale; neglecting hysteresis, we derived a vectorial Stefan-type problem, cf. (IV.8.1) and (IV.8.2). A different approach to ferromagnetism was proposed by Landau and Lifshitz in 1935. [9] According to this model, at a length scale of about 10^{-6} cm, a ferromagnetic body occupying a domain Ω can be represented as a continuum of elementary magnets, whose magnetization has prescribed magnitude $\mathcal{M}$:

$$|\vec{M}| = \mathcal{M} \quad \text{in } \Omega. \tag{4.1}$$

The Magnetic Energy Functional. Any stationary configuration is assumed to minimize a magnetic (free) energy functional, which consists of the following contributions.

(i) Exchange Energy. The ferromagnetic behaviour is due to a force of quantum origin, which locally tends to align the magnetization field. Following Heisenberg, this is represented by the *exchange energy*

$$\mathcal{E}_{\text{ex}}(\vec{M}) := \frac{1}{2} \sum_{i,j=1}^{3} a_{ij} \int_{\Omega} \frac{\partial \vec{M}}{\partial x_i} \cdot \frac{\partial \vec{M}}{\partial x_j} dx, \tag{4.2}$$

where $\{a_{ij}\}$ is a symmetric, positive definite 3×3-tensor.

(ii) Anisotropy Energy. This depends on the crystal structure of the ferromagnet. For instance, for a uniaxial material it reads

$$\mathcal{E}_{\text{an}}(\vec{M}) := -b \int_{\Omega} (\vec{M} \cdot \hat{z})^2 dx, \tag{4.3}$$

where b is a positive constant, and $\hat{z}$ is a unit vector. The minimization of this energy accounts for the tendency of $\vec{M}$ to align to the z-direction.

(iii) Magnetic Field Energy. The magnetic field $\vec{H}$ can be split into the sum of the *applied field* $\vec{H}_{\text{app}}$ and the *demagnetizing field* $\vec{H}_{\text{dem}}(\vec{M})$. The former is prescribed, whereas the latter is determined by the *magnetostatic equations*

$$\nabla \cdot (\vec{H}_{\text{dem}} + 4\pi\vec{M}) = 0, \quad c\nabla \times \vec{H}_{\text{dem}} = 4\pi\vec{J} \quad \text{in } \mathbf{R}^3, \tag{4.4}$$

---

[8] This theory deals with a length scale of about 10^{-6} cm, which is not far from that of molecular phenomena and may be regarded as *microscopic*, at least for the purpose of classification.

[9] See, e.g., Brown [100, 101] and Landau and Lifshitz [338, 339].

in Gauss units. Here $\vec{J}$ is a prescribed density of electric current, and $\vec{M}$ is extended with value $\vec{0}$ outside Ω. The fields $\vec{H}$ and $\vec{M}$ are assumed to vanish at infinity. The magnetic energy stored in the field $\vec{H}$ is

$$\mathcal{E}_{\text{field}}(\vec{M}) := -\int_{\Omega} [\vec{H}_{\text{app}} + \frac{1}{2}\vec{H}_{\text{dem}}(\vec{M})] \cdot \vec{M}\,dx. \qquad (4.5)$$

Total Magnetic Energy. Neglecting other (in particular, mechanical) contributions, this is

$$\mathcal{E}_{\text{mag}}(\vec{M}) := \mathcal{E}_{\text{ex}}(\vec{M}) + \mathcal{E}_{\text{an}}(\vec{M}) + \mathcal{E}_{\text{field}}(\vec{M}). \qquad (4.6)$$

The Stationary Problem. The functional $\mathcal{E}_{\text{mag}}$ has at least one absolute minimizer in $H^1(\Omega; \mathbf{R}^3)$, under the nonconvex constraint (4.1). This can easily be checked, by setting

$$J(\vec{M}) := \begin{cases} \mathcal{E}_{\text{mag}}(\vec{M}) & \text{if (4.1) is fulfilled a.e.} \\ +\infty & \text{otherwise} \end{cases} \qquad \forall \vec{M} \in H^1(\Omega; \mathbf{R}^3),$$

and then applying Theorem XI.7.4, with $B_0 := L^2(\Omega; \mathbf{R}^3)$ and $B := H^1(\Omega; \mathbf{R}^3)$.

If the applied field $\vec{H}_{\text{app}}$ is not too large, *relative* minimizers can occur, due to the nonconvexity of the constraint (4.1). This *multi-stability* is a source of *hysteresis* in evolution. [10]

The minimization of $\mathcal{E}_{\text{mag}}$ accounts for the splitting of the body into small uniformly magnetized regions *(Weiss domains)*, separated by thin transition layers *(Bloch* and *Néel walls)*. An appropriate scaling consists of assuming that the exchange tensor $\{a_{ij}\}$ and the anisotropy coefficient are, respectively, proportional to a parameter $\eta > 0$ and to $1/\eta$. Passing to the Γ-limit [11] as η vanishes, one gets $\vec{M} = \pm \mathcal{M}\hat{z}$ a.e. in Ω, so that two phases can be distinguished. Moreover, a space interaction energy proportional to the perimeter of the interface occurs in the limit free energy. [12] The analogy with solid–liquid systems is evident: in that case the mesoscopic model (with $|\chi| = 1$) is derived from the microscopic *phase field* model (with $|\chi| \leq 1$) via a Γ-limit; see Sect. VI.5.

The Landau-Lifshitz Equation. Landau and Lifshitz [338] proposed the following equation to describe ferromagnetic processes:

$$\frac{\partial \vec{M}}{\partial t} = \lambda_1 \vec{M} \times \vec{H}^e - \lambda_2 \vec{M} \times (\vec{M} \times \vec{H}^e); \qquad (4.7)$$

---

[10] This is illustrated by the classical Stoner-Wohlfarth model; see, e.g., Landau and Lifshitz [339; Sect. 37].

[11] In the sense of De Giorgi; see, e.g., De Giorgi and Franzoni [180], De Giorgi [179], and Dal Maso [163]. See also Sect. VI.5.

[12] See Anzellotti, Baldo and V. [23].

here the *effective magnetic field* $\vec{H}^e$ is defined as [13]

$$\vec{H}^e := -\frac{\partial}{\partial \vec{M}} \mathcal{E}_{\text{mag}}(\vec{M}) = \sum_{i,j=1}^{3} a_{ij} \frac{\partial^2 \vec{M}}{\partial x_i \partial x_j} + bM_z \hat{z} + \vec{H}, \qquad (4.8)$$

and λ_1, λ_2 are constants, $\lambda_2 > 0$.

The equation (4.7) is obviously consistent with the constraint (4.1). It is the most simple dynamics for a magnet that is subject to a magnetic field $\vec{H}^e$ and is only free to rotate. This equation must be coupled with the Maxwell equations (I.3.3) and (I.3.4), and with the constitutive laws (I.3.5). Existence of a solution of the corresponding initial and boundary value problem can be proved by means of approximation by the *Faedo-Galerkin method,* a priori estimates corresponding to the energy integral, and passage to the limit via compactness and monotonicity techniques. [14]

IX.5 Some Comparisons

In Sects. IV.1 and IV.4 we dealt with a macroscopic model of solid–liquid systems (i.e., the classical Stefan problem), in Sect. VI.5 with a microscopic model of two-phase systems (i.e., the phase field model), in Sect. IV.8 with a macroscopic model of ferromagnetism, in this chapter with a mesoscopic model of solid–liquid systems and with a microscopic model of ferromagnetism (i.e., micromagnetics). In this section we briefly compare these models.

Comparison Among Fine Scale Models. [15] Micromagnetics shares some properties with either of the two fine scale models of solid–liquid systems, as schematically illustrated in Table 1.

Comparison Between Macroscopic and Fine Scale Models. Let us consider the macroscopic models first. For both solid–liquid and ferromagnetic systems, a (respectively, scalar and vectorial) problem of the form (II.1.1), (II.1.2) is derived, for which well-posedness can be proved; see Chap. II. If α is a cyclically maximal monotone graph, the constitutive law $w \in \alpha(u)$ is equivalent to the minimization of a *convex* free energy functional. This functional does not account for any space interaction, as it does not contain any space derivative. If α is multi-valued, the

---

[13] The functional $\mathcal{E}_{\text{field}}$ is *Fréchet differentiable,* in the sense of Sect. XI.4, when coupled with the magnetostatic equations (4.4); see Anzellotti, Baldo and V. [23]. $\partial \mathcal{E}_{\text{mag}}/\partial \vec{M}$ is the Fréchet differential.

[14] See V. [552].

[15] For convenience, we group mesoscopic and microscopic models under the label of *fine scale models.*

system can exhibit a sharp interface *(free boundary)*, and a strong formulation can be derived, under regularity assumptions.

Mesosc. solid–liquid	Micromagnetics	Phase field
Mesoscopic	Microscopic	Microscopic
$\|\chi\| = 1$	$\|\vec{M}\| = \mathcal{M}$	$\|\chi\| \leq 1$
Sharp interface	Diffuse interface	Diffuse interface
Nonconvex constraint	Nonconvex constraint	Double-well
$\sigma \int_\Omega \|\nabla \chi\|$	$\frac{1}{2} \sum_{i,j=1}^{3} a_{ij} \int_\Omega \frac{\partial \vec{M}}{\partial x_i} \cdot \frac{\partial \vec{M}}{\partial x_j} dx$	$\frac{\varepsilon a}{2} \int_\Omega \|\nabla \chi\|^2 dx$
$\chi \in BV(\Omega)$	$\vec{M} \in H^1(\Omega; \mathbf{R}^3)$	$\chi \in H^1(\Omega)$

Table 1. Comparison among some properties of
(i) the mesoscopic model of solid–liquid systems, cf. Sect. VI.1,
(ii) micromagnetics, cf. Sect. IX.4,
(iii) the phase field model, cf. Sect. VI.5.
The lines from the second to the seventh, respectively, concern: the length scale, the constraint acting on the phase function, the interface structure, the source of the nonconvexity, the space interaction term occurring in the free energy, and the space regularity of the phase function.

The mesoscopic model of solid–liquid systems and micromagnetics exhibit the following relevant differences with respect to the corresponding macroscopic models:

(i) the presence of a *nonconvex constraint:* $|\chi| = 1$ for solid–liquid systems, and $|\vec{M}| = \mathcal{M}$ for ferromagnets;

(ii) the occurrence of a *space interaction term* in the free energy functional: the perimeter contribution $\sigma \int_\Omega |\nabla \chi|$ for (isotropic) solid–liquid systems, [16] and the exchange energy $\frac{1}{2} \sum_{i,j=1}^{3} a_{ij} \int_\Omega \frac{\partial \vec{M}}{\partial x_i} \cdot \frac{\partial \vec{M}}{\partial x_j} dx$ for ferromagnets. This term gives rise to a *fine structure,* whose length scale is determined by σ and $\{a_{ij}\}$, respectively.

So in both cases the stationary mesoscopic model is *nonconvex*. This is the source of *metastability,* and yields *hysteresis* in evolution. [17] However, the functional is convex with respect to the gradient of the phase variable, and this yields the existence of a minimizer. In more abstract terms, the nonconvexity of the constraint is *compensated* by compactness and convexity with respect to higher order derivatives. See Table 2.

---

[16] We assumed isotropy for the sake of simplicity; however, our analysis can be extended to the anisotropic setting. As a matter of fact, crystals, like ferromagnets, are never isotropic.

[17] The connection between multi-stability and hysteresis is illustrated, e.g., in V. [564; Sect. I.3].

Comparison Between Fine Scale Models of Solid–Liquid and Ferromagnetic Systems. A fundamental difference between the two phenomena is that they are respectively characterized by scalar and vectorial state variables: θ, u, χ and $\vec{H}, \vec{B}, \vec{M}$, respectively.

The constraint $|\chi| = 1$ is equivalent to $\chi = \pm 1$, whereas the condition $|\vec{M}| = 1$ corresponds to a continuum of values of $\vec{M}$. The form of the space interaction contribution to the respective free energies is strictly related to the different nature of these constraints.

	Macroscopic Scale	Fine Scale
Phase constraint	$|\chi| \leq 1$, $|\vec{M}| \leq \mathcal{M}$	$|\chi| = 1$, $|\vec{M}| = \mathcal{M}$
Convexity	Yes	No
Metastability and hysteresis	No	Yes
Space interaction in free energy	No	Yes
Fine structure	No	Yes

Table 2. Comparison of macroscopic models versus the mesoscopic model of solid–liquid systems and micromagnetics (the two latter ones are here labelled as *fine scale* models).

	Mesosc. solid–liquid	Micromagnetics
Algebraic structure	Scalar	Vectorial
Nonconvex constraint	$|\chi| = 1$	$|\vec{M}| = \mathcal{M}$
Space interaction term	$\sigma \int_\Omega |\nabla \chi|$	$\frac{1}{2} \sum_{i,j=1}^{3} a_{ij} \int_\Omega \frac{\partial \vec{M}}{\partial x_i} \cdot \frac{\partial \vec{M}}{\partial x_j} dx$
Phase regularity	$\chi \in BV(\Omega)$	$\vec{M} \in H^1(\Omega; \mathbf{R}^3)$
Evolution law	Mean curvature flow	Landau–Lifshitz equation

Table 3. Comparison between the mesoscopic model of solid–liquid systems and micromagnetics.

In the scalar setting, a term proportional to $\int_\Omega |\nabla \chi|^p dx$ $(p > 1)$ would force χ to be constant. Actually, the perimeter is the only *isotropic* seminorm that contains first order derivatives, while it has nontrivial minima that are consistent with the constraint $|\chi| = 1$. On the other hand, the unit sphere of $\mathbf{R}^3$ is a smooth manifold, course, $p = 2$ is the simplest choice, since the corresponding Fréchet derivative, $-\Delta \vec{M}$, is linear.

As for evolution, the Landau-Lifshitz equation is a simple dynamics over the sphere of equation $|\vec{M}| = 1$. In the scalar setting, the *mean curvature flow* [19] seems to be the most natural dynamics consistent with the nonconvex constraint $|\chi| = 1$. See Table 3.

IX.5 Comments

In this chapter we modified the mesoscopic problem studied in Chap. VIII, and outlined some alternative formulations. The model of Sect. IX.1 is also studied in V. [566, 567]; that of Sect. IX.2 was introduced in V. [559]. That of Sect. IX.3 is here proposed for the first time, and will be developed in more detail elsewhere; a similar model seems appropriate to represent ferromagnetic hysteresis.

We conclude by mentioning an open question. Micromagnetics is a microscopic theory, because of the smallness of the exchange coefficients $\{a_{ij}\}$. Then it seems natural to consider the possibility of introducing a mesoscopic model, by replacing the exchange energy (4.2) with a perimeter-type contribution of the form, say, $\int_\Omega |A \cdot \nabla \vec{M}| dx$, where $(A \cdot \nabla \vec{M})_{i\ell} := \{\sum_{j=1}^3 A_{ij} \cdot \frac{\partial M_\ell}{\partial x_j}\}$. Does a model of this sort make sense? Can an energy such as this be derived via a Γ-limit? For a strongly anisotropic uniaxial material, this guess is supported by the already mentioned result of Anzellotti, Baldo and V. [23].

---

[19] See Sects. VII.6 and VII.7.

Chapter X. Compactness by Strict Convexity

Outline

Let K be a closed convex subset of $\mathbf{R}^M$. If $u_n \to u$ weakly in $L^1\left(\Omega; \bar{\mathbf{R}}^M\right)$, $u_n(x) \in K$ for any n, and $u(x)$ is an extremal point of K for a.a. $x \in \Omega$, then the convergence is strong (essentially because asymptotically $u_n(x)$ cannot oscillate about $u(x)$).

This is applied to show that the weak convergence and the convergence of the values attained by a strictly convex functional imply strong convergence.

Prerequisites. Integration theory as well as basic properties of Banach spaces are used. Some results of convex analysis are also applied; see Sect. X.4.

X.1 Extremality and Compactness

In this section we present the main result of this chapter.

At first we review some simple definitions and properties. Let K be a closed subset of $\mathbf{R}^M$ ($M \geq 1$). K is *convex* iff $\xi = \lambda\xi' + (1-\lambda)\xi'' \in K$ for any $\xi', \xi'' \in K$ and any $0 < \lambda < 1$. ξ is said to be an *extremal* point of K whenever it can be represented as indicated only if $\xi = \xi' = \xi''$. ξ is said to be an *exposed* point of K iff there exists a *supporting hyperplane* [1] H of $\mathbf{R}^M$ such that $K \cap H = \{\xi\}$. All exposed points are extremal, but the converse may fail; for instance, let $K := \{(x,y) \in \mathbf{R}^2 : (x^+)^2 \leq y\}$ and $\xi := (0,0)$.

For any subset A of $\mathbf{R}^M$, by $\overline{\mathrm{co}}(A)$ we denote its *closed convex hull*, namely, the smallest closed convex set that contains A.

We deal with a subset Ω of $\mathbf{R}^N$ ($N \geq 1$), endowed with the ordinary N-dimensional Lebesgue measure. However, our results hold unchanged in any complete σ-finite measure space. [2]

A multi-valued mapping $K : \Omega \to 2^{\mathbf{R}^M}$ is said to be measurable iff there exists a countable family of measurable functions $\{k_m : \Omega \to \mathbf{R}^M\}$ such that, for a.a. $x \in \Omega$, $k_m(x) \in K(x)$ for any m, and $K(x)$ is the closure of $\bigcup_{m \in \mathbf{N}} k_m(x)$. [3]

---

[1] A closed $(M-1)$-dimensional hyperplane H of $\mathbf{R}^M$ is said *to support* a set $K \subset \mathbf{R}^M$ at a point $y \in K$ iff $y \in H$ and K is contained in one of the two closed half-spaces delimited by H.

[2] A measure space is said to be σ-*finite* iff any measurable set can be represented as a countable union of sets with finite measure.

[3] See, e.g., Castaing and Valadier [125; Sect. III.2].

Lemma 1.1 *Let x be an extremal point of a closed and convex subset K of $\mathbf{R}^M$. Then for any $\varepsilon > 0$, $x \notin \overline{co}\left(K \setminus B_\varepsilon(x)\right)$.*

Proof. Obviously, it is not restrictive to assume that $x = 0$.

Let us set $S_\varepsilon := \{y \in K : |y| = \varepsilon\}$, and $\delta := \min\{|y| : y \in \overline{co}(S_\varepsilon)\}$. This minimum is attained, since $\overline{co}(S_\varepsilon)$ is compact; $\delta > 0$, by the hypothesis of extremality. To prove our result, it suffices to show that $|y| \geq \delta$, for any $y \in \overline{co}\left(K \setminus B_\varepsilon(0)\right)$.

By contradiction, let $\tilde{y} \in \overline{co}\left(K \setminus B_\varepsilon(0)\right)$ be such that $|\tilde{y}| < \delta$, and consider any (two-dimensional) plane Π that contains the origin 0 and $\tilde{y}$. So $S_\varepsilon \cap \Pi$ is an arc AB, as in Fig. 1. As K is convex, it is easy to see that $K \cap \Pi$ is included in the angle $A0B$; hence $\tilde{y}$ cannot stay in the triangle $A0B$. □

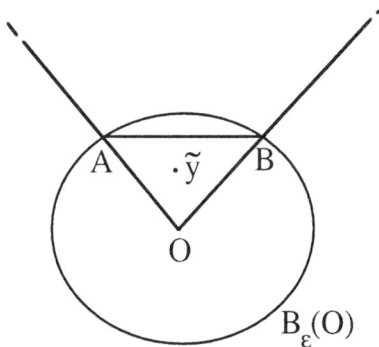

Figure 1. *Auxiliary picture for the proof of Lemma 1.1.*

Theorem 1.2 [4] *Let $K : \Omega \to 2^{\mathbf{R}^M}$ be a measurable multi-valued mapping, and $K(x)$ be closed and convex for a.a. $x \in \Omega$. Assume that*

$$u_n \to u \quad \text{weakly in } L^1\left(\Omega; \mathbf{R}^M\right), \tag{1.1}$$

$$u_n(x) \in K(x) \quad \text{for a.a. } x \in \Omega, \forall n, \tag{1.2}$$

$$u(x) \text{ is an extremal point of } K(x) \text{ for a.a. } x \in \Omega. \tag{1.3}$$

Then

$$u_n \to u \quad \text{strongly in } L^1\left(\Omega; \mathbf{R}^M\right). \tag{1.4}$$

(Hence $u_n \to u$ in measure and, possibly extracting a subsequence, a.e. in Ω.)

Proof. Possibly replacing K by $\tilde{K} := K - u$ and u_n by $\tilde{u}_n := u_n - u$, we can assume that $u = 0$ a.e. in Ω, without loss of generality.

---

[4] See V. [549].

Let us fix any $\varepsilon > 0$, set $A_{\varepsilon n} := \{x \in \Omega : |u_n(x)| > \varepsilon\}$ for any n, and

$$w_{\varepsilon n}(x) := \begin{cases} 0 & \text{in } A_{\varepsilon n}, \\ u_n(x) & \text{in } \Omega \setminus A_{\varepsilon n}, \end{cases} \qquad v_{\varepsilon n}(x) := \begin{cases} u_n(x) & \text{in } A_{\varepsilon n}, \\ 0 & \text{in } \Omega \setminus A_{\varepsilon n}, \end{cases}$$

for a.a. $x \in \Omega$ and for any n. Hence $w_{\varepsilon n} + v_{\varepsilon n} = u_n$ for any n. By (1.1) and Corollary XI.3.7, there exist w_ε and a subsequence $w_{\varepsilon n'}$ such that $w_{\varepsilon n'} \to w_\varepsilon$ weakly in $L^1(\Omega; \mathbf{R}^M)$ as $n' \to \infty$. Hence $v_{\varepsilon n'} \to v_\varepsilon := u - w_\varepsilon$ weakly in $L^1(\Omega; \mathbf{R}^M)$.

Note that $w_\varepsilon, v_\varepsilon \in K(x)$, as $w_{\varepsilon n'}, v_{\varepsilon n'}, 0 \in K(x)$ for any n'. Moreover $w_\varepsilon/2 + v_\varepsilon/2 = u/2 = 0$ a.e. in Ω; hence $w_\varepsilon = v_\varepsilon = 0$, as the origin 0 is an extremal point of $K(x)$ by (1.3). Therefore the whole sequences $\{w_{\varepsilon n}\}$ and $\{v_{\varepsilon n}\}$ vanish weakly in $L^1(\Omega; \mathbf{R}^M)$ as $n \to \infty$.

By the previous lemma, for a.a. $x \in \Omega$, $0 \notin \overline{\text{co}}(K(x) \setminus B_\varepsilon(0))$, hence there exists an $(M-1)$-dimensional hyperplane in $\mathbf{R}^M$ that strictly separates 0 from $K(x) \setminus B_\varepsilon(0)$; this hyperplane depends measurably on x, as does $K(x)$. Therefore, setting $\delta_\varepsilon(x) := \inf\{|y|^2 : y \in \overline{\text{co}}(K(x) \setminus B_\varepsilon(0))\}$ for a.a. $x \in \Omega$, there exists $z_\varepsilon \in L^\infty(\Omega; \mathbf{R}^M)$ such that $\int_\Omega z_\varepsilon \cdot v_{\varepsilon n} dx = \int_{A_{\varepsilon n}} z_\varepsilon \cdot v_{\varepsilon n} dx \geq \int_{A_{\varepsilon n}} \delta_\varepsilon(x) dx$ for any n.

By (1.1), $\int_\Omega z_\varepsilon \cdot v_{\varepsilon n} dx \to 0$ as $n \to \infty$, hence $\int_{A_{\varepsilon n}} \delta_\varepsilon(x) dx \to 0$. For any finite measure set $\tilde{\Omega} \subset \Omega$, we claim (and show in the following) that $|A_{\varepsilon n} \cap \tilde{\Omega}| \to 0$ as $n \to \infty$, since $\delta_\varepsilon > 0$ a.e. in Ω. As this occurs for any $\varepsilon > 0$, we infer that $|u_n| \to 0$ in measure in $\tilde{\Omega}$. Finally, this yields (1.4), by Theorem XI.3.9.

We are left with the proof of the claim, which is just an exercise in measure theory. For any $m \in \mathbf{N}$, let us set $C_m := \{x \in \tilde{\Omega} : 1/(m+1) < \delta_\varepsilon(x) \leq 1/m\}$, and note that $A_{\varepsilon n} \cap \tilde{\Omega} = \bigcup_m (A_{\varepsilon n} \cap C_m)$ for any n. For any m, $|A_{\varepsilon n} \cap C_m| \leq (m+1) \int_{A_{\varepsilon n}} \delta_\varepsilon(x) dx \to 0$ as $n \to \infty$, whence $|A_{\varepsilon n} \cap \tilde{\Omega}| \to 0$. □

In some cases, the weak convergence in $L^1(\Omega; \mathbf{R}^M)$ can be derived via a compactness result of the type of Dunford-Pettis, cf. Theorem XI.3.6. More frequently in applications, *a priori estimates* yield either weak convergence in $L^p(\Omega; \mathbf{R}^M)$ for some $p \in [1, +\infty[$ or weak star convergence in $L^\infty(\Omega; \mathbf{R}^M)$, cf. Theorem XI.3.1. The next result, which is a straightforward consequence of Theorem 1.2 and Proposition XI.3.10, shows that in those cases the extremality of the limit allows us to improve the convergence.

Corollary 1.3 *Let K be as in Theorem 1.2, and $1 \leq q < p < +\infty$. If (1.2), (1.3) are satisfied and*

$$u_n \to u \qquad \text{weakly in } L^p_{\text{loc}}(\Omega; \mathbf{R}^M), \tag{1.5}$$

then

$$u_n \to u \qquad \text{strongly in } L^q_{\text{loc}}(\Omega; \mathbf{R}^M). \tag{1.6}$$

The Infinite Dimensional Setting. If $\mathbf{R}^M$ is replaced by an infinite dimensional Banach space B, Theorem 1.2 does not hold.

As a counterexample, let us take $B = \ell^2 \times \mathbf{R}$ (ℓ^2:= Hilbert space of square summable real sequences), which is a Hilbert space endowed with the product structure. Let $\{e_n\}$ be the canonical basis of ℓ^2, and set $K := \overline{\text{co}}\{(e_n, 1/n)\}$. $(e_n, 1/n) \to (0,0)$ weakly in B and not strongly, nevertheless $(0,0)$ is an extremal (even exposed) point of K, as $K \cap (\ell^2 \times \{0\}) = \{(0,0)\}$.

Exercises.

1.1 Prove Theorem 1.2 under the stronger assumption that $u(x)$ is an *exposed* point of $K(x)$ for a.a. $x \in \Omega$.

Hint. For a.a. $x \in \Omega$, there exists a unit vector $\nu(x) \in \mathbf{R}^M$ such that $[z - u(x)] \cdot \nu(x) \geq 0$ for any $z \in K(x)$. Show that $(u_n - u) \cdot \nu(x) \to 0$ strongly in $L^1(\Omega)$, by (1.1). Hence, possibly extracting a subsequence, the latter sequence also converges a.e. in Ω. Show that this implies that $u_n \to u$ a.e. in Ω. Hence $u_n \to u$ in measure in any subset of Ω with finite measure, and this yields (1.4) by Theorem XI.3.9. Show that (1.4) then holds for the whole sequence.

1.2 Check that in Theorem 1.2 the assumption (1.2) can be replaced by the requirement that $K(x) = \overline{\text{co}} \left(\bigcap_{n \in \mathbf{N}} \{u_n(x)\} \right)$ for a.a. $x \in \Omega$.

1.3 Discuss the possibility of removing the conditions of closedness and convexity in the definition of extremality, and the possible validity of Theorem 1.2 under these reduced assumptions.

1.4 In Corollary 1.3, can one conclude that u_n converges strongly in $L^p_{\text{loc}}(\Omega; \mathbf{R}^M)$?

1.5 Prove the following statement.

Proposition 1.4 [5] *Let K be a closed and convex subset of $\mathbf{R}^M$. Let $u : \Omega \to K$ be measurable, essentially bounded, and such that $u(x)$ is not extremal for K, for a.a. x in a measurable set $\tilde{\Omega} \subset \Omega$ of positive measure.*

Then there exists a sequence $\{u_n\}$ of measurable functions $\Omega \to K$ such that $u_n \to u$ weakly star in $L^\infty(\Omega; \mathbf{R}^M)$, but not strongly in $L^1_{\text{loc}}(\Omega; \mathbf{R}^M)$.

Hint. Consider the following example: set $v_n(x) := \text{sign}(\sin(nx))$ for any $x \in]0, 2\pi[$ and any $n \in \mathbf{N}$. We have $v_n \to 0$ weakly star in $L^\infty(0, 2\pi)$, but not strongly in $L^1(0, 2\pi)$. Show that a similar oscillatory behaviour can be constructed in any set $\tilde{\Omega}$ as previously.

1.6 With reference to the counterexample at the end of this section, show that $K \cap (\ell^2 \times \{0\}) = \{(0,0)\}$.

---

[5] See V. [549; p. 444–445].

X.2 Strictly Convex Functionals

In this section we apply Theorem 1.2 to improve the convergence of the approximating sequence, a posteriori of passage to the limit in strictly convex functionals. Let [6]

$$\varphi : \mathbf{R}^M \to \mathbf{R} \cup \{+\infty\} \text{ be proper, strictly convex, lower semicontinuous} \quad (2.1)$$

and define the convex functional

$$\Phi(v) := \begin{cases} \int_\Omega \varphi(v) dx & \text{if } \varphi(v) \in L^1(\Omega), \\ +\infty & \text{otherwise,} \end{cases} \quad \forall v \in L^1_{\text{loc}}(\Omega; \mathbf{R}^M). \quad (2.2)$$

We often consider the restriction of this functional to $L^1(\Omega; \mathbf{R}^M)$, which is proper iff

$$|\Omega| = +\infty \implies \varphi(0) = 0. \quad (2.3)$$

Lemma 2.1 *If (2.1) through (2.3) are fulfilled, and*

$$|\Omega| = +\infty \implies \exists L \in \mathbf{R}^M : \forall v \in \mathbf{R}^M, L \cdot v \leq \varphi(v), \quad (2.4)$$

then $\Phi : L^1(\Omega; \mathbf{R}^M) \to \mathbf{R} \cup \{+\infty\}$ is (proper, convex, and) lower semicontinuous.

Proof. By (2.3) and (2.4), there exist $w \in \mathbf{R}^M$ and $C \in \mathbf{R}$ such that $\varphi(v) \geq w \cdot v - C$ for any $v \in \mathbf{R}^M$, and $C = 0$ if $|\Omega| = +\infty$. The functional $L : L^1(\Omega; \mathbf{R}^M) \to \mathbf{R} : v \mapsto \int_\Omega (w \cdot v - C) dx$ is obviously linear and continuous.

The functional $\Phi - L : L^1(\Omega; \mathbf{R}^M) \to \mathbf{R} \cup \{+\infty\} : v \mapsto \Phi(v) - \int_\Omega (w \cdot v - C) dx$ is then nonnegative and convex; hence it is strongly (equivalently, weakly) lower semicontinuous, by Fatou's lemma. This yields our statement. □

Theorem 2.2 *Assume that (2.1) through (2.4) are fulfilled, and that*

$$u_n \to u \quad \text{weakly in } L^1(\Omega; \mathbf{R}^M), \quad (2.5)$$

$$\Phi(u_n) \to \Phi(u) \neq +\infty. \quad (2.6)$$

Then

$$u_n \to u \quad \text{strongly in } L^1(\Omega; \mathbf{R}^M), \quad (2.7)$$

$$\varphi(u_n) \to \varphi(u) \quad \text{strongly in } L^1(\Omega). \quad (2.8)$$

---

[6] By *proper* we mean that $\varphi \not\equiv +\infty$. *Convex* and *strictly convex* functions are defined in Sect. XI.4.

Proof. Let us set $K := \text{epi}(\varphi) \subset \mathbf{R}^{M+1}$, and note that $(u, \varphi(u))$ is an extremal point of K a.e. in Ω. By Lemma 2.3, we can then apply Theorem 1.2 taking $(u_n, \varphi(u_n)) \in \mathbf{R}^{M+1}$ in place of $u_n \in \mathbf{R}^M$. □

Lemma 2.3 *If (2.1) through (2.6) are fulfilled, then*

$$\varphi(u_n) \to \varphi(u) \qquad \text{weakly in } L^1(\Omega). \tag{2.9}$$

Proof. For any measurable set $A \subset \Omega$, applying Lemma 2.1 with either A or $\Omega \setminus A$ in place of Ω, we get

$$\liminf_{n \to \infty} \int_A \varphi(u_n) dx \geq \int_A \varphi(u) dx, \qquad \liminf_{n \to \infty} \int_{\Omega \setminus A} \varphi(u_n) dx \geq \int_{\Omega \setminus A} \varphi(u) dx.$$

Moreover we have

$$\limsup_{n \to \infty} \int_A \varphi(u_n) dx = \lim_{n \to \infty} \int_\Omega \varphi(u_n) dx - \liminf_{n \to \infty} \int_{\Omega \setminus A} \varphi(u_n) dx$$

$$\leq \int_\Omega \varphi(u) dx - \int_{\Omega \setminus A} \varphi(u) dx = \int_A \varphi(u) dx.$$

Therefore $\int_A \varphi(u_n) dx \to \int_A \varphi(u) dx$. As the linear space spanned by characteristic functions is dense in $L^\infty(\Omega)$, we can now apply Theorem XI.3.12. □

The next result generalizes a well-known property of $L^p(\Omega; \mathbf{R}^M)$ ($1 < p < +\infty$), and more generally of any *locally uniformly (strictly) convex* Banach space.

Theorem 2.4 *Assume that (2.1) through (2.4) are fulfilled, and*

$$u_n \to u \qquad \text{weakly in } L^1_{\text{loc}}(\Omega; \mathbf{R}^M), \tag{2.10}$$

$$\exists p \in [1, +\infty[, \exists L > 0, \exists M \geq 0 \, (M = 0 \text{ if } |\Omega| = +\infty) :$$
$$\forall \xi \in \mathbf{R}, \varphi(\xi) \geq L|\xi|^p - M. \tag{2.11}$$

Then

$$u_n \to u \qquad \text{strongly in } L^p(\Omega; \mathbf{R}^M). \tag{2.12}$$

Proof. [7] For any compact set $D \subset \Omega$, we can apply Theorem 2.2; hence, possibly extracting a subsequence, $u_n \to u$ a.e. in Ω. Moreover (2.11) yields

$$|u - u_n|^p \leq 2^p \left(|u|^p + |u_n|^p\right) \leq 2^p \left(\frac{\varphi(u) + \varphi(u_n)}{L} + 2M\right) \qquad \text{a.e. in } \Omega.$$

---

[7] Here is an alternative argument, which uses slightly less elementary tools. The sequence $\{|u_n|^p\}$ is weakly compact in $L^1(\Omega; \mathbf{R}^M)$, by (2.11) and Theorem XI.3.6. By Theorem 2.2, a subsequence converges a.e. in Ω, hence in measure. It then suffices to apply Theorem XI.3.9. Notice that this also entails $\varphi(u_n) \to \varphi(u)$ a.e. in Ω, whence (2.8) by (2.9) and Theorem XI.3.9.

By the following lemma, then we get (2.12) for the extracted subsequence. As the limit is independent of the subsequence, this holds for the whole sequence. □

Lemma 2.5 *Let $\{v_n\}$, $\{w_n\}$ be sequences of measurable functions $\Omega \to \mathbf{R}$ such that $|v_n| \leq w_n$ for any n, $v_n \to v$ and $w_n \to w$ a.e. in Ω, $w \in L^1(\Omega)$, and $\int_\Omega w_n dx \to \int_\Omega w dx$. Then $v \in L^1(\Omega)$ and $\int_\Omega v_n dx \to \int_\Omega v dx$.*

Proof. By Fatou's lemma, we have

$$\liminf_{n\to\infty} \int_\Omega v_n dx = \liminf_{n\to\infty} \int_\Omega (w_n + v_n)dx - \lim_{n\to\infty} \int_\Omega w_n dx$$
$$\geq \int_\Omega (v+w)dx - \int_\Omega w dx = \int_\Omega v dx,$$

$$\limsup_{n\to\infty} \int_\Omega v_n dx = -\liminf_{n\to\infty} \int_\Omega (w_n - v_n)dx + \lim_{n\to\infty} \int_\Omega w_n dx$$
$$\leq -\int_\Omega (w-v)dx + \int_\Omega w dx = \int_\Omega v dx.$$

□

One can also deal with the convergence in the sense of distributions.

Theorem 2.6 [8] *Assume that $|\Omega| < +\infty$, that (2.1) is fulfilled, and*

$$\exists M, C : \forall v \in \mathbf{R}^M, \ |\varphi(v)| \leq M|v| + C. \tag{2.13}$$

Let $\{u_n\} \subset L^1(\Omega; \mathbf{R}^M)$ and $u \in L^1(\Omega; \mathbf{R}^M)$ be such that (2.6) is satisfied, and

$$u_n \to u \quad \text{in } \mathcal{D}'(\Omega; \mathbf{R}^M). \tag{2.14}$$

Then

$$u_n \to u \quad \text{strongly in } L^1_{\text{loc}}(\Omega; \mathbf{R}^M). \tag{2.15}$$

If, moreover,

$$\varphi(v) \to +\infty \quad \text{as } |v| \to +\infty, \tag{2.16}$$

then (2.7) holds.

The following counterexample shows that (2.7) may fail if (2.16) is not satisfied, and that in Theorem 1.2 the assumption (1.1) cannot be replaced by (2.14). Let us take

$$\Omega :=]0, 1[, \quad u_n(x) := \begin{cases} n^2 & \text{if } 0 < x < 1/n, \\ 0 & \text{if } 1/n < x < 1, \end{cases} \quad \varphi(v) := e^{-v} \ \forall v \in \mathbf{R}.$$

---

[8] See Brézis [93], where the result is proved.

Then $u_n \to 0$ in $\mathcal{D}'(]0, 1[)$ and strongly in $L^1(A)$ for any compact set $A \subset]0, 1[$, but not in $L^1(0, 1)$, although $\int_0^1 e^{-u_n} dx \to \int_0^1 e^0 dx = 1$.

Exercises.

2.1 Is the assumption (2.4) needed in Lemma 2.1?

Hint. Consider the function $\varphi(x) := -\sqrt{x}$ for any $x \geq 0$, $\varphi(x) := +\infty$ for any $x < 0$.

2.2 Detail the justification of the first sentence of the proof of Lemma 2.1.

2.3 Detail the justification of the first sentence of the proof of Lemma 2.3.

2.4 Prove Lemma 2.5 by using the classical *Vitali theorem,* which we state in a footnote in Sect. XI.3.

2.5 Prove the following result. Let $\{u_n\}$ and u be elements of $L^1(\Omega)$. If $u_n \to u$ a.e. in Ω and $\|u_n\|_{L^1(\Omega)} \to \|u\|_{L^1(\Omega)}$, then $u_n \to u$ strongly in $L^1(\Omega)$.

(A similar property does not hold in $L^p(\Omega)$ for any $p > 1$; see also Proposition XI.3.10.)

X.3 Applications

In this section we apply Theorem 1.2 to improve the convergence of an approximating sequence, a posteriori of passage to the limit in a nonlinear term. [9] This technique may be used in the study of certain nonlinear P.D.E.s.

The next statement provides both passage to the limit in a nonlinear term and strong convergence. We denote by ∂ the *subdifferential* operator, and by φ^* and Φ^* the *convex conjugates* of φ and Φ, respectively; cf. Sect. XI.4.

Theorem 3.1 *Assume that (2.1) through (2.4) are satisfied,*

$$u_n \to u \qquad \text{weakly in } L^1\left(\Omega; \mathbf{R}^M\right), \tag{3.1}$$

that $\varphi(u) \in L^1(\Omega)$, and there exists a sequence $\{w_n\}$ such that

$$w_n \in \partial\varphi(u_n) \qquad \text{a.e. in } \Omega, \forall n, \tag{3.2}$$

$$w_n \to w \qquad \text{weakly star in } L^\infty\left(\Omega; \mathbf{R}^M\right), \tag{3.3}$$

$$\limsup_{n \to \infty} \int_\Omega w_n u_n dx \leq \int_\Omega wu dx. \tag{3.4}$$

---

[9] Here we refer to the *approximation — a priori estimate — passage to the limit* procedure, that we outlined in Sect. I.2.

Then
$$w \in \partial\varphi(u) \quad \text{a.e. in } \Omega, \tag{3.5}$$

$$u_n \to u \quad \text{strongly in } L^1\left(\Omega; \mathbf{R}^M\right), \tag{3.6}$$

$$\Phi(u_n) \to \Phi(u). \tag{3.7}$$

Moreover, assume that φ^* is strictly convex, fulfills (2.11) for some $p \in [1,+\infty[$, and is such that $\varphi^*(w) \in L^1(\Omega)$. Then

$$w_n \to w \quad \text{strongly in } L^p\left(\Omega; \mathbf{R}^M\right), \tag{3.8}$$

$$\Phi^*(w_n) \to \Phi^*(w). \tag{3.9}$$

Proof. The inclusion (3.5) follows from Lemma XI.5.1.

By definition of subdifferential, (3.2) entails that $\varphi(u_n) < +\infty$ and $w_n(u_n - u) \geq \varphi(u_n) - \varphi(u)$ a.e. in Ω for any n. The left member of this inequality is integrable, and we have $\int_\Omega w_n(u_n - u)dx \geq \Phi(u_n) - \Phi(u)$. Passing to the superior limit, by (3.4) we get $\limsup_{n\to\infty} \Phi(u_n) \leq \Phi(u)$. On the other hand, Lemma 2.1 yields $\liminf_{n\to\infty} \Phi(u_n) \geq \Phi(u)$, so that $\Phi(u_n) \to \Phi(u)$. Theorem 2.2 then yields (3.6).

Similarly one can show that $\Phi^*(w_n) \to \Phi^*(w)$, as $\int_\Omega u_n(w_n - w)dx \geq \Phi^*(w_n) - \Phi^*(w)$. Theorem 2.4 then yields (3.8). □

Corollary 3.2 *If in Theorem 3.1 the condition (3.4) is replaced by*

$$\liminf_{n\to\infty} \int_\Omega w_n u_n \, dx \leq \int_\Omega wu \, dx, \tag{3.10}$$

then (3.6) through (3.9) hold for suitable subsequences $\{u_{n'}\}$ and $\{w_{n'}\}$.

Proof. By (3.10), there exists a subsequence $\{(u_{n'}, w_{n'})\}$ such that

$$\lim_{n'\to\infty} \int_\Omega w_{n'} u_{n'} \, dx \leq \int_\Omega wu \, dx.$$

Then it suffices to apply the previous theorem. □

Compactness by Strict Monotonicity. Here we prove a result based on the strict and maximal monotonicity of a graph α, which holds in infinite dimension as well. Let B be a (real) Banach space, denote its norm by $\|\cdot\|$, its dual by B', and the duality pairing by $\langle\cdot,\cdot\rangle$.

Let $\alpha : \text{Dom}(\alpha) \subset B \to 2^{B'}$ be maximal monotone. [10] We say that α is *strictly monotone* if

$$\forall u_i \in \text{Dom}(\alpha), \forall w_i \in \alpha(u_i)(i = 1, 2) \\ \langle w_1 - w_2, u_1 - u_2 \rangle = 0 \quad \Rightarrow \quad u_1 = u_2. \tag{3.11}$$

---

[10] See Sect. XI.5 for the definition and the main properties.

This entails that $\beta := \alpha^{-1} : \mathrm{Dom}(\beta) \subset B' \to 2^B$ is single valued (the converse does not hold). Hence it seems more convenient to state our result in terms of β rather than α. Notice that in the following statement we assume a stronger property than (3.11), cf. (3.13), which includes some uniformity.

Proposition 3.3 *(Compactness by Strict Monotonicity) Let B be a Banach space, and let either B be reflexive or B' be separable.* [11] *Assume that*

$$\beta : \mathrm{Dom}(\beta) \subset B' \to B \text{ is maximal monotone,} \tag{3.12}$$

$$\forall \eta \in \mathrm{Dom}(\beta), \forall \{\eta_n\} \subset \mathrm{Dom}(\beta), \\ \langle \beta(\eta_n) - \beta(\eta), \eta_n - \eta \rangle \to 0 \;\Rightarrow\; \beta(\eta_n) \to \beta(\eta) \text{ strongly in } B', \tag{3.13}$$

$$w_n \to w \qquad \text{weakly star in } L^\infty(\Omega; B') \left(= L^1(\Omega; B)'\right), \tag{3.14}$$

$$\beta(w_n) \to u \qquad \text{weakly in } L^1(\Omega; B), \tag{3.15}$$

$$\limsup_{n\to\infty} \int_\Omega \langle w_n, \beta(w_n) \rangle \, dx \leq \int_\Omega \langle w, u \rangle \, dx. \tag{3.16}$$

Then

$$\beta(w_n) \to \beta(w) \qquad \text{strongly in } L^1(\Omega; B). \tag{3.17}$$

Proof. By Lemma XI.5.1, we have $u = \beta(w)$ a.e. in Ω. Let us set

$$\psi(v, x) := \langle v - w(x), \beta(v) - u(x) \rangle \qquad \forall v \in \mathrm{Dom}(\beta), \text{ for a.a. } x \in \Omega.$$

By (3.14) through (3.16) we have $\limsup_{n\to\infty} \int_\Omega \psi(w_n(x), x) dx \leq 0$. As ψ is nonnegative, this means that $\psi(w_n, \cdot) \to 0$ strongly in $L^1(\Omega)$. Hence there exists a subsequence $\{w_{n'}\}$ such that $\psi(w_{n'}, \cdot) \to 0$ a.e. in Ω, and (3.13) yields

$$\beta(w_{n'}) \to u \qquad \text{strongly in } B, \text{ a.e. in } \Omega. \tag{3.18}$$

Therefore $\|\beta(w_{n'}) - u\| \to 0$ in measure in any subset of Ω with finite measure.

By Theorem XI.3.9, this yields (3.17) for the extracted subsequence. Finally, the whole sequence converges, since the limit is independent of the subsequence. □

By the simple argument used in Corollary 3.2, one can also prove the following result.

Corollary 3.4 *If in Proposition 3.3 the condition (3.16) is replaced by*

$$\liminf_{n\to\infty} \int_\Omega \langle w_n, \beta(w_n) \rangle \, dx \leq \int_\Omega \langle w, u \rangle \, dx, \tag{3.19}$$

---

[11] Hence $L^1(\Omega; B)' = L^\infty(\Omega; B')$. See, e.g., Kufner, John and Fučik [330; Sect. 2.22.5], Diestel and Uhl [195; Chap. III], and Dunford and Schwartz [202; Chap. V].

then (3.17) holds for a suitable subsequence $\{w_{n'}\}$.

The condition (3.13) cannot be dropped. A counterexample is simply constructed, if $B = \mathbf{R}^M$, $M = 2$, and α is a rotation of $\pi/2$ rad. Then the question arises of determining sufficient conditions for (3.13). For instance, (3.13) is fulfilled if $(\alpha :=)\beta^{-1}$ is strongly monotone, in the sense of (II.2.13).

Generalizations. The previous results can be extended in several ways.
(i) Some variants of Theorem 3.1 and Proposition 3.3 can be proved. For instance, with reference to the former statement, if $|\Omega| < +\infty$, one can assume that

$$u_n \to u \text{ weakly in } L^p(\Omega; \mathbf{R}^M), \quad w_n \to w \text{ weakly in } L^q(\Omega; \mathbf{R}^M),$$

with $p, q \in]1, +\infty[$ and $1/p + 1/q = 1$, in place of (3.3) and (3.6). In this case,

$$u_n \to u \text{ strongly in } L^{\tilde{p}}(\Omega; \mathbf{R}^M), \quad w_n \to w \text{ strongly in } L^{\tilde{q}}(\Omega; \mathbf{R}^M),$$

for any $\tilde{p} < p$ and any $\tilde{q} < q$. Proposition 3.3 can be extended similarly.
(ii) In Theorems 2.2, 2.4, and 3.1, the convex function φ may also depend on x; for instance, one may assume that $\varphi(v) = \varphi(v, x)$, with φ *convex normal integrand*. [12] Similarly, in Proposition 3.3 the function β may also depend on x.
(iii) The requirement that φ be strictly convex can be systematically replaced by the condition that $(u(x), \varphi(u(x)))$ be an extremal point of epi(φ) for a.a. $x \in \Omega$. This condition is weaker than the strict convexity. For instance, it is fulfilled by $\varphi(\xi) := \left[(|\xi| - 1)^+\right]^2$ and $|u| \geq 1$, but there exists no strictly convex function $\tilde{\varphi}$ such that $\tilde{\varphi}(\xi) = \varphi(\xi)$ if $|\xi| \geq 1$.
(iv) Derivatives may also occur in the functional Φ. For instance, the results of Sect. X.2 can be applied for $u_n = \nabla v_n$ with $v_n \in W^{1,1}(\Omega)$. In this case φ may also explicitly depend on x and $v_n(x)$, under appropriate conditions.

In similar cases the hypothesis of convexity may be replaced by that of *quasiconvexity*, cf. Evans and Gariepy [217]. □

Exercises.

3.1 Let $\beta_1, \beta_2 : D (\subset \mathbf{R}^M) \to \mathbf{R}^M$ be monotone functions. Show that if β_1 fulfills (3.13), then the same occurs for $\beta_1 + \beta_2$.

3.2 Prove that if $\varphi : \mathbf{R}^M \to \mathbf{R} \cup \{+\infty\}$ is strictly convex, lower semicontinuous and of class C^2 at the interior of its domain, then $\beta := (\partial \varphi)^{-1}$ fulfills (3.13).

3.3 State precisely and prove the result outlined in the preceding Remark (ii).

3.4 Illustrate by an example the preceding Remark (iv).

---

[12] See, e.g., Brézis [88], and Castaing and Valadier [125; Chap. VII].

3.5 Let B be a Banach space, and $\varphi : B \to \mathbf{R} \cup \{+\infty\}$ be strictly convex and lower semicontinuous. Show by a counterxample that $\beta := (\partial \varphi)^{-1}$ does not necessarily fulfill (3.13).

X.4 Comments

This chapter has been devoted to the illustration of a fairly general technique, which we called *compactness by strict convexity,* which can be used in several nonlinear problems; see, for example, Sect. III.4.

As far as this author knows, Olech [433], [434; Lemma 1, p. 300] was the first one who pointed out the relationship existing between extremality and strong convergence, albeit in a rather different form from that considered in this chapter. Along those lines see also Artstein and Rzezuchowski [27].

Theorem 1.2 was proved by V. [549], and then extended by Rzezuchowski [491]. As we saw, for functions with range in a Banach space, in general extremality and weak L^1-convergence do not yield strong convergence. Balder [48, 49, 50] showed that those conditions yield a different kind of convergence, intermediate between the weak and the strong convergence. His approach is based on *Young measures;* in this respect, see also Tartar [522].

Rzezuchowski [492] introduced the concept of *point of dentability,* and extended Theorem 1.2 to functions taking values in an infinite dimensional Banach space. Improvements were also introduced by Amrani, Castaing and Valadier [18, 19], and Valadier [541, 542, 543]. See also B.-L. Lin, P.-K. Lin and Troyanski [76].

Less concern was apparently devoted to results about strictly convex functionals like Theorem 2.2, with the exceptions of V. [549] and Brézis [93].

Chapter XI. Toolbox

Outline

Some definitions and fundamental results about function spaces, compactness, convexity, monotonicity, nonlinear semigroups of contractions, functional minimization, and geometric measure theory are here reviewed, mainly without proofs.

Prerequisites. Linear algebra, calculus, topology, Lebesgue's measure and integration theory, and basic notions about Banach and Hilbert spaces.

Readers who are not familiar with the topics of this chapter should also consult other texts. See the references quoted in the *Book Selection*.

XI.1 Some Function Spaces

Our toolbox contains several concepts and results that we used in this volume. Most of them are classic, and are presented without proofs. Some complements are also included.

In this section we review some spaces of functions acting from a Euclidean set to a Banach space. References on this subject can be found in the *Book Selection*.

Spaces of Continuous Functions. Let Ω be an open subset of $\mathbf{R}^N$ ($N \geq 1$) and B an either real or complex Banach space, whose topological dual we denote by B'.

We say that Ω is of class $C^{k,\nu}$ ($k \in \mathbf{N}_0 := \mathbf{N} \cup \{0\}, 0 < \nu \leq 1$) iff

(i) for any point y of its boundary Γ there exists a neighbourhood U of y such that $U \cap \Omega$ stays only on one side of $U \cap \Gamma$, and

(ii) possibly after a rotation of the axes, $U \cap \Gamma$ is the graph of a function $f : D\left(\subset \mathbf{R}^{N-1}\right) \to \mathbf{R}$ of class $C^{k,\nu}$.

Henceforth we assume that Ω is bounded and Γ of Lipschitz (i.e., $C^{0,1}$) class.

For any *multi-index* $\alpha = (\alpha_1, \ldots, \alpha_N) \in \mathbf{N}_0^N$, we set $|\alpha| := \alpha_1 + \ldots + \alpha_N$, $D^\alpha := \partial^{|\alpha|} / \partial x_1^{\alpha_1} \ldots \partial x_N^{\alpha_N}$. We define the space of *vector-valued* continuous functions [1]

$$C^0(\bar{\Omega}; B) := \left\{ v : \bar{\Omega} \to B \text{ strongly continuous} \right\},$$

---

[1] We denote the closure of a generic set A by $\bar{A}$.

equipped with the norm $\|v\| := \max_{x \in \Omega} \|v(x)\|_B$. For any $k \in \mathbf{N}$, we set

$$C^k(\bar\Omega; B) := \left\{ v \in C^0(\bar\Omega; B) : D^\alpha v \in C^0(\bar\Omega; B), \forall \alpha, |\alpha| \le k \right\};$$

here derivatives of functions $\bar\Omega \to B$ are meant as strong limits in B of the corresponding incremental ratios. For any $\nu \in]0, 1]$, we define the spaces of vector-valued *Hölder continuous* functions

$$C^{0,\nu}(\bar\Omega; B) := \left\{ v \in C^0(\bar\Omega; B) : \sup_{x_1, x_2 \in \bar\Omega, x_1 \ne x_2} \frac{\|v(x_1) - v(x_2)\|_B}{|x_1 - x_2|^\nu} < +\infty \right\},$$

$$C^{k,\nu}(\bar\Omega; B) := \left\{ v \in C^k(\bar\Omega; B) : D^\alpha v \in C^{0,\nu}(\bar\Omega; B), \forall \alpha, |\alpha| \le k \right\}.$$

We also set $C^{k,0}(\bar\Omega; B) := C^k(\bar\Omega; B)$. These are (either real or complex) Banach spaces, equipped with the respective *graph norms*. [2]

We also set $C^1_0(\Omega; B) := \{ v \in C^1(\bar\Omega; B) : v = 0 \text{ on } \Gamma \}$, which is a subspace of $C^1(\bar\Omega; B)$.

Lebesgue Spaces. Let us denote by $S(\Omega; B)$ the family of *simple functions* $\Omega \to B$, namely, functions with finite range such that the inverse of any element of B is measurable. We can then introduce the space [3] of *strongly measurable* functions $\Omega \to B$ [4]

$$M(\Omega; B) := \left\{ v : \Omega \to B : \exists \{v_n\}_{n \in \mathbf{N}} \subset S(\Omega; B) \text{ such that} \right.$$
$$\left. v_n \to v \text{ strongly in } B, \text{ a.e. in } \Omega \right\}.$$

As the range of any strongly measurable function is confined to a separable subspace, it is natural to assume that B is separable.

The *Lebesgue spaces* of vector-valued functions

$$L^p(\Omega; B) := \left\{ v \in M(\Omega; B) : \int_\Omega \|v\|_B^p \, dx < +\infty \right\} \qquad \forall p \in [1, +\infty[,$$

$$L^\infty(\Omega; B) := \left\{ v \in M(\Omega; B) : \operatorname*{ess\,sup}_\Omega \|v\|_B < +\infty \right\},$$

are (either real or complex) Banach spaces equipped with the norms

$$\|v\|_{L^p(\Omega; B)} := \left(\int_\Omega \|v\|_B^p \, dx \right)^{1/p}, \qquad \|v\|_{L^\infty(\Omega; B)} := \operatorname*{ess\,sup}_\Omega \|v\|_B,$$

---

[2] If B_1 and B_2 are Banach spaces (equipped with the norms $\|\cdot\|_{B_1}, \|\cdot\|_{B_2}$, resp.) and $\Lambda : \operatorname{Dom}(\Lambda) \subset B_1 \to B_2$ is a linear unbounded operator, then $\{ v \in B_1 : \Lambda v \in B_2 \} := \operatorname{Dom}(\Lambda)$ is a Banach space equipped with the *graph norm* $\|v\| := \|v\|_{B_1} + \|\Lambda v\|_{B_2}$.

[3] By *space* we mean a linear space over either $\mathbf{R}$ or $\mathbf{C}$, with linear operations defined in an obvious way.

[4] As usual, *a.e.* stands for "almost everywhere", and *a.a.* for "almost any".

respectively. For any $p \in [1,+\infty]$, $L^p(\Omega;B)$ consists of classes of functions, induced by the equivalence relation $u \sim v$ iff $u = v$ a.e. in Ω. Nevertheless, we write $X \subset C^0(\bar{\Omega};B)$ whenever $X \subset L^p(\Omega;B)$ and representatives of X can be selected in $C^0(\bar{\Omega};B)$. We define [5]

$$L^p_{\text{loc}}(\Omega;B) := \{v \in M(\Omega;B) : v|_D \in L^p(D;B), \forall D \subset\subset \Omega\} \quad \forall p \in [1,+\infty]$$

($v|_D :=$ restriction of v to D). A sequence $\{u_n\}$ is said to converge weakly (strongly, resp.) in $L^p_{\text{loc}}(\Omega;B)$ iff for any $D \subset\subset \Omega$ the restriction of $\{u_n\}$ to D converges weakly (strongly, resp.) in $L^p(D;B)$. This is a *Fréchet space* (i.e., a complete metric and linear space, in which the linear operations are continuous).

One can similarly define other *local* spaces.

With abuse of notation, for any set $A \subset B$ we write, for example, $L^p(\Omega;A)$ in place of $\{v \in L^p(\Omega;B) : v \in A \text{ a.e. in } \Omega\}$.

Distributions. For any $K \subset\subset \Omega$, let us denote by $\mathcal{D}_K(\Omega)$ the space of infinitely differentiable functions $f : \Omega \to \mathbf{R}$ (or $\mathbf{C}$) whose support is included in K. This is a Fréchet space equipped with the family of seminorms $|f|_{K,m} := \sum_{|\alpha| \leq m} \sup_K |D^\alpha f|$ ($m \in \mathbf{N}_0$).

We then introduce the space of **test functions** $\mathcal{D}(\Omega) := \bigcup_{K \subset\subset \Omega} \mathcal{D}_K(\Omega)$. This is a *locally convex topological space* equipped with the *inductive limit* topology. [6] This is the finest topology on $\mathcal{D}(\Omega)$ that makes the injection $\mathcal{D}_K(\Omega) \to \mathcal{D}(\Omega)$ continuous for any $K \subset\subset \Omega$. Then $\mathcal{D}(\Omega)$ is a complete and nonmetrizable locally convex topological space.

For instance, set $f(x) := \exp\left[\left(|x|^2 - 1\right)^{-1}\right]$ if $|x| < 1$, $f(x) := 0$ if $|x| \geq 1$. For any $x_0 \in \Omega$ and any ε smaller than the distance of x_0 from $\mathbf{R}^N \setminus \Omega$, we have $f\left((\cdot - x_0)/\varepsilon\right) \in \mathcal{D}(\Omega)$.

Proposition 1.1 $v_n \to 0$ in $\mathcal{D}(\Omega)$ iff there exists $K \subset\subset \Omega$ that includes the supports of all the v_ns, and $|D^\alpha v_n| \to 0$ uniformly in K for any $\alpha \in \mathbf{N}_0^N$.

Definitions. The elements of $\mathcal{D}'(\Omega)$, that is, the linear functionals $\mathcal{D}(\Omega) \to \mathbf{R}$ (or $\mathbf{C}$) that are continuous with respect to the inductive limit topology of $\mathcal{D}(\Omega)$, are called **distributions**.

Any $f \in L^1_{\text{loc}}(\Omega)$ is identified with the distribution $T_f : v \mapsto \int_\Omega fv\,dx$. For any $x_0 \in \Omega$, the *Dirac measure* $\delta_{x_0} : v \mapsto v(x_0)$ is also a distribution.

A sequence $\{T_n\}$ is said to converge to T in $\mathcal{D}'(\Omega)$ iff

$$\langle T_n, v \rangle := T_n(v) \to \langle T, v \rangle \quad \forall v \in \mathcal{D}(\Omega). \tag{1.1}$$

---

[5] With standard notation, by $D \subset\subset \Omega$ we mean that D is a bounded open subset of $\mathbf{R}^N$ and $\bar{D} \subset \Omega$. However, dealing with Banach spaces, $A \subset\subset B$ means that $A \subset B$ with continuous injection, and that any bounded subset of A is relatively strongly compact in B.

[6] For this concept and for other results about distributions, see, e.g., Gel'fand et al. [261], Hörmander [300], Horváth [301], Schwartz [497], Treves [535], and Yosida [589].

Proposition 1.2 *Let T be a linear functional $\mathcal{D}(\Omega) \to \mathbf{R}$ (or $\mathbf{C}$). Then $T \in \mathcal{D}'(\Omega)$ iff either*

$$v_n \to 0 \text{ in } \mathcal{D}(\Omega) \quad \Longrightarrow \quad \langle T, v_n \rangle \to 0, \tag{1.2}$$

or

$$\forall K \subset\subset \Omega, \exists m \in \mathbf{N}_0, \exists C > 0 : \forall v \in \mathcal{D}_K(\Omega),$$
$$|\langle T, v \rangle| \leq C \sum_{|\alpha| \leq m} \sup_K |D^\alpha v|. \tag{1.3}$$

Derivatives are defined in $\mathcal{D}'(\Omega)$ through the *integration by parts* formula

$$\langle D^\alpha T, v \rangle := (-1)^{|\alpha|} \langle T, D^\alpha v \rangle \qquad \forall T \in \mathcal{D}'(\Omega), \forall v \in \mathcal{D}(\Omega), \tag{1.4}$$

and are linear and continuous operators in $\mathcal{D}'(\Omega)$.

Distributions taking values in Banach spaces can be defined by a *similar* construction. [7]

Space of Measures. If in (1.3) the integer m can be chosen independent of K, and m is the minimum of such values, then the distribution T is said to be of order m. In particular the distributions of zero order are called *Radon measures,* and the corresponding subspace of $\mathcal{D}'(\Omega)$ can be identified with the topological dual of $C_c^0(\Omega)$. [8] The latter is the Fréchet space of continuous (real-valued) functions with compact support, equipped with the family of seminorms $|f|_{K,0} := \sup_K |f|$ with $K \subset\subset \Omega$.

We also define the space of (real-valued) functions having **bounded total variation:**

$$BV(\Omega) := \left\{ u \in L^1(\Omega) : \int_\Omega |\nabla u| < +\infty \right\},$$

where

$$\int_\Omega |\nabla u| := \sup \left\{ \int_\Omega u \nabla \cdot \vec{\eta} : \vec{\eta} \in C_0^1(\Omega)^N, |\vec{\eta}| \leq 1 \text{ in } \Omega \right\} \tag{1.5}$$

This is a Banach space equipped with the graph norm.

---

[7] See, e.g., Schwartz [498].
[8] See, e.g., Brézis [92; Chap. IV] and Edwards [206; Chap. 4].

XI.2 Sobolev Spaces

In this section we review the definitions of Sobolev spaces of vector-valued functions, as well as the classical embedding results and some trace theorems.

We still assume that Ω is a bounded open set of Lipschitz class, and that B is an either real or complex Banach space. We set

$$W^{k,p}(\Omega; B) := \{v \in L^p(\Omega; B) : D^\alpha v \in L^p(\Omega; B), \forall \alpha, |\alpha| \leq k\}$$
$$\forall k \in \mathbf{N}, \forall p \in [1, +\infty],$$

where D^α is the derivative in the sense of distributions $\Omega \to B$.

For any $p \in [1, +\infty[$, $W^{k,p}(\Omega; B)$ coincides with the completion of $C^\infty(\bar{\Omega}; B)$ with respect to the norm $\|v\|_{W^{k,p}(\Omega; B)} := \left(\sum_{|\alpha| \leq k} \|D^\alpha v\|^p_{L^p(\Omega; B)}\right)^{1/p}$.

We set

$$W^{k+\nu, p}(\Omega; B) := \Big\{v \in W^{k,p}(\Omega; B) :$$

$$\sum_{|\alpha|=k} \iint_{\Omega^2} \frac{\|D^\alpha v(x_1) - D^\alpha v(x_2)\|^p_B}{|x_1 - x_2|^{(N+\nu p)}} dx_1 dx_2 < +\infty\Big\}$$

$$\forall k \in \mathbf{N}_0, \forall \nu \in]0, 1[, \forall p \in [1, +\infty[,$$

$$W^{k+\nu, \infty}(\Omega; B) := C^{k,\nu}(\bar{\Omega}; B) \qquad \forall k \in \mathbf{N}_0, \forall \nu \in]0, 1[,$$

and, denoting by $C^\infty_c(\Omega; B)$ the space of infinitely differentiable functions $\Omega \to B$ with compact support,

$$W^{s,p}_0(\Omega; B) := \text{ closure of } C^\infty_c(\Omega; B) \text{ in } W^{s,p}(\Omega; B) \quad \forall s > 0, \forall p \in [1, +\infty[.$$

These are (either real or complex) Banach spaces equipped with the respective graph norms.

If either B is reflexive or B' is separable, we have $L^{p'}(\Omega; B') = L^p(\Omega; B)'$ for any $p \in [1, +\infty[$, where $p' := p/(p-1)$ for any $p \in]1, +\infty[$ and $1' := \infty$. [9]

We then set

$$W^{-s,p'}(\Omega; B') := \left(W^{s,p}_0(\Omega; B)\right)' \qquad \forall s > 0, \forall p \in [1, +\infty[.$$

We also set $H^s(\Omega; B) := W^{s,2}(\Omega; B)$ for any $s \in \mathbf{R}$; this is a Hilbert space if so is B.

Henceforth in this section we only deal with scalar-valued Sobolev spaces.

---

[9] See, e.g., Kufner, John and Fučik [330; Sect. 2.22.5], Diestel and Uhl [195; Chap. III], and Dunford and Schwartz [202; Chap. V].

We remind the reader of the classical Sobolev and Morrey embeddings.

Theorem 2.1 [10] *Let Ω be an open bounded subset of $\mathbf{R}^N$ of Lipschitz class, $r, s \in \mathbf{R}$, $0 \leq r < s$, and $k \in \mathbf{N}_0$.*
(i) Let $1 < p \leq q < \infty$ and $0 < \alpha < 1$.

$$\text{If} \quad r - \tfrac{N}{q} \leq s - \tfrac{N}{p} \quad \text{then} \quad W^{s,p}(\Omega) \subset W^{r,q}(\Omega), \tag{2.1}$$

$$\text{if} \quad k + \alpha \leq s - \tfrac{N}{p} \quad \text{then} \quad W^{s,p}(\Omega) \subset C^{k,\alpha}(\bar{\Omega}), \tag{2.2}$$

with continuous injections.
(ii) If $p = 1$, then (2.1) is satisfied for $q = \infty$ (i.e., $W^{s,1}(\Omega) \subset W^{s-N,\infty}(\Omega)$ if $s \geq N$), and (2.2) for $\alpha = 0$ or 1. If $p = 1$ and $q < \infty$, then (2.1) holds if either $r, s \in \mathbf{N}$, or $s \notin \mathbf{N}$, or $s - r \geq 1$.

For instance, $W^{N,1}(\Omega) \subset C^0(\bar{\Omega})$ with continuous injection. $W^{1,N}(\Omega)$ is not included in $L^\infty(\Omega)$; however, $W^{1,N}(\Omega) \subset W^{1-\varepsilon,N}(\Omega) \subset L^q(\Omega)$ for any $q \in [N, +\infty[$, with $\varepsilon \in]0, 1[$ depending on q.

The statement (2.1) is based on the following classical *Sobolev inequality*:

$$\begin{aligned} &\forall p \in [1, N[, \exists C > 0 : \text{ setting } p^* := \tfrac{Np}{N-p}, \\ &\|u\|_{L^{p^*}(\mathbf{R}^N)} \leq C \|\nabla u\|_{L^{p^*}(\mathbf{R}^N; \mathbf{R}^N)}. \end{aligned} \tag{2.3}$$

Notice that (2.2) represents the limit case $q = \infty$ in (2.1), as $C^{k,\alpha}(\bar{\Omega}) =: W^{k+\alpha,\infty}(\Omega)$ if $\alpha \in]0, 1[$; here $\alpha \neq 0, 1$ iff $r \notin \mathbf{N}$. Thus (2.1) also holds for $p \geq 1$, $q = \infty$ and $r \notin \mathbf{N}$. Therefore exceptions only occur in the following cases:
(i) in (2.1) if $p > 1$, $q = \infty$, $r \in \mathbf{N}$ (i.e. in (2.2) if $p > 1$, $\alpha = 0$ or 1);
(ii) in (2.1) if $p = 1$, $q < \infty$, $r \notin \mathbf{N}$, $s \in \mathbf{N}$.

The preceding result also holds for Ω unbounded (which requires only few modifications in the definition of the preceding spaces) and possibly fulfilling weaker regularity conditions.

For any $h \in \mathbf{R}^N$, by δ_h we denote the increment operator $v \mapsto v(\cdot + h) - v$, and set $D + h := \{x + h : x \in D\}$.

Proposition 2.2 [11] *(i) $w \in H^1(\Omega)$ iff $w \in L^2(\Omega)$ and*

$$\exists C > 0 : \forall D \subset\subset \Omega, \forall h \in \mathbf{R}^N, \quad D + h \subset \Omega \Rightarrow \|\delta_h w\|_{L^2(D)} \leq C|h|. \tag{2.4}$$

---

[10] See, e.g., Adams [3; Chap. V], Alt [11; Chaps. A5, 8], Brézis [92; Sect. IX.3], Brezzi and Gilardi [96; Sect. 2.7], Kufner, John and Fučik [330; Chap. 5], Lions [348; Chap. II], Maz'ja [381; Chaps. 1, 5], Morrey [404; Sect. 1.3.5], Nečas [414; Sect. 2.3], [415; Sect. 2.2], Sobolev [511; Chap. 1], and Wloka [580; Sect. 6].
[11] See, e.g., Brézis [92; Sect. IX.1].

The optimal constant is $C = \|\nabla w\|_{L^2(\Omega;\mathbf{R}^N)}$.

(ii) $w \in BV(\Omega)$ *iff* $w \in L^1(\Omega)$ *and*

$$\exists C > 0 : \forall D \subset\subset \Omega, \forall h \in \mathbf{R}^N, \quad D + h \subset \Omega \Rightarrow \|\delta_h w\|_{L^1(D)} \leq C|h|. \quad (2.5)$$

The optimal constant is $C = \int_\Omega |\nabla w|$.

Traces. For any $s > 0$ and $p \in [1, +\infty]$, if $\Gamma := \partial\Omega$ is of class $C^{[s],1}$ ($[s] :=$ integer part of s), the Sobolev space $W^{s,p}(\Gamma)$ can also be defined via the local Cartesian representation of Γ. [12]

Theorem 2.3 *(Traces in Sobolev Spaces) Let* $1 \leq p \leq +\infty$, $s > 1/p$ *and* Γ *be of class* $C^{[s],1}$. *Then:*

(i) *Assume that* $s - 1/p \notin \mathbf{N}$ *if* $2 < p < +\infty$. *Then there exists a unique linear and continuous "trace" operator* $\gamma_0 : W^{s,p}(\Omega) \to W^{s-1/p,p}(\Gamma)$, *such that* $\gamma_0 v = v_{|\Gamma}$ *for any* $v \in W^{s,p}(\Omega) \cap C^\infty(\bar\Omega)$. *(This conclusion also holds if* $p = 1$ *and* $s \geq 1$.)

(ii) *Assume that* $s - 1/p \notin \mathbf{N}$ *if* $1 \leq p < 2$. *Then there exists a linear and continuous "lift" operator* $\mathcal{R} : W^{s-1/p,p}(\Gamma) \to W^{s,p}(\Omega)$ *such that* $\gamma_0 \mathcal{R} v = v$ *for any* $v \in W^{s-1/p,p}(\Gamma)$. *(This conclusion also holds if* $p = s = 1$.)

Obviously $\mathcal{R}$ is nonunique. Higher order trace operators representing normal derivatives can also be introduced. Let $k \in \mathbf{N}$, Γ be of class C^k, denote the outward unit normal vector by $\vec\nu$, and set $D_{\vec\nu}^k := \left(\sum_{i=1}^N \nu_i D_i\right)^k$.

Theorem 2.4 *(Traces of Normal Derivatives) Assume that* $1 \leq p \leq +\infty$, $k \in \mathbf{N}$, $s > k + 1/p$, $s - 1/p \notin \mathbf{N}$, *and* Γ *is of class* $C^{[s],1}$. *Then:*

(i) *For* $j = 0, \ldots, k$ *there exists a unique linear and continuous operator* $\gamma_j : W^{s,p}(\Omega) \to W^{s-j-1/p,p}(\Gamma)$, *such that* $\gamma_j v = \left(D_{\vec\nu}^k v\right)\big|_\Gamma$ *for any* $v \in W^{s,p}(\Omega) \cap C^\infty(\bar\Omega)$.

(ii) *There exists a linear and continuous operator*

$$\mathcal{R} : X := W^{s-1/p,p}(\Gamma) \times \cdots \times W^{s-k-1/p,p}(\Gamma) \to W^{s,p}(\Omega)$$

such that $\gamma_j \mathcal{R} v = v_j$ *for any* $v := (v_0, \ldots, v_k) \in X$ *and* $j = 0, \ldots, k$.

Theorem 2.5 *(Characterization of* $W_0^{s,p}(\Omega)$) [13] *Assume that* $1 < p < +\infty$, $k \in \mathbf{N}$, $s > k + 1/p$, $s - 1/p \notin \mathbf{N}$, *and* Γ *is of class* $C^{[s],1}$. *Then:*

$$W_0^{s,p}(\Omega) = \{v \in W^{s,p}(\Omega) : \gamma_j u = 0 \text{ for } j = 0, \ldots, k\}. \quad (2.6)$$

---

[12] For this construction and the two following theorem, see, e.g., Adams [3; Chap. VII], Alt [11; p. 168], Baiocchi and Capelo [47; Chap. 5], Brezzi and Gilardi [96; Sects. 2.8, 2.9], Kufner, John and Fučik [330; Chap. 6], Lions [348; Chap. III], Lions and Magenes [356; vol. I, Chap. 2], Nečas [414; Sects. 2.4, 2.5], [415; Sect. 2.4], and Wloka [580; Sect. 8].

[13] See, e.g., Baiocchi and Capelo [47; Sect. 5.4], Lions [348; Chap. III], and Nečas [414; Sect. 2.4].

For instance, $H_0^1(\Omega) = \{v \in H^1(\Omega) : \gamma_0 v = 0 \text{ a.e. on } \Gamma\}$.
In the next statement we equip $BV(\Omega)$ with the metric

$$d(u,v) := \int_\Omega |u-v|dx + \Big| \|\nabla u\|_{C_c^0(\Omega;\mathbf{R}^N)'} - \|\nabla v\|_{C_c^0(\Omega;\mathbf{R}^N)'} \Big| \qquad (2.7)$$

$$\forall u, v \in BV(\Omega).$$

This induces a topology that is intermediate between the strong and weak star topologies. [14]

Theorem 2.6 *(Traces in $BV(\Omega)$)* [15] *Assume that Γ is of Lipschitz class, and let $BV(\Omega)$ be equipped with the metric d. Then there exists a unique linear and continuous operator $\gamma_0 : BV(\Omega) \to L^1(\Gamma)$, such that $\gamma_0 v$ equals the restriction of v to Γ for any $v \in C^0(\bar{\Omega}) \cap BV(\Omega)$.*

Interpolation of Sobolev Spaces. We only recall a simple example of interpolation of function spaces. [16]

Theorem 2.7 *(Interpolation of Vector-Valued Sobolev Spaces) For any $r_i, s_i \in \mathbf{R}$ ($i = 1, 2$) and $\lambda \in [0,1]$, setting $r = \lambda r_1 + (1-\lambda) r_2$ and $s = \lambda s_1 + (1-\lambda) s_2$, we have*

$$H^{r_1}(0,T; H^{s_1}(\Omega)) \cap H^{r_2}(0,T; H^{s_2}(\Omega)) \subset H^r(0,T; H^s(\Omega)) \qquad (2.8)$$

with continuous injection.

---

[14] Incidentally we notice that $X := C^\infty(\Omega) \cap BV(\Omega)$ is dense in $BV(\Omega)$ with respect to the topology induced by the metric (2.7), but not with respect to the strong topology of $BV(\Omega)$, cf. Anzellotti and Giaquinta [24]. In fact the closure of X in the latter topology coincides with $W^{1,1}(\Omega)$.

The metric (2.7) is not induced by any norm, since it is not invariant for translations. However, in several variational problems the convergence of the norms is a more natural property than the convergence in norm. For instance, this occurs for any sequence that minimizes a functional equal to the sum of a norm and a continuous term.

[15] See Anzellotti and Giaquinta [24], Giusti [268; Chap. 2], Evans and Gariepy [218; Sect. 5.3], and Ziemer [597; Sect. 5.10].

[16] For this theory see, e.g., Bergh and Löfström [72], Lions and Magenes [356], and Triebel [536].

XI.3 Compactness

In this section we review some classical compactness results for function spaces.

A subset of a topological space is said to be *relatively compact* iff its closure is compact. On the other hand, a mapping between two Banach spaces is also said to be *compact* iff it maps bounded sets onto relatively strongly compact sets. [17]

Let us review some classical properties of Banach spaces.

Theorem 3.1 [18] *(i) (Weak Star Compactness) Any bounded subset of the dual of a Banach space is relatively weakly star compact.*

(ii) (Weak Star Sequential Compactness) Any bounded subset of the dual of a separable Banach space is relatively weakly star sequentially compact.

(iii) (Weak Sequential Compactness) Any bounded subset of a reflexive Banach space is relatively weakly sequentially compact.

The next theorem plays a key role in the derivation of several compactness results based on control of the increments.

Theorem 3.2 *(Ascoli-Arzelà)* [19] *Let K be a compact metric space equipped with the metric d. $\mathcal{F} \subset C^0(K)$ is relatively strongly compact iff it is bounded and*

$$\sup_{f \in \mathcal{F}} \max_{d(x,y) \leq h} |f(x) - f(y)| \to 0 \quad \text{as } h \to 0 \text{ (equicontinuity)}. \tag{3.1}$$

The argument of the "if" part is based on the *total boundedness* of K, the relative compactness of bounded subsets of $\mathbf{R}$, and a diagonalization procedure. If B is a Banach space, this result can easily be extended to $\mathcal{F} \subset C^0(K; B)$, provided that $\{f(x) : f \in \mathcal{F}\}$ is relatively strongly compact in B for any $x \in K$.

Theorem 3.3 *(Fréchet-Riesz-Kolmogorov)* [20] *Let Ω be a measurable subset of $\mathbf{R}^N$ and $p \in [1, +\infty[$. $\mathcal{F} \subset L^p(\Omega)$ is relatively strongly compact iff it is bounded and (setting functions equal to 0 outside Ω)*

$$\sup_{f \in \mathcal{F}} \|f(\cdot + h) - f\|_{L^p(\Omega)} \to 0 \quad \text{as } h \to 0, \tag{3.2}$$

---

[17] Compact sets are *essentially* finite dimensional, in the sense illustrated by the following result, see, e.g., Diestel [194; p. 3].

Theorem. Let K be a (strongly) compact subset of a normed space B. Then there is a sequence $\{x_n\} \subset B$ such that $K \subset \overline{co}\{x_n\}$ *(closed convex hull)* and $x_n \to 0$ strongly in B.

[18] See, e.g., Brézis [92; Sect. 3.4], Dunford and Schwartz [202; vol. I, V.4 – V.6], and Mukherjea and Pothoven [405; vol. II, Sect. 6.4].

[19] See, e.g., Alt [11; p. 69], Dunford and Schwartz [202; vol. I, IV.6], Edwards [206; Sect. 0.4], Kufner, John and Fučik [330; Sect. 1.6], Mukherjea and Pothoven [405; vol. I, Sect. 1.6], Reed and Simon [463; Sect. 1.5], and Yosida [589; p. 85].

[20] See, e.g., Alt [11; p. 70], Brézis [92; Sect. IV.5], Edwards [206; 4.20], Kufner, John and Fučik [330; Sect. 2.13], and Yosida [589; Sect. X.1].

$$\sup_{f \in \mathcal{F}} \int_{\Omega \setminus \tilde{\Omega}} |f(x)|^p dx \to 0 \quad as \ \tilde{\Omega} \to \Omega, \ with \ |\tilde{\Omega}| < +\infty. \tag{3.3}$$

The proof is based on regularization and application of the Ascoli-Arzelà theorem. It is then easy to see that this result can be extended to $\mathcal{F} \subset L^p(\mathbf{R}^N; B)$, where B is a Banach space, provided that $\{f(x) : f \in \mathcal{F}\}$ is relatively strongly compact in B for a.a. $x \in \mathbf{R}^N$.

Theorem 3.4 [21] *Let Ω be an open bounded subset of $\mathbf{R}^N$ of Lipschitz class, $0 \leq r < s$, $1 \leq p \leq q < +\infty$, $k_i \in \mathbf{N}_0$, and $0 \leq \alpha_i \leq 1$ ($i = 1, 2$). Then:*

(i) (Rellich-Kondrachov) If either in (2.1) or in (2.2) the inequality is strict, then the respective injection is compact.

(ii) If $k_1 + \nu_1 < k_2 + \nu_2$, then $C^{k_2,\nu_2}(\bar{\Omega}) \subset C^{k_1,\nu_1}(\bar{\Omega})$ with continuous and compact injection.

Inclusion and compactness results can also be derived for $BV(\Omega)$, for any open bounded set Ω. Indeed $W^{1,1}(\Omega) \subset BV(\Omega) \subset W^{s,1}(\Omega)$, for any $s < 1$, with continuous injection. [22]

The theory of space interpolation provides further compactness results; here is an example. We recall that $B_0 \subset\subset B$ means that any bounded subset of B_0 is relatively strongly compact in B.

Theorem 3.5 *Let B, B_0, B_1 be Banach spaces, B_0, B_1 be reflexive, $B_0 \subset\subset B \subset B_1$ with continuous injections, and $1 < p_i < +\infty$ ($i = 0, 1$). Then:*

(i) (Ehrling)

$$\forall \varepsilon > 0, \exists C_\varepsilon > 0 : \forall v \in B_0, \ \|v\|_B \leq \varepsilon \|v\|_{B_0} + C_\varepsilon \|v\|_{B_1}. \tag{3.4}$$

(ii) (Lions-Aubin) [23]

$$L^{p_0}(0, T; B_0) \cap W^{1,p_1}(0, T; B_1) \subset L^{p_0}(0, T; B), \tag{3.5}$$

with compact injection.

Some Results of Integration Theory. We consider a measurable set $\Omega \subset \mathbf{R}^N$, and denote the ordinary N-dimensional Lebesgue measure by $|\cdot|$. However, the

---

[21] See, e.g., Alt [11; p. 159], Adams [3; Chaps. V, VI], Brézis [92; Sect. IX.3], Kufner, John and Fučik [330; Chap. 5], Lions [348; Chap. II], Maz'ja [381; Sect. 1.4], Morrey [404; Sect. 1.3.4], Nečas [414; Sect. 2.6], [415; Sect. 2.3], Sobolev [511; Chap. 1], and Wloka [580; Sect. 7].

[22] For the compactness of the inclusion $BV(\Omega) \subset L^1(\Omega)$, see also, e.g., Giusti [268; p. 17], Evans and Gariepy [218; p. 188], and Ziemer [597; Sect. 5.3].

[23] See Aubin [38], Lions [351; Sect. 1.5], and Simon [506].

following results hold in a more general setting, for instance, for vector-valued functions defined in a σ-finite measure space.

Parts (ii) and (iii) of Theorem 3.1 provide simple criteria for weak compactness in $L^p(\Omega)$, for any $1 < p \leq +\infty$. The case of $L^1(\Omega)$ is more interesting.

Theorem 3.6 *(Weak Compactness in $L^1(\Omega)$) A set $\mathcal{F} \subset L^1(\Omega)$ is relatively weakly compact iff any of the following conditions is fulfilled*
(i) $\mathcal{F}$ is relatively sequentially weakly compact; [24]
(ii) (Dunford-Pettis) [25] *$\mathcal{F}$ is bounded and*

$$\sup_{f \in \mathcal{F}} \int_A |f(x)| dx \to 0 \quad \text{as } |A| \to 0 \text{ (equiintegrability)}, \tag{3.6}$$

$$\sup_{f \in \mathcal{F}} \int_{\Omega \setminus \tilde{\Omega}} |f(x)| dx \to 0 \quad \text{as } \tilde{\Omega} \to \Omega, \text{ with } |\tilde{\Omega}| < +\infty; \tag{3.7}$$

(iii) (De la Vallée Poussin) (3.7) is fulfilled and there exists a Borel function $\psi : \mathbf{R}^+ \to \mathbf{R}^+$ such that

$$\frac{\psi(t)}{t} \to +\infty \quad \text{as } t \to +\infty, \quad \max_{f \in \mathcal{F}} \int_\Omega \psi(|f(x)|) dx < +\infty; \tag{3.8}$$

(iv) (3.7) is fulfilled and

$$\sup_{f \in \mathcal{F}} \int_{\{x \in \Omega : |f(x)| > m\}} |f(x)| dx \to 0 \quad \text{as } m \to +\infty \tag{3.9}$$

(uniform integrability);

(v) $\mathcal{F}$ is bounded and

$$\sup_{f \in \mathcal{F}} \int_{A_k} |f(x)| dx \to 0 \text{ as } k \to \infty, \text{ for any nonincreasing} \tag{3.10}$$

sequence $\{A_k\}$ of measurable subsets of Ω such that $\left| \bigcap_{k=1}^\infty A_k \right| = 0$.

---

[24] This follows from the following classical result, see, e.g., Beauzamy [62; p. 61], Day [177; Sect. III.2], Diestel [194; Chap. III], Dunford and Schwartz [202; vol. I, Sect. V.6], Edwards [206; Sect. 8.12], and Holmes [298; Sect. III.18].

Theorem. (Eberlein-Šmulian) In any Banach space, a set is relatively weakly compact iff it is relatively *sequentially* weakly compact. Hence it is weakly compact iff it is *sequentially* weakly compact. (Sometimes this theorem is stated in a different (weaker) form, see, e.g., Brézis [92; Sect. III.6], Yosida [589; p. 141], and Zeidler [591; vol. I, p. 782].)

[25] See, e.g., Beauzamy [62; Sect. 6.2], Brézis [92; Chap. IV], Diestel and Uhl [195; p. 76], Dunford and Schwartz [202; vol. I, Sect. IV.8], Edwards [206; Sect. 4.21.1], and Kufner, John and Fučik [330; Sect. 2.14].

Therefore $\mathcal{F}$ is relatively weakly compact in $L^1(\Omega)$ iff so is $\{|f| : f \in \mathcal{F}\}$. More generally we have the following simple result.

Corollary 3.7 *Let $\mathcal{F}, \mathcal{G} \subset L^1(\Omega)$ be such that*

$$\forall g \in \mathcal{G}, \exists f \in \mathcal{F} : \text{ for a.a. } x \in \Omega, |g(x)| \leq |f(x)|. \tag{3.11}$$

If $\mathcal{F}$ is relatively weakly compact in $L^1(\Omega)$, then this property also holds for $\mathcal{G}$.

To deal with relatively *strong* compactness in L^p-spaces, we need some further concepts of convergence.

Definitions. A sequence $\{f_n\}$ of measurable functions $\Omega \to \mathbf{R}$ (or $\mathbf{C}$) is said to converge to a (measurable) function f *in measure* in Ω iff

$$|\{x \in \Omega : |f_n - f| \leq \varepsilon\}| \to 0 \quad \text{as } n \to \infty, \forall \varepsilon > 0. \tag{3.12}$$

$\{f_n\}$ is said to converge to f *almost uniformly* in Ω iff

$$\forall \varepsilon > 0, \exists \Omega_\varepsilon \subset \Omega : |\Omega \setminus \Omega_\varepsilon| \leq \varepsilon, \sup_{\Omega_\varepsilon} |f_n - f| \to 0 \text{ as } n \to \infty. \tag{3.13}$$

In the next statement we gather some classical properties.

Theorem 3.8 [26] *(i) (Egoroff theorem) Almost uniform convergence implies convergence a.e.. The converse holds in any finite measure set.*

(ii) Almost uniform convergence implies convergence in measure. The converse holds for a subsequence.

(iii) Convergence a.e. entails convergence in measure in any subset of finite measure. Conversely, convergence in measure implies convergence a.e. for a suitable subsequence.

(iv) For any $p \in [1, +\infty]$, strong convergence in $L^p(\Omega)$ implies convergence in measure.

The next statement tells us that the weakest of these convergences joined with an appropriate property of weak compactness yields strong convergence.

Theorem 3.9 [27] *Let $1 \leq p < +\infty$ and $\{f_n\}$ be a sequence in $L^p(\Omega)$. Then $f \in L^p(\Omega)$ and $f_n \to f$ strongly in $L^p(\Omega)$ iff $f_n \to f$ in measure in any set $\tilde{\Omega} \subset \Omega$ of finite measure and $\{|f_n|^p\}$ is relatively weakly compact in $L^1(\Omega)$.* [28]

A simple counterexample shows that, for any $p \in]1, +\infty[$, the latter assumption cannot be replaced by the relative weak compactness (namely, the boundedness)

---

[26] See, e.g., Friedman [251; Chap. 2], Hewitt and Stromberg [290; Sect. III.11], Mukherjea and Pothoven [405; vol. I, Sect. 3.1], Kolmogorov and Fomin [327; Chap. VII], and Rudin [489; Chap. 3].

[27] See Dunford and Schwartz [202; vol. I, pp. 122, 295] and Edwards [206; Sect. 4.21.5].

[28] As we saw, for $p = 1$ the latter condition is fulfilled iff $\{f_n\}$ is relatively weakly compact in $L^1(\Omega)$.

of $\{f_n\}$ in $L^p(\Omega)$: just set $f_n(x) = n^{1/p}$ if $0 < x < 1/n$ and $f_n(x) = 0$ if $1/n < x < 1$.

Several conclusions can be drawn by coupling this result with Theorem 3.6. [29]

Although the inclusion among L^p-spaces is not compact, [30] the following result shows that the convergence in measure yields that property.

Proposition 3.10 *Let $|\Omega| < +\infty$ and $1 \leq q < p < +\infty$. If the sequence $\{f_n\}$ is bounded in $L^p(\Omega)$ and $f_n \to f$ in measure in Ω, then $f_n \to f$ strongly in $L^q(\Omega)$.*

Proof. By Egoroff's theorem, a suitable subsequence $\{f_{n'}\}$ converges almost uniformly. That is, for any $\varepsilon > 0$, there exists $\Omega_\varepsilon \subset \Omega$ such that $|\Omega \setminus \Omega_\varepsilon| \leq \varepsilon$ and $f_{n'} \to f$ uniformly in Ω_ε. Hence, by the Schwarz-Hölder inequality,

$$\int_\Omega |f - f_{n'}|^q dx \leq \int_{\Omega \setminus \Omega_\varepsilon} |f - f_{n'}|^q dx + \sup_{\Omega_\varepsilon} |f - f_{n'}|^q |\Omega_\varepsilon|$$

$$\leq \|f - f_{n'}\|^q_{L^p(\Omega \setminus \Omega_\varepsilon)} |\Omega \setminus \Omega_\varepsilon|^{(p-q)/p} + \sup_{\Omega_\varepsilon} |f - f_{n'}|^q |\Omega_\varepsilon| \to 0$$

as $n \to \infty$ and $\varepsilon \to 0$. That is, $f_{n'} \to f$ strongly in $L^q(\Omega)$.

Since such a converging sequence can be extracted from any subsequence of $\{f_n\}$, we conclude that the whole sequence converges strongly in $L^q(\Omega)$. □

The following more general result can be proved by a similar argument.

Proposition 3.11 *Let $|\Omega_i| < +\infty$ ($i = 1, 2$) and $1 \leq q < p < +\infty$, $1 \leq s < r < +\infty$. If the sequence $\{f_n\}$ is bounded in $L^p(\Omega_1; L^r(\Omega_2))$ and $f_n \to f$ in measure in $\Omega_1 \times \Omega_2$, then $f_n \to f$ strongly in $L^q(\Omega_1; L^s(\Omega_2))$.*

Some of the previous results can be extended to functions taking values in a Banach space B, equipped with the norm $\|\cdot\|$.

Here is another theorem, which is used in Sect. X.2.

Theorem 3.12 [31] *Let B be a Banach space and X a strongly dense subset of B'. If $\{u_n \in B\}$ is a bounded sequence and*

$$\langle f, u_n \rangle \to \langle f, u \rangle \qquad \forall f \in X, \tag{3.14}$$

---

[29] These also include the following classical result, see, e.g., Alt [11; p. 46], Dunford and Schwartz [202; vol. I, p. 150], and Kufner, John and Fučik [330; Sect. 2.1].

Theorem. (Vitali) Let $f_n \to f$ a.e. in Ω and $1 \leq p < +\infty$. $f \in L^p(\Omega)$ and $f_n \to f$ strongly in $L^p(\Omega)$ iff the sequence $\{g_n := |f_n|^p\}$ is bounded in $L^1(\Omega)$ and fulfills (3.6), (3.7).

[30] Incidentally we notice the following result.

Theorem. (Zolezzi [598]) Let Ω be of finite measure. If $u_n \to u$ weakly in $L^\infty(\Omega)$, then $u_n \to u$ strongly in $L^p(\Omega)$ for any $p \in [1, +\infty[$.

[31] See, e.g., Dunford-Schwartz [202; vol. I, p. 291] and Kufner, John and Fučik [330; Sect. 0.15].

then $u_n \to u$ weakly in B.

Finally, we note a result of *compensated compactness*. (Here $\nabla \cdot := \text{div}$ and $\nabla \times := \text{curl}$.)

Theorem 3.13 *(Div-Curl Lemma)* [33] *If*

$$u_n \to u, \quad w_n \to w \quad \text{weakly in } L^2\left(\Omega; \mathbf{R}^3\right), \tag{3.15}$$

$$\|\nabla \cdot u_n\|_{L^2(\Omega)}, \|\nabla \times w_n\|_{L^2(\Omega;\mathbf{R}^3)} \leq \text{Constant}, \tag{3.16}$$

then

$$\int_\Omega u_n \cdot w_n \varphi dx \to \int_\Omega u \cdot w \varphi dx \quad \forall \varphi \in C_c^0(\Omega). \tag{3.17}$$

XI.4 Convexity

In this section we review the *Legendre transformation*, some properties of convex and lower semicontinuous functions in Banach spaces, the *Legendre-Fenchel transformation*, the concept of subdifferential, the *Gâteaux differential*, and a generalization of the subdifferential. What follows is mainly a selection of results from Ekeland and Temam [207; Chap. I]. Further references can be found in the *Book Selection*.

The Legendre Transformation. [33] Let $u : D\left(\subset \mathbf{R}^N\right) \to \mathbf{R}$ be of class C^1 and set $S := \{(x, u(x)) : x \in D\}$. The tangent hyperplane to S at a generic point $(x, u(x)) \in S$ is the set of all points $(\tilde{x}, \tilde{u})$ such that $\tilde{u} = \nabla u(x) \cdot (\tilde{x} - x) + u(x)$. That is,

$$\tilde{u} = \xi \cdot \tilde{x} - w, \tag{4.1}$$

where ξ and w are functions of x:

$$\begin{cases} \xi = u_x(x) \; (:= \nabla u(x)), \\ w = \xi \cdot x - u(x). \end{cases} \tag{4.2}$$

If the Hessian matrix of u is nonsingular at some point, then locally $(4.2)_1$ can be inverted: $x = x(\xi)$. Let us assume that this condition holds globally. The equation

---

[32] Cf. Murat [409] and Tartar [522]. See also, e.g., Dacorogna [160], Evans [216; Chap. 5], Struwe [517; Chap. I.3], and Tartar [523].

[33] See, e.g., Giaquinta and Hildebrandt [264; Chap. 7].

$(4.2)_2$ then reads $w(\xi) = \xi \cdot x(\xi) - u(x(\xi))$, whence $w_\xi = x + \xi \cdot x_\xi - u_x \cdot x_\xi = x$. Hence (4.2) is equivalent to

$$\begin{cases} x = w_\xi(\xi), \\ u = \xi \cdot x - w(\xi). \end{cases} \quad (4.3)$$

In conclusion, S is the hull of the hyperplanes of the form (4.1), each one being characterized by (w, ξ). The function of $w = w(\xi) := \xi \cdot x - u$ defined by (4.3) represents this hull, and so provides a *dual representation* of S. This function is called the **Legendre transform** of the function u.

Notice that if u is of class C^1 and strictly convex, then $(4.2)_1$ can be inverted globally, and $w(\xi) = \sup_y \{\xi \cdot y - u(y)\}$. In fact, $\xi \cdot x - u(x) = \sup_y \{\xi \cdot y - u(y)\}$ iff $(\partial/\partial y)[\xi \cdot y - u(y)]|_{y=x} = 0$; that is, $\xi = u_x(x)$. This is at the basis of the *Legendre-Fenchel polar transformation* we introduce in this section.

Convex and Lower Semicontinuous Functions. Although most of the following results hold in locally convex topological spaces, here we deal with a real Banach space B with topological dual B'. We denote the extended real line $\mathbf{R} \cup \{-\infty, +\infty\}$ by $\tilde{\mathbf{R}}$. For any function $F : B \to \tilde{\mathbf{R}}$, the sets $\text{Dom}(F) := \{v \in B : F(v) < +\infty\}$ and $\text{epi}(F) := \{(v, a) \in B \times \mathbf{R} : F(v) \leq a\}$ are, respectively, called the *effective domain* and the *epigraph* of F.

For any set $K \subset B$, we also define the *indicator function* of K:

$$I_K : B \to \tilde{\mathbf{R}} : v \mapsto \begin{cases} 0 & \text{if } v \in K, \\ +\infty & \text{otherwise.} \end{cases}$$

Definitions. Any set $K \subset B$ is said to be **convex** iff $\lambda v_1 + (1-\lambda)v_2 \in K$, for any $v_1, v_2 \in K$ and any $\lambda \in [0, 1]$. By convention, the empty set is also included in the class of convex sets.

Any function $F : B \to \tilde{\mathbf{R}}$ is said to be **convex** iff

$$F(\lambda v_1 + (1-\lambda)v_2) \leq \lambda F(v_1) + (1-\lambda)F(v_2) \quad \forall v_1, v_2 \in B, \forall \lambda \in [0,1]. \quad (4.4)$$

Here it is assumed that

$$(+\infty) + (-\infty) = (-\infty) + (+\infty) = +\infty. \quad (4.5)$$

Whenever the inequality (4.4) is strict for any $\lambda \in]0, 1[$ and $v_1 \neq v_2$, the function F is said to be **strictly convex**.

Any function $F : B \to \tilde{\mathbf{R}}$ is said to be strongly (weakly, resp.) **lower semicontinuous** iff for any $a \in \mathbf{R}$ the set $\{v \in B : F(v) \leq a\}$ is strongly (weakly, resp.) closed. Notice that F is strongly lower semicontinuous iff it is strongly *sequentially* lower semicontinuous; that is, $\liminf_{v \to u} F(v) \geq F(u)$ for any $u \in B$. [34]

---

[34] This holds since B endowed with the strong topology is a metric space.

Any convex function $F : B \to \tilde{\mathbf{R}}$ is said to be *proper* iff $-\infty \notin F(B)$ and $F(B) \neq \{+\infty\}$.

Proposition 4.1 *Any function $F : B \to \tilde{\mathbf{R}}$ is convex (lower semicontinuous, resp.) iff epi(F) is convex (closed, resp.).*

Any set $K \subset B$ is convex (closed, resp.) iff I_K is convex (lower semicontinuous, resp.).

The *Hahn-Banach theorem* yields the following result.

Proposition 4.2 *Any convex set $K \subset B$ is strongly closed iff it is weakly closed.*

Any convex function $F : B \to \tilde{\mathbf{R}}$ is strongly lower semicontinuous iff it is weakly lower semicontinuous.

On account of this result, one often speaks of convex lower semicontinuous functions, without specifying the topology. (However, if B is a dual space, this does not imply lower semicontinuity with respect to the weak star topology.)

Proposition 4.3 *If $\{F_i : B \to \tilde{\mathbf{R}}\}_{i \in I}$ is a family of convex (lower semicontinuous, resp.) functions, then their upper hull $F : v \mapsto \sup_{i \in I} F_i(v)$ is convex (lower semicontinuous, resp.).*

Let $\Gamma(B)$ be the class of the functions $F : B \to \tilde{\mathbf{R}}$ that are the upper hull of a family of continuous affine functions $B \to \mathbf{R}$.

Proposition 4.4 *$\Gamma(B)$ consists of the class $\Gamma_0(B)$ of convex lower semicontinuous and proper functions, and of the two functions which are identically equal to either $+\infty$ or $-\infty$.*

For instance, the function $F : \mathbf{R} \to \tilde{\mathbf{R}}$, $F(x) := -\infty \; \forall x \leq 0$, $F(x) := +\infty$ $\forall x > 0$ is convex and lower semicontinuous, but $F \notin \Gamma(B)$.

The Legendre-Fenchel Transformation. Let $F : B \to \tilde{\mathbf{R}}$. The function

$$F^* : B' \to \tilde{\mathbf{R}} : u^* \mapsto \sup_{u \subset B} \{\langle u^*, u \rangle - F(u)\} \tag{4.6}$$

is called the (convex) **conjugate** (or **polar**) **function** of F. In turn, the conjugate function of F^*

$$F^{**} : B \to \tilde{\mathbf{R}} : u \mapsto \sup_{u^* \in B'} \{\langle u^*, u \rangle - F^*(u^*)\} \tag{4.7}$$

is called the **biconjugate** (or **bipolar**) **function** of F.

(Here the duality between B' and B is considered, rather than that between B' and B''.)

Theorem 4.5 *For any* $F : B \to \tilde{\mathbf{R}}$, $F^* \in \Gamma(B')$; $F^{**} \leq F$; $F^{**} = F$ *iff* $F \in \Gamma(B)$; $(F^*)^{**} = F^*$.

The conjugacy transformation is a bijection between $\Gamma(B)$ and $\Gamma(B')$, as well as between $\Gamma_0(B)$ and $\Gamma_0(B')$.

Definition. Let $F : B \to \tilde{\mathbf{R}}$ be convex. We define its **subdifferential** $\partial F : \text{Dom}(F) \subset B \to 2^{B'}$ (the power set) as follows:

$$\partial F(u) := \{u^* \in B' : \langle u^*, u - v\rangle \geq F(u) - F(v), \forall v \in B\} \quad (4.8)$$
$$\forall u \in B, F(u) \in \mathbf{R},$$

cf. Fig. 1. $\partial F^*(u^*)(\subset B)$ is similarly defined.

Note that $\partial F(u) = \emptyset$ is not excluded, so that one can take the subdifferential even of either nonconvex or non-lower-semicontinuous functions at any point of their domain.

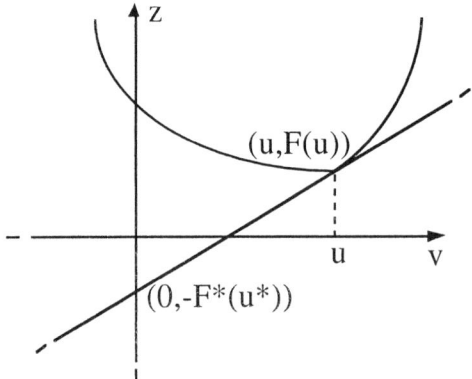

Figure 1. The straight line represents a *supporting hyperplane* to the epigraph of F. It is characterized by $z = \langle u^*, v - u\rangle + F(u)$, or equivalently by $z = \langle u^*, v\rangle - F^*(u^*)$, where $u^* \in \partial F(u)$.

Theorem 4.6 *Let* $F : B \to \tilde{\mathbf{R}}$. *Then for any* $u \in B$ *and any* $u^* \in B'$:
(i) $F(u) + F^*(u^*) \geq \langle u^*, u\rangle$; $u^* \in \partial F(u)$ iff $F(u) + F^*(u^*) = \langle u^*, u\rangle$;
(ii) $u^* \in \partial F(u)$ only if $u \in \partial F^*(u^*)$; the converse holds if $F(u) = F^{**}(u)$, in particular $\partial F^* = (\partial F)^{-1}$ if $F \in \Gamma_0(B)$;
(iii) $\partial F(u)$ is convex and weakly star closed in B'; $\partial F^*(u^*)$ is convex and weakly closed in B;
(iv) $F(u) = \inf F \in \mathbf{R}$ iff $0 \in \partial F(u)$;
(v) $\partial F(u) \neq \emptyset$ only if $F(u) = F^{**}(u)$;
(vi) $F(u) = F^{**}(u)$ only if $\partial F(u) = \partial F^{**}(u)$ (possibly $= \emptyset$).

The two latter properties cannot be inverted in general.

Theorem 4.7 *Let $F_1, F_2 : B \to \tilde{\mathbf{R}}$. Then*

$$\partial F_1(u) + \partial F_2(u) \subset \partial (F_1 + F_2)(u) \qquad \forall u \in \text{Dom}(F_1) \cap \text{Dom}(F_2). \quad (4.9)$$

The opposite inclusion holds if F_1, F_2 are both convex and lower semicontinuous, and either F_1 or F_2 is continuous at some point $u_0 \in \text{Dom}(F_1) \cap \text{Dom}(F_2)$.

The latter assumption cannot be dropped. As a counterexample, take $B = \mathbf{R}$, $F_1(x) := +\infty$ for any $x < 0$, $F_1(x) := -\sqrt{x}$ for any $x \geq 0$, $F_2(x) := F_1(-x)$ for any $x \in \mathbf{R}$. Then $(F_1 + F_2)(0) = 0$ and $(F_1 + F_2)(x) = +\infty$ for any $x \neq 0$; hence $\partial (F_1 + F_2)(0) = \mathbf{R}$, whereas $\partial F_1(0) + \partial F_2(0) = \emptyset + \emptyset = \emptyset$.

Theorem 4.8 *Let $F : B \to \tilde{\mathbf{R}}$ be convex, proper and bounded from above in an open subset of B. Then the interior of $\text{Dom}(F)$ is nonempty, there F is locally Lipschitz continuous, and $\partial F \neq \emptyset$.*

Definitions. $F : B \to \tilde{\mathbf{R}}$ is said to be (strongly) **Gâteaux differentiable** at $u \in B$ iff

$$\exists u^* \in B' : \forall v \in B, \quad \frac{F(u + \lambda v) - F(u)}{\lambda} \to \langle u^*, v \rangle \quad \text{as } \lambda \to 0. \quad (4.10)$$

Such u^* is necessarily unique; it is called the (strong) *Gâteaux differential* of F at u, and is denoted by $F'(u)$.

$F : B \to \tilde{\mathbf{R}}$ is said to be **Fréchet differentiable** at $u \in B$ iff

$$\exists u^* \in B' : \frac{1}{\|v\|} \left[F(u + v) - F(u) - \langle u^*, v \rangle \right] \to 0 \quad \text{as } \|v\| \to 0. \quad (4.11)$$

Such u^* is necessarily unique, it is called the *Fréchet differential* of F at u, and is also denoted by $F'(u)$.

If F is *Fréchet differentiable*, then it is also *Gâteaux differentiable* and the two differentials coincide, but the converse may fail even for a finite dimensional B.

Proposition 4.9 *Let $F : B \to \tilde{\mathbf{R}}$ be convex, and $u \in \text{Dom}(F)$. If F is Gâteaux differentiable at u, then $\partial F(u) = \{F'(u)\}$. Conversely, if F is continuous at u and $\partial F(u)$ is a singleton, then it is Gâteaux differentiable at u and $\partial F(u) = \{F'(u)\}$.*

Proposition 4.10 *Let U be a convex subset of B, and $F : U \to \mathbf{R}$ be Gâteaux differentiable at any point. Then F is convex iff $F' : B \to B'$ is monotone; that is,*

$$\langle u_1 - u_2, F'(u_1) - F'(u_2) \rangle \geq 0 \qquad \forall u_1, u_2 \in B. \quad (4.12)$$

Proposition 4.11 *Let B' be a separable Banach space. Assume that $F : B \to \tilde{\mathbf{R}}$ is convex and lower semicontinuous, and*

$$u \in H^1(0, T; B), \qquad w \in L^2(0, T; B'), \quad (4.13)$$

$$w \in \partial F(u) \qquad \text{a.e. in }]0,T[. \tag{4.14}$$

Then

$$F(u) \in W^{1,1}(0,T), \qquad \frac{d}{dt}F(u) = \left\langle w, \frac{du}{dt} \right\rangle \qquad \text{a.e. in }]0,T[. \tag{4.15}$$

Proof. Note that $L^2(0,T;B)' = L^2(0,T;B')$, as B' is separable. [35]

For any $[t',t''] \subset [0,T]$, let us consider a generic finite (increasingly ordered) family $\{t_n\}_{n=0,\ldots,m} \subset [t',t'']$ of continuity points of w. We have

$$\langle w(t_{n-1}), u(t_n) - u(t_{n-1}) \rangle \leq F(u(t_n)) - F(u(t_{n-1})) \leq \langle w(t_n), u(t_n) - u(t_{n-1}) \rangle$$

$$\text{for } n = 1, \ldots, m.$$

Summing for $n = 1,\ldots,m$ and taking the limit as $\max_n\{t_n - t_{n-1}\} \to 0$, we get $F(u(t'')) - F(u(t')) = \int_{t'}^{t''} \langle w, du/dt \rangle \, dt$. This is equivalent to (4.15). □

Another Concept of Subdifferential. Let $F : B \to \tilde{\mathbf{R}}$. We define the operator $\partial^- : \mathrm{Dom}(F) \subset B \to 2^{B'}$ as follows [36]

$$\partial^- F(u) := \left\{ u^* \in B' : \liminf_{v \to u} \frac{F(v) - F(u) - \langle u^*, v - u \rangle}{\|v - u\|_B} \geq 0 \right\} \tag{4.16}$$

$$\forall u \in B, F(u) \in \mathbf{R};$$

here v is assumed to converge to u strongly in B.

Proposition 4.12 *Let $F : B \to \tilde{\mathbf{R}}$. Then $\partial F(u) \subset \partial^- F(u)$ for any $u \in B$; the equality holds if $F(u) = F^{**}(u)$.*

The operator ∂^- also extends the Fréchet differential (where it exists), and for convex functions also the Gâteaux differential (where it exists).

Let us now briefly interpret the condition $\partial^- F(u) \ni 0$. This holds if u is either a stationary or a minimum point for the functional F (in a neighbourhood of u); or if F is the sum of two functionals, $F = F_1 + F_2$, such that u is a minimum point of F_1 and a stationary point of F_2. The preceding condition is even more general: for instance, u may be a minimum point of F with respect to some directions of B, and be just stationary with respect to others. It is not excluded that u be a (stationary) *maximum* point for F.

---

[35] See, e.g., Diestel and Uhl [195; Chap. III], and Kufner, John and Fučik [330; Sect. 2.22].
[36] See De Giorgi, Marino and Tosques [181].

XI.5 Monotonicity

In this section we review the concept of maximal monotone operator, on the basis of Brézis [90]; see also Brézis [87] and Pazy [447]. Further references can be found in the *Book Selection*.

The Banach Setting. As the inverse of a nondecreasing real function may be multi-valued, we deal with possibly multi-valued monotone functions or operators. (In this context, by *monotone* one means nondecreasing).

We denote by B a (real) Banach space and by B' its topological dual. It is convenient to identify an operator $A : B \to 2^{B'}$ (the power set) with its graph $\{(u, w) \in B \times B' : w \in A(u)\}$. The terms *operator* and *graph* are used equivalently.

Definitions. Let $A : B \to 2^{B'}$. The sets $\mathrm{Dom}(A) := \{v \in B : A(v) \neq \emptyset\}$ and $R(A) := \bigcup_{v \in B} A(v)$ are, respectively, called the *domain* and the *range* of A.

A is said to be **monotone** iff

$$_{B'}\langle u^* - v^*, u - v \rangle_B \geq 0 \qquad \forall u, v \in B, \forall u^* \in A(u), \forall v^* \in A(v). \tag{5.1}$$

A monotone operator A is said to be **maximal monotone** iff it is not properly included in any other monotone operator $B \to 2^{B'}$, or equivalently iff

$$\begin{gathered} u \in \mathrm{Dom}(A), u^* \in A(u) \iff \\ _{B'}\langle u^* - v^*, u - v \rangle_B \geq 0 \qquad \forall v \in \mathrm{Dom}(A), \forall v^* \in A(v). \end{gathered} \tag{5.2}$$

By Zorn's lemma, any monotone operator can be extended to a maximal monotone operator.

The inverse of a monotone operator is defined as the operator that has the inverse graph: $w \in A^{-1}(u)$ iff $u \in A(w)$. This definition is at variance with that of inverse of a function; for example, $A \circ A^{-1}$ may be different from the identity. A is maximal monotone iff A^{-1} is maximal monotone.

Composing two (maximal) monotone operators does not necessarily yield a monotone operator. For instance, let us set

$$\mathrm{sign}(x) := \{-1\} \text{ if } x < 0, \quad \mathrm{sign}(0) := [-1, 1], \quad \mathrm{sign}(x) := \{1\} \text{ if } x > 0; \tag{5.3}$$

set also $A : \mathbf{R} \to 2^{\mathbf{R}} : x \mapsto x + \mathrm{sign}(x)$, The graph of the operator $A \circ A^{-1}$ includes the whole square $[-1, 1]^2$, hence it is not monotone.

Maximal Monotone Operators and Subdifferentials. If $F : B \to \mathbf{R} \cup \{+\infty\}$ is a proper convex function, its *subdifferential* ∂F is a monotone operator. If F is also lower semicontinuous, then ∂F is maximal monotone.

Not any maximal monotone operator can be represented as the subdifferential of a lower semicontinuous convex function. This fails even for linear operators; as a counterexample, consider the operator $A_\varphi : \mathbf{R}^2 \to \mathbf{R}^2$ representing rotation by an angle $\varphi \in [-\pi/2, \pi/2]$. Indeed subdifferentials of convex functions are characterized by *cyclical monotonicity,* which reads

$$\forall M \in \mathbf{N}, \forall \{u_i \in B\}_{i=1,\ldots,M}, \forall \{u_i^* \in A(u_i)\}_{i=1,\ldots,M},$$
$$\sum_{i=1}^{M} {}_{B'}\langle u_{i+1}^*, u_{i+1} - u_i\rangle_B \geq 0 \quad \text{(here } u_{M+1} := u_1, u_{M+1}^* := u_1^*\text{)}. \tag{5.4}$$

The following simple result is often useful to pass to the limit in (nonlinear) maximal monotone operators.

Lemma 5.1 *Let* $\alpha : B \to 2^{B'}$ *be a maximal monotone operator. If*

$$w_n \in \alpha(u_n) \qquad \forall n \in \mathbf{N}, \tag{5.5}$$

$$u_n \to u \quad \text{weakly in } B, \qquad w_n \to w \quad \text{weakly in } B', \tag{5.6}$$

$$\liminf_{n \to \infty} {}_{B'}\langle w_n, u_n\rangle_B \leq {}_{B'}\langle w, u\rangle_B, \tag{5.7}$$

then $w \in \alpha(u)$.

Proof. (5.5) is equivalent to

$${}_{B'}\langle w_n - \eta, u_n - \xi\rangle_B \geq 0 \qquad \forall \xi \in B, \forall \eta \in \alpha(\xi), \forall n. \tag{5.8}$$

By (5.6) and (5.7), this inequality is preserved in the limit, whence $w \in \alpha(u)$. □

Theorem 5.2 [37] *Let B be reflexive, $A : B \to 2^{B'}$ be maximal monotone, and*

$$\frac{{}_{B'}\langle w, v\rangle_B}{\|v\|_B} \to +\infty \qquad \text{as } w \in A(v), \|v\|_B \to \infty. \tag{5.9}$$

Then for any $f \in B'$ there exists $u \in B$ such that $A(u) \ni f$.

The next result can be compared with Theorem 4.7.

Theorem 5.3 [38] *Let B be reflexive, $A, \tilde{A} : B \to 2^{B'}$ be maximal monotone and such that*

$$\text{int}(\text{Dom}(A)) \cap \text{Dom}(\tilde{A}) \neq \emptyset. \tag{5.10}$$

---

[37] See Barbu [53; p. 48].
[38] See Barbu [53; p. 46].

Then $A + \tilde{A}$ is maximal monotone.

The Hilbert Setting. Henceforth in this section we deal with a real Hilbert space H, equipped with the scalar product $(\cdot, \cdot)$ and the norm $\|\cdot\|$.

Theorem 5.4 *(Minty)* [39] *A monotone operator $A : H \to 2^H$ is maximal monotone iff $R(I + \lambda A) = H$ for some (equivalently, for any) $\lambda > 0$.*

Cauchy Problem. Let $A : H \to 2^H$ be a maximal monotone operator, $T > 0$, $f \in L^1(0, T; H)$, $u^0 \in \overline{\mathrm{Dom}(A)}$ (the strong closure of $\mathrm{Dom}(A)$), and consider the following initial value problem

$$\begin{cases} \dfrac{du}{dt} + A(u) \ni f & \text{in }]0, T[, \\ u(0) = u^0. \end{cases} \quad (5.11)$$

A function $u : [0, T] \to H$ is said to be a **strong solution** of $(5.11)_1$ iff it is continuous in $[0, T]$, it is absolutely continuous in any $[a, b] \subset]0, T[$ (hence strongly differentiable a.e. in $]0, T[$, as H is a Hilbert space), $u \in \mathrm{Dom}(A)$ a.e. in $]0, T[$, and $(5.11)_1$ is fulfilled a.e. in $]0, T[$.

$u : [0, T] \to H$ is said to be a **weak solution** of $(5.11)_1$ iff there exists a sequence $\{(u_n, f_n)\}$ such that u_n is a strong solution of $(5.11)_1$ with f_n in place of f for any n, and

$$\begin{aligned} u_n &\to u & \text{strongly in } C^0([0, T]; H), \\ f_n &\to f & \text{strongly in } L^1(0, T; H). \end{aligned} \quad (5.12)$$

$u : [0, T] \to H$ is said to be an **integral solution** of $(5.11)_1$ iff it is continuous in $[0, T]$; $u(t) \in \overline{\mathrm{Dom}(A)}$ for any $t \in [0, T]$; and

$$\|u(t_2) - v\|^2 \leq \|u(t_1) - v\|^2 + 2 \int_{t_1}^{t_2} (u(\tau) - v, f(\tau) - z) d\tau \quad (5.13)$$

$$\forall v \in \mathrm{Dom}(A), \forall z \in A(v), \forall [t_1, t_2] \subset [0, T].$$

These definitions are extended to the preceding Cauchy problem assuming $(5.11)_2$ to hold. The concepts of weak and integral solution are equivalent. Any strong solution is also a weak solution; conversely it can be proved that a weak solution is a strong solution, whenever it is absolutely continuous in any $[a, b] \subset]0, T[$.

In the remainder of this section, we select some results from Brézis [90; Chap. III], in view of possible application to the problems that we studied in Chap. II.

Theorem 5.5 *(i) For any $f \in L^1(0, T; H)$ and any $u^0 \in \overline{\mathrm{Dom}(A)}$, there exists one and only one weak solution of (5.11).*

---

[39] See Minty [392].

(ii) If $u^0 \in \text{Dom}(A)$ and $f \in BV(0, T; H)$, then the weak solution of (5.11) is also a strong solution and is Lipschitz continuous.

(iii) For any $f_i \in L^1(0, T; H)$ and any $u_i^0 \in \overline{\text{Dom}}(A)$ ($i = 1, 2$), denoting by u_i the corresponding weak solution of (5.11),

$$\|u_1(t) - u_2(t)\| \leq \|u_1^0 - u_2^0\| + \int_0^t \|f_1 - f_2\| d\tau \qquad \forall t > 0. \qquad (5.14)$$

Remark about the Proof. A first a priori estimate for u in $L^\infty(0, T; H)$ is *formally* obtained multiplying the equation by u, then integrating in $]0, t[$ for any $t > 0$. However, this does not suffice to prove (i), which is actually derived by using a suitable approximation, and then showing the Cauchy property for the corresponding sequence of approximate solutions.

The statement (ii) follows from a uniform estimate that formally can be obtained by taking the time differential ratio of (5.11) (or rather of an approximate equation), multiplying it by the time differential ratio of u, and integrating in t.

The (nonstrict) contraction property (5.14) for strong solutions can be derived writing the equation for u_1 and u_2, taking the difference between these identities, multiplying it by $u_1 - u_2$, and integrating in t. A simple passage to the limit then yields it for weak solutions, too. □

Remark. The estimation procedure that we pointed out for part (ii) of the latter theorem also yields the following result. Assume that $f \in BV(0, T; X)$, $u^0 \in \text{Dom}(A) \subset X$, where X is a reflexive Banach space with norm $\|\cdot\|_X$, and

$$\exists a > 0 : \forall u, v \in \text{Dom}(A), \forall \hat{u} \in A(u), \forall \hat{v} \in A(v),$$
$$(\hat{u} - \hat{v}, u - v) \geq a\|u - v\|_X^2. \qquad (5.15)$$

Then $u \in H^1(0, T; X)$. □

Semigroups. Let us now take $f = 0$ and set $T = +\infty$. By (5.14), for any $t > 0$ the operator $u^0 \mapsto u(t)$ is a (nonstrict) contraction in $\text{Dom}(A)$, hence it can be extended to $\overline{\text{Dom}}(A)$ by continuity. This extended mapping $t \mapsto S(t)$ is a *semigroup of (nonlinear) contractions* in H, that is:
  (i) $S(0) = I$ (identity in $\overline{\text{Dom}}(A)$), and $S(t_1 + t_2) = S(t_1) \circ S(t_2)$ for any $t_1, t_2 \geq 0$,
  (ii) $S(t)v \to v$ strongly in H as $t \to 0^+$ for any $v \in \overline{\text{Dom}}(A)$,
  (iii) $\|S(t)v_1 - S(t)v_2\| \leq \|v_1 - v_2\|$ for any $v_1, v_2 \in \overline{\text{Dom}}(A)$ and any $t \geq 0$.

Regularity Results. Several properties can be derived whenever A is a subdifferential.

Theorem 5.6 *Assume that $f \in L^2(0, T; H)$, $u^0 \in \overline{\text{Dom}}(A)$ and*

$$A = \partial \varphi, \quad \text{where}$$
$$\varphi : H \to \mathbf{R} \cup \{+\infty\} \text{ is proper, convex, and lower semicontinuous.} \qquad (5.16)$$

Then any weak solution u of (5.11) is a strong solution, and

$$\sqrt{t}\frac{du}{dt} \in L^2(0,T;H), \qquad \varphi(u) \in L^1(0,T) \cap W^{1,1}(\delta,T) \quad \forall \delta > 0. \tag{5.17}$$

Moreover, if $u^0 \in \mathrm{Dom}(A)$, then

$$\frac{du}{dt} \in L^2(0,T;H), \qquad \varphi(u) \in W^{1,1}(0,T). \tag{5.18}$$

Note that, by multiplying the equation by tdu/dt (du/dt, resp.) and integrating in time, a priori estimates corresponding to (5.17) ((5.18), resp.) are *formally* obtained.

Proposition 5.7 *Assume that $f \in W^{1,1}(0,T;H)$, $u^0 \in \overline{\mathrm{Dom}}(A)$ and (5.16) is fulfilled. Then the weak solution u of (5.11) is such that $tdu/dt \in L^\infty(0,T;H)$. Moreover, if $u^0 \in \mathrm{Dom}(A)$, then $du/dt \in L^\infty(0,T;H)$.*

These properties are formally derived differentiating the equation in t, and then multiplying it either by $t^2 du/dt$, or by du/dt if $u^0 \in \mathrm{Dom}(A)$.

Asymptotic Behaviour. [40] We still assume that A is a maximal monotone operator in H, but now take $T = +\infty$.

Theorem 5.8 *Assume that $f \in L^1_{\mathrm{loc}}([0,+\infty[;H)$, and* [41]

$$\operatorname*{ess\,sup}_{]t,+\infty[} \|f - f_\infty\| \to 0 \quad as\ t \to +\infty, \tag{5.19}$$

$$\exists a > 0 : \forall u,v \in \mathrm{Dom}(A), \forall \hat{u} \in A(u), \forall \hat{v} \in A(v), \\ (\hat{u} - \hat{v}, u - v) \geq a\|u - v\|^2. \tag{5.20}$$

Let u be the solution of (5.11) in $[0,+\infty[$ and set $u_\infty := A^{-1}(f_\infty)$. [42] *Then*

$$\operatorname*{ess\,sup}_{]t,+\infty[} \|u - u_\infty\| \to 0 \quad as\ t \to +\infty. \tag{5.21}$$

Moreover, if $f \in W^{1,1}(0,+\infty;H)$ and $u^0 \in \mathrm{Dom}(A)$, then

$$\operatorname*{ess\,sup}_{]t,+\infty[} \left\|\frac{du}{dt}\right\| \to 0 \quad as\ t \to +\infty. \tag{5.22}$$

---

[40] For the following results we still refer to Brézis [87], and also to Brézis [91], Bruck [103], Dafermos and Slemrod [162], and Pazy [446, 447].

[41] The condition (5.19) is obviously equivalent to the existence of a function $\hat{f} :]0,+\infty[\to H$ such that $f = \hat{f}$ a.e. in $]0,+\infty[$ and $\hat{f}(t) \to f_\infty$ strongly in H as $t \to +\infty$.

[42] By Theorem 5.4 and (5.20) there exists a (necessarily unique) $u_\infty \in H$ such that $A(u_\infty) = f_\infty$.

Theorem 5.9 *Assume that (5.16) is fulfilled and*

$$\frac{df}{dt} \in L^1(0, +\infty; H), \tag{5.23}$$

so that $f(t)$ strongly converges in H to some f_∞, as $t \to +\infty$. Assume that $f_\infty \in R(\partial\varphi)$. Let u be a weak solution of (5.11) in $]0, +\infty[$. Then (5.22) holds. Moreover, if $t\,df/dt \in L^1(0, +\infty; H)$, then

$$\operatorname*{ess\,sup}_{t>0}\, t \left\|\frac{du}{dt}(t)\right\| < +\infty. \tag{5.24}$$

Theorem 5.10 *Assume that (5.16) is fulfilled, and*

$$\exists f_\infty \in R(\partial\varphi) : f - f_\infty \in L^1(0, +\infty; H). \tag{5.25}$$

$$\{x \in H : \varphi(x) + |x|^2 \le C\} \text{ is strongly compact, } \forall C > 0. \tag{5.26}$$

Let u be a weak solution of (5.11) in $]0, +\infty[$. Then (5.21) is fulfilled and $A(u_\infty) \ni f_\infty$.

Periodic Solution. We still assume that A is a maximal monotone operator in H and $f \in L^1(0, T; H)$. We then consider the periodic problem consisting of $(5.11)_1$ coupled with the periodicity condition

$$u(0) = u(T). \tag{5.27}$$

The extension of the definitions of strong and weak solutions is straightforward.

Theorem 5.11 *If $f \in BV(0, T; H)$ and (5.20) is fulfilled, then there exists one and only one strong periodic solution. Moreover, this is Lipschitz continuous in $[0, T]$.*

Theorem 5.12 *If $f \in L^1(0, T; H)$ and A is coercive, in the sense that*

$$\exists u_0 \in H : \frac{(\hat{u}, u - u_0)}{|u|} \to +\infty \quad \text{as } \hat{u} \in A(u), |u| \to +\infty, \tag{5.28}$$

then the periodic problem has a weak solution.

Theorem 5.13 *If $f \in L^2(0, T; H)$ and (5.16) and (5.28) are fulfilled, then the periodic problem has a strong solution $u \in H^1(0, T; H)$.*

XI.6 Accretiveness

In this section we review some results of the classical theory of (nonlinear) m-accretive operators in Banach spaces, and the associated semigroups of nonlinear contractions. We refer the reader to the surveys of Crandall [145, 146, 147], Evans [215], and Pazy [449], in addition to the monographs we quoted in the *Book Selection*. See also, for example, Baillon [43], Bénilan [63, 65], Cioranescu [136], Crandall and Liggett [149], Crandall and Pazy [151, 152], Kato [313], and Pazy [448].

Definitions. Let B be a Banach space with norm $\|\cdot\|_B$. An operator $A : B \to 2^B$ is said to be **accretive** iff

$$\forall u_i \in \mathrm{Dom}(A), \forall v_i \in A(u_i)(i=1,2), \forall \lambda > 0,$$
$$\|u_1 - u_2\| \leq \|u_1 - u_2 + \lambda(v_1 - v_2)\|. \tag{6.1}$$

A is said to be **m-accretive** iff it is accretive and $R(I + \lambda A) = H$ for some $\lambda > 0$ (equivalently, for any $\lambda > 0$).

If B is a Hilbert space, B' can be identified with B, so that monotonicity and accretiveness can be compared. By Minty's Theorem 5.4, the operator A is maximal monotone iff it is m-accretive. One can also introduce the concept of *maximal accretive* operator in an obvious way, but in general maximal accretiveness is weaker than m-accretiveness, and the latter is more suited for the Cauchy problem that now we introduce.

Semi-Inner Product. In Hilbert spaces one disposes of the scalar product and of the *Riesz operator* (cf. Sect. XI.9). In Banach spaces one can generalize these concepts, defining the *semi-inner product* $\langle \cdot, \cdot \rangle_s : B^2 \to \mathbf{R}$ by

$$\langle u, v \rangle_s := \lim_{\lambda \searrow 0} \frac{\|u + \lambda v\|^2 - \|u\|^2}{2\lambda} \qquad \forall u, v \in B, \tag{6.2}$$

and the (normalized, possibly multi-valued) *duality mapping* $F : B \to B'$ by

$$F(u) := \left\{ z \in B' : {}_{B'}\langle z, u \rangle_B = \|u\|^2 = \|z\|_{B'}^2 \right\}. \tag{6.3}$$

Notice that $u \mapsto \langle u, v \rangle_s$ and F are linear iff B is a Hilbert space. One can show that
$$\langle u, v \rangle_s = \sup\{ {}_{B'}\langle z, v \rangle_B : z \in F(u)\} \qquad \forall u, v \in B, \tag{6.4}$$

and that an operator $A : \mathrm{Dom}(A) \subset B \to B$ is *accretive* iff

$$\langle u_1 - u_2, w_1 - w_2 \rangle_s \left(= \sup\{ {}_{B'}\langle z, w_1 - w_2 \rangle_B : z \in F(u_1 - u_2)\}\right) \geq 0$$
$$\forall u_i \in B, \forall w_i \in A(u_i) \ (i=1,2). \tag{6.5}$$

Cauchy Problem. Let $T > 0$, $f \in L^1(0, T; B)$, $u^0 \in \overline{\mathrm{Dom}}(A)$ (the strong closure of $\mathrm{Dom}(A)$), and consider the initial value problem

$$\begin{cases} \dfrac{du}{dt} + A(u) \ni f & \text{in }]0, T[, \\ u(0) = u^0. \end{cases} \tag{6.6}$$

A function $u : [0, T] \to B$ is said to be a **strong solution** of $(6.6)_1$ iff it is continuous in $[0, T]$, it is absolutely continuous on any $[a, b] \subset]0, T[$ and strongly differentiable a.e. in $]0, T[$, $u \in \mathrm{Dom}(A)$ a.e. in $]0, T[$, and $(6.6)_1$ is fulfilled a.e. in $]0, T[$.

The concept of **weak solution** of $(6.6)_1$ is defined as for $(5.11)_1$.

$u : [0, T] \to B$ is said to be an **integral solution** of $(6.6)_1$ iff it is continuous on $[0, T]$, $u(t) \in \overline{\mathrm{Dom}}(A)$ for any $t \in [0, T]$, and

$$\|u(t_2) - v\|^2 \leq \|u(t_1) - v\|^2 + 2 \int_{t_1}^{t_2} \langle u(\tau) - v, f(\tau) - z \rangle_s \, d\tau \tag{6.7}$$

$$\forall v \in \mathrm{Dom}(A), \forall z \in A(v), \forall [t_1, t_2] \subset [0, T].$$

The concepts of weak and integral solution are equivalent, and any strong solution is also a weak solution. These definitions are easily extended to the periodic problem, which corresponds to $(6.6)_1$ coupled with the periodicity condition

$$u(0) = u(T). \tag{6.8}$$

The following theorem can be compared with analogous results for maximal monotone operators in Hilbert spaces, which we stated in the previous section.

Theorem 6.1 *Let B be a Banach space and $A : B \to 2^B$ be an m-accretive operator. Then:*

(i) If $f \in L^1(0, T; B)$ and $u^0 \in \overline{\mathrm{Dom}}(A)$, then (6.6) has one and only one weak solution.

(ii) If u_i is the weak solution of (6.6) corresponding to $u_i^0 \in \overline{\mathrm{Dom}}(A)$ and $f_i \in L^1(0, T; B)$ $(i = 1, 2)$, then

$$\|u_1(t) - u_2(t)\| \leq \|u_1^0 - u_2^0\| + \int_0^t \|f_1(s) - f_2(s)\| ds \quad \forall t \in [0, T]. \tag{6.9}$$

(iii) If the weak solution of (6.6) is strongly differentiable in $]0, T[$, then it is a strong solution.

(iv) If $f \in BV(0, T; B)$ and $u^0 \in \mathrm{Dom}(A)$, then the weak solution of (6.6) is Lipschitz continuous in $[0, T]$.

(v) For any $n \in \mathbf{N}$, let A_n be an m-accretive operator in B, $u_n^0 \in \overline{\mathrm{Dom}}(A_n)$,

$f_n \in L^1(0, T; B)$, and denote by u_n the weak solution of the corresponding problem (6.6). Assume that

$$u_n^0 \to u^0 \quad \text{strongly in } B, \qquad f_n \to f \quad \text{strongly in } L^1(0, T; B), \qquad (6.10)$$

$$(I + A_n)^{-1}(v) \to (I + A)^{-1}(v) \qquad \text{strongly in } B, \forall v \in B. \qquad (6.11)$$

Then, denoting by u the weak solution of (6.6),

$$u_n \to u \qquad \text{strongly in } B, \text{ uniformly in } [0, T].$$

(vi) Let $f \equiv 0$. For any $t > 0$ and any $u^0 \in \text{Dom}(A)$, let u be the weak solution of (6.6). The operator $u^0 \mapsto u(t)$ can be extended by continuity into an operator $S(t)$ acting in $\overline{\text{Dom}}(A)$. The mapping $(0 \leq)t \mapsto S(t)$ is a continuous semigroup of (nonstrict) contractions over $\overline{\text{Dom}}(A)$.

(vii) If $A - aI$ is accretive for some constant $a > 0$, then the periodic problem $(6.6)_1$, (6.8) has one and only one weak solution.

A Banach space B is said to have the *Radon-Nikodým property* iff any Lipschitz continuous function $[0, 1] \to B$ is strongly differentiable a.e. in $]0, T[$. If either B is reflexive, or it is the dual of a Banach space and is separable (e.g., ℓ^1 but neither $L^1(\Omega)$ nor $L^\infty(\Omega)$), then it has that property. [43]

If B has the *Radon-Nikodým property*, $f \in BV(0, T; B)$ and $u^0 \in \text{Dom}(A)$, then the weak solution of (6.6) is a strong solution, by parts (iii) and (iv) of the preceding theorem.

If the operator A is replaced by $\tilde{A} := A + F$, where $F : B \to B$ is a Lipschitz continuous operator with Lipschitz-constant ω, $S(t)$ is a *continuous semigroup of ω-contractions*. This means that, in the definition of semigroup which we gave in the previous section, the contraction condition (iii) is replaced by

(iii)' $\quad \|S(t)v_1 - S(t)v_2\| \leq e^{\omega t}\|u^0 - v^0\| \qquad \forall v_1, v_2 \in \overline{\text{Dom}}(A), \forall t \geq 0.$

It is easy to prove the following result.

Proposition 6.2 *Assume that $f - f_\infty \in L^1(0, +\infty; B)$, and $A^{-1}(f_\infty) \neq \emptyset$. Then any weak solution of $(6.8)_1$ in $[0, +\infty[$ is uniformly bounded.*

Definitions. B is called a **Banach lattice** iff it is a Banach space and a *lattice* (i.e., an ordered set such that any finite subset admits infimum and supremum), and, setting $|u| := \sup\{u, -u\}$ and $u \leq v$ if $u = \inf\{u, v\}$, it satisfies the following conditions:

---

[43] See, e.g., Bénilan [64], Kufner, John and Fučik [330; Sect. 2.22.5], and Diestel and Uhl [195; Chap. III].

(i) if $u \leq v$ then $u + w \leq v + w$,
(ii) if $u \leq v$ and $\alpha > 0$, then $\alpha u \leq \alpha v$,
(iii) if $u \leq v$ then $-v \leq -u$,
(iv) if $|u| \leq |v|$ then $\|u\| \leq \|v\|$.

For any $u \in B$, we set $u^+ := \sup\{u, 0\}$ and $u^- := \sup\{-u, 0\}$. If B is a Banach lattice, an operator $A : B \to 2^B$ is said to be **T-accretive** iff

$$\forall u_i \in \text{Dom}(A), \forall v_i \in A(u_i) (i = 1, 2), \forall \lambda > 0,$$
$$\|(u_1 - u_2)^+\| \leq \|[u_1 - u_2 + \lambda(v_1 - v_2)]^+\|. \tag{6.12}$$

If for any $u, v \in B$

$$\|u^+\| \leq \|v^+\|, \|u^-\| \leq \|v^-\| \quad \text{imply} \quad \|u\| \leq \|v\|, \tag{6.13}$$

then any T-accretive operator in B is also accretive.

Theorem 6.3 *If B is a Banach lattice and A is m- and T-accretive, then the weak solution of problem (6.6) depends monotonically on the data. That is, if u_i is the solution corresponding to u_i^0, f_i ($i = 1, 2$) and $u_1^0 \leq u_2^0$, $f_1 \leq f_2$, then $u_1 \leq u_2$.*

XI.7 Minimization

In this section we briefly deal with the minimization of functionals defined on Banach spaces.

Compactness and Lower Semicontinuity. A family of sets is said to have the *finite intersection property* iff the intersection of any finite subfamily is nonempty. A subset K of a topological space S is said to be *compact* iff any family of closed subsets of K with the finite intersection property has nonempty intersection in K. K is said to be *sequentially compact* iff any sequence of points of K has a subsequence that converges to some point of K. In general these two compactness properties are not correlated, but in metric spaces they are equivalent.

A functional $J : S \to \mathbf{R} \cup \{+\infty\}$ is *lower semicontinuous* iff the *sublevel set* $S_a := \{x \in S : J(x) \leq a\}$ is closed for any $a \in \mathbf{R}$. J is *sequentially lower semicontinuous* iff $\liminf J(v_n) \geq J(v)$ whenever $v_n \to v$. In nonmetric spaces the latter condition is weaker than the previous one, whereas in metric spaces they are equivalent.

Theorem 7.1 *Let S be a topological space and $J : S \to \mathbf{R} \cup \{+\infty\}$. Assume that either*
  (i) there exists $\tilde{a} \in \mathbf{R}$ such that $S_{\tilde{a}}$ is nonempty and compact,

(ii) J is lower semicontinuous,

or

(iii) there exists $\tilde{a} \in \mathbf{R}$ such that $S_{\tilde{a}}$ is nonempty and sequentially compact,
(iv) J is sequentially lower semicontinuous.
Then there exists $x \in S$ such that $J(x) = \inf J$.

Proof. If (i) and (ii) are satisfied, we can use the *level method* to prove existence of a minimizer. The family $\mathcal{F} := \{S_a : a \leq \tilde{a}, S_a \neq \emptyset\}$ consists of closed subsets of $S_{\tilde{a}}$ by (ii), and the intersection of any finite subfamily is nonempty. Hence by (i) the intersection of the whole family, $\bigcap \mathcal{F}$, is nonempty. It is easy to see that $J(x) = \inf J$ for any $x \in \bigcap \mathcal{F}$.

Now we assume that (iii) and (iv) are satisfied, and use the so-called *direct method* to show existence of a minimizer. If $\inf J = \tilde{a}$, then $J(y) = \inf J$ for any $y \in S_{\tilde{a}}$. If instead $\inf J < \tilde{a}$, let $\{x_n\} \subset S_{\tilde{a}}$ be such that $J(x_n) \to \inf J$. By (iii), there exists $x \in S$ such that, possibly extracting a subsequence, $x_n \to x$. Hence $J(x) = \inf J$, by (iv). □

Corollary 7.2 *Let B be a Banach space and $J : B' \to \mathbf{R} \cup \{+\infty\}$, $J \not\equiv +\infty$ (i.e., J is not identically equal to $+\infty$). Assume that:*
(i) J is (strongly) coercive; that is, $J(v_n) \to +\infty$ whenever $\|v_n\|_{B'} \to +\infty$;
(ii) J is weakly star lower semicontinuous.
Then there exists $x \in B'$ such that $J(x) = \inf J$.

Proof. Let $x_0 \in B'$ be such that $J(x_0) < +\infty$. By (i), $\{x \in B' : J(x) \leq J(x_0)\}$ is bounded; by (ii), this set is weakly star closed; hence it is compact by Theorem 3.1(i). The preceding theorem can then be applied. □

If B is separable, by Theorem 3.1(ii), in Corollary 7.2 it suffices to assume that J is *sequentially* weakly star lower semicontinuous. Similarly, by Theorem 3.1(iii), if B is reflexive it suffices to assume that J is *sequentially* weakly lower semicontinuous.

By Proposition 4.2, then we have the following classical statement.

Theorem 7.3 *Let B be a reflexive Banach space and $J : B \to \mathbf{R} \cup \{+\infty\}$, $J \not\equiv +\infty$. Assume that:*
(i) J is coercive,
(ii) J is convex and lower semicontinuous.
Then there exists $x \in B$ such that $J(x) = \inf J$ (which is unique if J is strictly convex).

Here is another simple result, which can be useful for applications.

Theorem 7.4 *Let B, B_0 be Banach spaces and $J : B \to \mathbf{R} \cup \{+\infty\}$. Assume that:*

(i) $B \subset B_0$ with compact injection,
(ii) J is coercive with respect to the metric of B,
(iii) J is lower semicontinuous with respect to the strong topology of B_0.
Then there exists $x \in B$ such that $J(x) = \inf J$.

Proof. We can assume that $J \not\equiv +\infty$, otherwise the result would be obvious. We can also extend J setting $J(x) := +\infty$ for any $x \in B_0 \setminus B$.

Let $\{x_n\} \subset B$ and $J(x_n) \to \inf J(< +\infty)$. By (ii), this sequence is uniformly bounded in B; hence, by (i), there exists $x \in B_0$ such that, possibly extracting a subsequence, $x_n \to x$ strongly in B_0. Therefore $J(x) = \inf J$ by (iii), and $x \in B$ since $J(x) < +\infty$. □

The preceding results are easily extended to minimization over a nonempty set K, just by replacing J by $J + I_K$, where I_K is the indicator function of K (i.e., $I_K = 0$ in K, $I_K = +\infty$ outside). In fact if K is closed (convex, resp.) and J is lower semicontinuous (convex, resp.), then $J + I_K$ is also lower semicontinuous (convex, resp.).

XI.8 Geometric Measure Theory

In this section we review some well-known results about *almost minimal boundaries*. References can be found in the *Book Selection*.

The Perimeter Functional. Let Ω be a domain of $\mathbf{R}^N$ ($N \geq 2$). For any $E, F \subset \Omega$ we denote by $E \triangle F$ their symmetric difference, and by $d(x, E)$ the distance of a point x from E. Let us set $\chi_E := 1$ in E, $\chi_E := -1$ in $\Omega \setminus E$.

For any measurable function $\chi : \Omega \to \mathbf{R}$, let us define the *total variation functional* $\int_\Omega |\nabla \chi|$ as in (1.5), and set

$$P(\chi) := \begin{cases} \frac{1}{2} \int_\Omega |\nabla \chi| \ (\leq +\infty) & \text{if } |\chi| = 1 \text{ a.e. in } \Omega, \\ +\infty & \text{otherwise.} \end{cases} \quad (8.1)$$

If $|\chi| = 1$ a.e. in Ω, $P(\chi)$ is the *perimeter* in Ω of the set $\Omega^+ := \{x \in \Omega : \chi(x) = 1\}$ in the sense of Caccioppoli.[44] If the boundary $\partial \Omega^+$ of Ω^+ in Ω is of Lipschitz class, this perimeter coincides with the $(N-1)$-dimensional *Hausdorff measure* of $\partial \Omega^+$.

---

[44] See, e.g., Giusti [268; Chap. 1] and Evans and Gariepy [218; Chap. 5].

XI.8 Geometric Measure Theory

Definitions. For any $\gamma \in]0, 1]$, a set $E \subset \Omega$ of finite perimeter is said to have a locally γ-**almost minimal boundary** in Ω iff, denoting by $B_r(x)$ the closed ball of center x and radius r,

$$\exists C > 0 : \forall x \in \Omega, \forall r < d(x, \partial\Omega), \forall F \subset \Omega, \\ \text{if } E \triangle F \subset B_r(x) \text{ then } P(\chi_E) \leq P(\chi_F) + Cr^{N-1+2\gamma}. \tag{8.2}$$

The **reduced boundary**, $\partial^* E$, of a set $E \subset \Omega$ is the set of the $x \in \Omega$ such that the following *approximate inner normal vector* exists: [45]

$$\vec{\nu}_E(x) := \lim_{r \to 0} \frac{\int_{B_r(x)} \nabla \chi_E}{\int_{B_r(x)} |\nabla \chi_E|}. \tag{8.3}$$

The closure of the reduced boundary, denoted by $\partial_e E$, is called the **essential boundary** of E, and can be characterized as

$$\partial_e E = \{x \in \Omega : \forall r > 0, |B_r(x) \cap E| > 0, |B_r(x) \cap (\Omega \setminus E)| > 0\},$$

where by $|A|$ we denote the ordinary N-dimensional Lebesgue measure of the set A. Denoting the topological boundary by ∂E, obviously we have $\partial^* E \subset \partial_e E \subset \partial E$.

Almost Minimal Boundaries. Here is a classical result of geometric measure theory.

Theorem 8.1 (Almgren) [46] *Let $0 < \gamma \leq 1$ and $E \subset \Omega$ have a locally γ-almost minimal boundary. Then $\partial^* E$ is an $(N-1)$-dimensional manifold of class $C^{1,\gamma}$.* [47] *Moreover, for any $n > N - 8$, the Hausdorff n-dimensional measure of $\partial_e E \setminus \partial^* E$ vanishes. (In particular $\partial^* E = \partial_e E$ if $N < 8$.)*

For any $\eta \in L^1(\Omega)$, let us set

$$\Phi_\eta(\chi) := \begin{cases} P(\chi) - \int_\Omega \eta \chi \, dx & \forall \chi \in \text{Dom}(P), \\ +\infty & \forall \chi \in L^1(\Omega) \setminus \text{Dom}(P). \end{cases}$$

This functional can be compared with that defined in (VI.1.7).

---

[45] See, e.g., Colombini, De Giorgi and Piccinini [144; Chap. I]. Here $\int_{B_r(x)} \nabla \chi_E$ represents the total mass of the vector measure $\nabla \chi_E$. One can show that $|\vec{\nu}_E(x)| = 1$ for any $x \in \partial^* E$.

[46] See Almgren [8].

[47] That is, locally $\partial^* E$ is the Cartesian graph of a function of class $C^{1,\gamma}$.

Proposition 8.2 [48] *For any $\eta \in L^1(\Omega)$, there exists an (in general not unique) $\chi \in \mathrm{Dom}(P)$ that minimizes Φ_η.*

Proof. We can apply Proposition 7.4 with $B = BV(\Omega)$ and $B_0 = L^1(\Omega)$. The assumptions (i) and (iii) are fulfilled; see, for example, Giusti [268; pp. 7, 17] and Evans and Gariepy [218; pp. 183, 188]. □

Proposition 8.3 *(Regularity) Let $\eta \in L^p(\Omega)$ $(p > N)$, $\chi \in \mathrm{Dom}(P)$ and* [49]

$$\liminf \frac{\Phi_\eta(v) - \Phi_\eta(\chi)}{\|v - \chi\|_{L^1(\Omega)}} \geq 0 \quad \text{as } v \to \chi \text{ strongly in } L^1(\Omega). \tag{8.4}$$

Then $\partial^* \Omega^+ := \partial^* \{x \in \Omega : \chi(x) = 1\}$ is a manifold of class $C^{1,(p-N)/2p}$.

Proof. Let us set $\omega_N := \Gamma\left(\frac{1}{2}\right)^N / \Gamma(N/2 + 1)$, the measure of the unit ball of $\mathbf{R}^N$.

For any ball $B_r(x) \subset \Omega$ and any $v \in \mathrm{Dom}(P)$ such that $v = \chi$ in $\Omega \setminus B_r(x)$, we have

$$\begin{aligned}
P(\chi) - P(v) &\leq \int_\Omega \eta(\chi - v) dx + o\left(\|\chi - v\|_{L^1(\Omega)}\right) \\
&= \int_{B_r(x)} \eta(\chi - v) dx + o\left(\int_{B_r(x)} |\chi - v| dx\right) \\
&\leq 2\omega_N^{1-1/p} \|\eta\|_{L^p(B_r(x))} r^{N-1+(p-N)/p} + o\left(r^N\right),
\end{aligned} \tag{8.5}$$

where $o(\xi)/\xi \to 0$ as $\xi \to 0$. Thus $\partial^* \Omega^+$ is a locally $(p - N)/2p$-almost minimal manifold, and the thesis follows from Theorem 8.1. □

The following statement yields a property of uniform regularity.

Theorem 8.4 [50] *Let the sequence $\{\eta_n\}$ be uniformly bounded in $L^p(\Omega)$ $(p > N)$, and χ_{E_n} be a minimizer of the functional Φ_{η_n} for any n. Assume that $\chi_{E_n} \to \chi$ strongly in $L^1(\Omega)$. Then the following holds:*

(i) if a sequence $\{x_n\}$ is such that $x_n \in \partial^* E_n$ for any n and $x_n \to x \in \Omega$, then $x \in \partial^* E$;

(ii) locally the $\partial^* E_n$s are Cartesian graphs of functions φ_ns defined on a same

---

[48] Incidentally, we notice that the converse holds as well, see Barozzi, Gonzalez and Tamanini [60].

Theorem. For any $\chi \in \mathrm{Dom}(P)$, there exists $\eta \in L^1(\Omega)$ such that χ minimizes Φ_η.

[49] The equation (8.4) also reads $\partial^- \Phi_\eta(\chi) \ni 0$ in $L^\infty(\Omega)$. We refer to (4.16) for the definition of the operator ∂^- and for the interpretation of this condition, which holds in particular for any relative minimizer of Φ_η.

[50] See Massari and Pepe [380; Theorem 3] and Tamanini [519, 520].

(smooth) set $G \subset \mathbf{R}^{N-1}$ (possibly after a suitable rotation of the axes). These functions are uniformly bounded in $C^{1,(p-N)/2p}(\bar{G})$.

Theorem 8.5 [51] *Let $0 < \gamma \leq 1$, and $E \subset \Omega$ have a locally γ-almost minimal boundary in Ω. If C is as in (8.2) then*

$$\frac{\omega_{N-1}}{2N\omega_N} \leq \frac{|E \cap B_r(x)|}{\omega_N r^N} \leq 1 - \frac{\omega_{N-1}}{2N\omega_N}$$

$$\forall x \in \partial_e E, \forall r \in \left]0, \left(\frac{\gamma \omega_{N-1}}{C(N-1+2\gamma)}\right)^{1/(2\gamma)}\right]. \tag{8.6}$$

XI.9 Other Results

Here we gather some further results, which we used in this volume.

Theorem 9.1 (*Gronwall's Lemma*) [52] *Let $0 < T \leq +\infty$, and $\varphi, \alpha, \beta : [0, T[\to \mathbf{R}$ be continuous functions, with α nondecreasing and $\beta \geq 0$. If*

$$\varphi(t) \leq \alpha(t) + \int_0^t \beta(\tau)\varphi(\tau)d\tau \qquad \forall t \in [0, T[, \tag{9.1}$$

then

$$\varphi(t) \leq \alpha(t) \exp\left(\int_0^t \beta(\tau)d\tau\right) \qquad \forall t \in [0, T[. \tag{9.2}$$

(It is easy to see that the assumption (9.1) can be replaced by the weaker condition $\varphi(t) \leq \alpha(t) + \int_0^t \beta(\tau) \max_{[0,\tau]} |\varphi| d\tau$ for any $t \in [0, T[$.)

Theorem 9.2 (*Lax-Milgram*) [53] *Let B be a Banach space, and $A : B \to B'$ be linear, continuous, and coercive, in the sense that there exists $c > 0$ such that ${}_{B'}\langle Av, v\rangle_B \geq c\|v\|_B^2$ for any $v \in B$. Then A is onto B'.*

Theorem 9.3 (*Generalized Poincaré Inequality*) *Assume that Ω is of Lipschitz class and connected. Let $\Gamma_1 \subset \partial \Omega$ have positive $(N-1)$-Hausdorff measure. Then there exists a constant $C > 0$ such that*

$$\int_\Omega |v|^2 dx \leq C \left(\int_\Omega |\nabla v|^2 dx + \int_{\Gamma_1} |\gamma_0 v|^2 d\sigma\right) \qquad \forall v \in H^1(\Omega). \tag{9.3}$$

---

[51] See Tamanini [519; Lemma 4].

[52] See, e.g., Walter [572; Sect. I.1].

[53] See, e.g., Baiocchi and Capelo [47; Sect. 3.1].

Proof. By contradiction let us assume that for any $n \in \mathbf{N}$ there exists $v_n \in H^1(\Omega)$ such that

$$\int_\Omega |v_n|^2 dx \geq n \left(\int_\Omega |\nabla v_n|^2 dx + \int_{\Gamma_1} |\gamma_0 v_n|^2 d\sigma \right). \tag{9.4}$$

Possibly dividing this inequality by $\|v_n\|^2_{L^2(\Omega)}$, we can assume that $\|v_n\|_{L^2(\Omega)} = 1$ for any n. By (9.4) the sequence $\{v_n\}$ is then uniformly bounded in $H^1(\Omega)$; hence there exists v such that, possibly extracting a subsequence, $v_n \to v$ weakly in $H^1(\Omega)$. The equation (9.4) also yields

$$\nabla v_n \to 0 \text{ strongly in } L^2(\Omega; \mathbf{R}^N), \quad \gamma_0 v_n \to 0 \text{ strongly in } L^2(\Gamma_1).$$

Hence $\nabla v = 0$ a.e. in Ω and $\gamma_0 v = 0$ a.e. on Γ_1. Therefore $v = 0$ in Ω as this set is connected, contradicting the fact that the v_n are normalized in $L^2(\Omega)$. □

Theorem 9.4 (*Schauder fixed point theorem*) [54] *Let M be a nonempty convex subset of a normed space N, $T : M \to N$ be strongly continuous, and $T(M)$ relatively strongly compact. Then there exists $x \in M$ such that $T(x) = x$.*

Transposition. Let E, F be Banach spaces and $L : E \to F$ be a linear and continuous operator. The transposed (or adjoint) $L^* : F' \to E'$ is defined as:

$$_{E'}\langle L^* f', e \rangle_E := {}_{F'}\langle f', Le \rangle_F \quad \forall e \in E, \forall f' \in F'. \tag{9.5}$$

Theorem 9.5 [55] *Under the preceding hypotheses,*
  (i) L^* is injective iff $L(E) \subset F$ with continuous and dense injection;
  (ii) L is injective if $L^*(F') \subset E'$ with continuous and dense injection. The converse also holds if E is reflexive.

Hilbert Triplets. Let V, B be Banach spaces such that $V \subset B$ with continuous and dense injection (i.e., V is a linear subspace of B and is dense with respect to the strong topology of B). By the latter result, B' can be identified with a linear subspace of V'; that is, $B' \subset V'$ with continuous injection (we are identifying functionals with their restrictions). The latter injection is also dense if V is reflexive.

Now let $B(=H)$ be a Hilbert space equipped with the scalar product $(\cdot, \cdot)$, and let $\mathcal{R} : H \to H'$ be the *Riesz operator*; that is, $_{H'}\langle \mathcal{R}u, v \rangle_H = (u, v)_H$ for any $u, v \in H$. If $V \subset H$ with continuous and dense injection, then we have $\mathcal{R}(V) \subset \mathcal{R}(H) = H' \subset V'$ with continuous injections. Omitting the operator $\mathcal{R}$, we get the *Hilbert triplet*

$$V \subset H = H' \subset V' \quad \text{with continuous and dense injections.} \tag{9.6}$$

---

[54] See, e.g., Baiocchi and Capelo [47; Sect. 9.3.2] and Smart [509; Chap. 4].
[55] See, e.g., Baiocchi and Capelo [47; Sect. 4.1] and Wloka [580; Sect. 17].

Book Selection

> *Il faut, Nathanael, que tu brûles en toi tous les livres.*
> (André Gide, Les Nourritures Terrestres)

Here we just list some references, for the benefit of the reader interested in deepening her/his knowledge either of the mathematical tools or of the topics dealt with in this volume. Due to the impressive amount of literature concerning those subjects, we confine ourselves to a restricted selection of monographs. Several choices are arbitrary, and even depend on availability at the library of the author's university. Admittedly, English written works are privileged.

Functional Analysis and Function Spaces. Adams [3], Alt [11], Aubin [37], Aubin [40], Bergh and Löfström [72], Brézis [92], Choquet-Bruhat [134], Day [177], Diestel [193, 194], Diestel and Uhl [195], Dunford and Schwartz [202], Edwards [206], Gel'fand et al. [261], Hille and Phillips [292], Holmes [298], Hörmander [300], Horváth [301], Kelley, Namioka et al. [315], Köthe [329], Kufner, John and Fučík [330], Lions and Magenes [356], Maz'ja [381], Morrey [404], Reed and Simon [463], Rudin [490], Schwartz [497, 498], Treves [535], Triebel [536], Yosida [589], Ziemer [596].

Analysis of P.D.E.s in Function Spaces. Brezzi and Gilardi [96], Carroll and Showalter [124], Dautray and Lions [176], Friedman [248], Fučík [257], Gilbarg and Trudinger [266], Grisvard [274], Ladyženskaja [333], Ladyženskaja, Solonnikov and Ural'ceva [334], Ladyženskaja and Ural'ceva [335], Lions [348, 351], Lions and Magenes [356], Mizohata [394], Nečas [414, 415], Renardy and Rogers [464], Sobolev [511], Wloka [580], Zeidler [591, 592, 593].

Semigroups of Contractions. Barbu [53], Bénilan [64], Brézis [90], Da Prato [174], Goldstein [270], Hille and Phillips [292], Lunardi [368], Miyadera [393], Pavel [444], Pazy [450], Vrabie [571].

Convex Analysis and Theory of Monotone Operators. Aubin and Ekeland [39], Barbu [53], Aubin [37; Chaps. 2, 10], Barbu and Precupanu [56], Brézis [87, 90], Browder [99], Castaing and Valadier [125], Cioranescu [136], Clarke [137, 138], Dunford and Schwartz [202; Chap. V], Ekeland and Temam [207], Gajewski, Gröger and Zacharias [260], Ghizzetti [263], Giles [267], Hiriart-Urruty and Lemarechal [293], Holmes [297], Ioffe and Tikhomirov [308], Kluge [326], Kusraev and Kutateladze [332], Moreau [401], Pascali and Sburlan [443], Phelps [453], Roberts and Varberg [467], Rockafellar [468, 469], Sewell [501], Showalter [504], Smith [510], Tikhomirov [533].

Variational Inequalities and Variational Methods. Attouch [36], Baiocchi and Capelo [47], Barbu [53], Buttazzo [104], Dacorogna [160, 161], Dal Maso [163], Damlamian and Kenmochi [171], Duvaut and Lions [205], Evans [216], Friedman [253], Giaquinta and Hildebrandt [264], Kinderlehrer and Stampacchia [324], Lions [349], Naumann [413], Nečas and Hlaváček [416], Panagiotopoulos [440, 441], Rodrigues [471], Showalter [504], Struwe [517].

Geometric Measure Theory. Colombini, De Giorgi and Piccinini [144], Evans and Gariepy [218], Federer [231], Finn [235], Giusti [268], Maz'ja [381; Chap. 6], Simon [507], Vol'pert and Hudjaev [570], Ziemer [596].

Physics of Phase Transitions. Brice [97], Chalmers [129], Christian [135], Doremus [200], Flemings [240], Kurz and Fisher [331], Pamplin [439], Skripov [508], Turnbull [537], Ubbelohde [539], Woodruff [583]. See also Hohenberg and Halperin [294].

For the *physics of surfaces* see also Abraham [1], Adamson [4], Aveyard and Haydon [42], Lüth [370], Murr [410].

For *general thermodynamics* see, for example, Astarita [29], Callen [119], Kittel and Krömer [325], Lupis [369], Müller [406], Woods [584].

For *nonequilibrium thermodynamics* see, for example, Defay, Prigogine, Bellemans and Everett [178], De Groot [182], De Groot and Mazur [183], Glansdorff and Prigogine [269], Lavenda [343], Nicholis and Prigogine [419], Prigogine [455].

Mathematics of Phase Transitions. Avdonin [41], Brokate and Sprekels [98], Datzeff [175], Gurtin [281], Hill [291], Meirmanov [388], Romano [477], Rubinstein [479], Tarzia [527], Yamaguchi and Nogi [588].

See also Cannon [120; Chaps. 17, 18], Crank [154], Elliott and Ockendon [212], Friedman [248; Chap. 8], Ladyženskaja, Solonnikov and Ural'ceva [334; Sect. V.9], Lions [351; Sect. 2.3], Naumann [413; Chap. 4].

Other Free Boundary Problems. Baiocchi and Capelo [47], Chipot [132], Crank [154], Damlamian and Kenmochi [170], Diaz [185], Duvaut and Lions [205], Elliott and Ockendon [212], Friedman [253], Kinderlehrer and Stampacchia [324], Monachov [398], Naumann [413], Nečas and Hlaváček [416], Radkevich and Melikulov [461], Rodrigues [471], Rubinstein and Martuzans [487].

Proceedings of Meetings on Free Boundary Problems. Albrecht, Collatz and Hoffmann [6], Antontsev, Hoffmann and Khludnev [22], Bossavit, Damlamian and Frémond [82], Brown and Davis [102] Chadam and Rasmussen [126, 127, 128], Delfour and Sabidussi [184], Diaz, Herrero, Liñán and Vázquez [186], Fasano and Primicerio [225], Fosdick, Dunn and Slemrod [242], Friedman and Spruck [256], Gurtin and McFadden [284], Hoffmann and Sprekels [295, 296], Kawarada, Kenmochi and Yanagihara [314], Kenmochi, Niezgódka and Strzelecki [320], Magenes [372], Neittanmäki [417], Ni, Peletier and Vázquez [418], Niezgódka [420], Ockendon and Hodgkins [432], Rodrigues [473], Visintin [565], Wilson, Solomon and Boggs [578], Wrobel and Brebbia [585, 586], Wrobel, Brebbia and Sarler [587].

Bibliography

1. F.F. Abraham: *Homogeneous Nucleation Theory.* Academic Press, New York 1974
2. E. Acerbi, N. Fusco: *Semicontinuity problems in the calculus of variations.* Arch. Rational Mech. Anal. **86** (1984) 125–145
3. R. Adams: *Sobolev Spaces.* Academic Press, New York 1975
4. A.W. Adamson: *Physical Chemistry of Surfaces.* Wiley, New York 1990
5. E.L. Albasiny: *The solution of nonlinear heat conduction problems on the pilot.* ACE, Proc. Inst. Electric. Engrg. **103** Part B, Suppl. 1 (1956) 158–162
6. J. Albrecht, L. Collatz, K.-H. Hoffmann (eds.): *Numerical Treatment of Free Boundary Value Problems.* Birkhäuser, Basel 1982
7. S.M. Allen, J.W. Cahn: *A microscopic theory for antiphase motion and its application to antiphase domain coarsening.* Acta Metall. **27** (1979) 1085–1095
8. F. Almgren: *Existence and Regularity Almost Everywhere of Elliptic Variational Problems with Constraints.* Memoirs A.M.S. **165** (1976)
9. F. Almgren, J.E. Taylor, L. Wang: *Curvature-driven flows: a variational approach.* S.I.A.M. J. Control and Optimization **31** (1993) 387–437
10. F. Almgren, L. Wang: *Mathematical existence of crystal growth with Gibbs-Thomson curvature effects.* J. Geometric Analysis (to appear)
11. H.W. Alt: *Lineare Funktionalanalysis.* Springer, Berlin 1985
12. H.W. Alt, S. Luckhaus: *Quasilinear elliptic-parabolic differential equations.* Math. Z. **183** (1983) 311–341
13. H.W. Alt, I. Pawlow: *Dynamics of non-isothermal phase separation.* In: Free Boundary Value Problems (K.-H. Hoffmann, J. Sprekels, eds.). Birkäuser, Boston 1990, pp. 1–26
14. H.W. Alt, I. Pawlow: *A mathematical model of dynamics of non-isothermal phase separation.* Physica D **59** (1992) 389–416
15. H.W. Alt, I. Pawlow: *Existence of solutions for non-isothermal phase separation.* Adv. Math. Sci. Appl. **1** (1992) 319–409
16. H.W. Alt, I. Pawlow: *On the entropy principle of phase transition models with a conserved order parameter.* Adv. Math. Sci. Appl. **6** (1996) 291–376
17. H. Amann: *Linear and Quasilinear Parabolic Equations.* Vol. 1. Birkhäuser, Basel 1995
18. A. Amrani, C. Castaing, M. Valadier: *Convergence forte dans L^1 impliquée par la convergence faible. Méthodes de troncature.* C.R. Acad. Sci. Paris, Série I **314** (1992) 37–40
19. A. Amrani, C. Castaing, M. Valadier: *Méthodes de troncature appliquées à des problèmes de convergence faible ou forte dans L^1.* Arch. Rational Mech. Anal. **117** (1992) 167–191
20. D. Andreucci, A. Fasano, R. Gianni, M. Primicerio, R. Ricci: *Diffusion driven crystallization in polymers.* Preprint, Firenze (1995)
21. D. Andreucci, A. Fasano, M. Primicerio, R. Ricci: *Mathematical problems in polymer crystallization.* Preprint, Firenze (1995)
22. S.N. Antontsev, K.-H. Hoffmann, A.M. Khludnev (eds.): *Free Boundary Problems in Continuum Mechanics.* Birkhäuser, Basel 1992
23. G. Anzellotti, S. Baldo, A. Visintin: *Asymptotic behavior of the Landau-Lifshitz model of ferromagnetism.* Appl. Math. Optim. **23** (1991) 171–192
24. G. Anzellotti, M. Giaquinta: *Funzioni BV e tracce.* Rend. Sem. Mat. Univ. Padova **60** (1978) 1–21
25. T. Arai: *On the existence of the solution for $\partial\varphi(u'(t)) + \partial\psi(u(t)) \ni f(t)$.* J. Fac. Sci. Univ. Tokyo. Sec. IA Math. **26** (1979) 75–96

26. D.G. Aronson: *The porous media equation.* In: Nonlinear Diffusion Problems (A. Fasano, M. Primicerio, eds.). Springer, Berlin 1986, pp. 1–46
27. Z. Artstein, T. Rzezuchowski: *A note on Olech's lemma.* Studia Mathematica **98** (1991) 91–94
28. G. Astarita: *A class of free boundary problems arising in the analysis of transport phenomena in polymers.* In: Free Boundary Problems: Theory and Applications (A. Fasano, M. Primicerio, eds.). Pitman, Boston 1983, pp. 606–612
29. G. Astarita: *Thermodynamics. An Advanced Textbook for Chemical Engineers.* Plenum Press, New York 1989
30. G. Astarita, G.C. Sarti: *A class of mathematical models for sorption of swelling solvents in glassy polymers.* Polymer Eng. Science **18** (1978) 388–395
31. I. Athanassopoulos, L.A. Caffarelli, S. Salsa: *Caloric functions in Lipschitz domains and the regularity of solutions to phase transition problems.* Ann. Math. **143** (1996) 413–434
32. I. Athanassopoulos, L.A. Caffarelli, S. Salsa: *Regularity of the free boundary in phase transition problems.* Acta Math. **176** (1996)
33. I. Athanassopoulos, L.A. Caffarelli, S. Salsa: *Degenerate phase transition problems of parabolic type: smoothness of the front.* Preprint, Princeton 1996
34. D.R. Atthey: *A finite difference scheme for melting problems.* J. Inst. Math. Appl. **13** (1974) 353–366
35. D.R. Atthey: *A finite difference scheme for melting problems based on the method of weak solutions.* In: Moving Boundary Problems in Heat Flow and Diffusion (J.R. Ockendon, W.R. Hodgkins, eds.). Clarendon Press, Oxford 1975, pp. 182–191
36. H. Attouch: *Variational Convergence for Functions and Operators.* Pitman, Boston 1984
37. J.-P. Aubin: *Applied Functional Analysis.* Wiley, New York 1979
38. J.-P. Aubin: *Un théorème de compacité.* C.R. Acad. Sci. Paris, Série I **256** (1963) 5042–5044
39. J.-P. Aubin, I. Ekeland: *Applied Nonlinear Analysis.* Wiley, New York 1984
40. T. Aubin: *Nonlinear Analysis on Manifolds: Monge-Ampère Equations.* Springer, New York 1982
41. N.A. Avdonin: *Mathematical Description of Processes of Crystallization.* Zinatne, Riga 1980 (Russian)
42. R. Aveyard, D.A. Haydon: *An Introduction to the Principles of Surface Chemistry.* Cambridge University Press, Cambridge 1973
43. J.B. Baillon: *Générateurs et semi-groupes dans les espaces de Banach uniformément lisses.* J. Funct. Anal. **29** (1978) 199–213
44. C. Baiocchi: *Sur un problème à frontière libre traduisant le filtrage de liquides à travers le milieux poreux.* C.R. Acad. Sci. Paris, Série A **273** (1971) 1215–1217
45. C. Baiocchi: *Su un problema di frontiera libera connesso a questioni di idraulica.* Ann. Mat. Pura Appl. **92** (1972) 107–127
46. C. Baiocchi: *Problèmes à frontière libre et inéquations variationnelles.* C.R. Acad. Sci. Paris, Série A **283** (1976) 29–32
47. C. Baiocchi, A. Capelo: *Variational and Quasivariational Inequalities, Applications to Free Boundary Problems.* Wiley, Chichester 1983
48. E.J. Balder: *On weak convergence implying strong convergence in L_1-spaces.* Bull. Austral. Math. Soc. **33** (1986) 363–368
49. E.J. Balder: *On equivalence of strong and weak convergence in L_1-spaces under extreme point conditions.* Israel J. Math. **75** (1991) 21–47
50. E.J. Balder: *On weak convergence implying strong convergence under extremal conditions.* J. Math. Anal. Appl. **163** (1992) 147–156
51. J. Ball: *Convexity conditions and existence theorems in nonlinear elasticity.* Arch. Rational Mech. Anal. **63** (1977) 337–403
52. V. Barbu: *Existence theorems for a class of two point boundary problems.* J. Differential Equations **17** (1975) 236–257
53. V. Barbu: *Nonlinear Semigroups and Differential Equations in Banach Spaces.* Noordhoff, Leyden 1976

54. V. Barbu: *Existence for nonlinear Volterra equations in Hilbert spaces.* S.I.A.M. J. Math. Anal. **10** (1979) 552–569
55. V. Barbu: *Optimal Control of Variational Inequalities.* Pitman, Boston 1984
56. V. Barbu, T. Precupanu: *Convexity and Optimization in Banach Spaces.* Editura Academiei, Bucuresti 1978
57. G.I. Barenblatt: *Dimensional Analysis.* Gordon and Breach, New York 1987
58. G. Barles: *Remark on a flame propagation model.* Rapport I.N.R.I.A. **464** (1985)
59. G. Barles: *Solutions de Viscosité des Équations de Hamilton-Jacobi.* Springer, Paris 1994
60. E. Barozzi, E. Gonzalez, I. Tamanini: *The mean curvature of a set of finite perimeter.* Proceedings A.M.S. **99** (1987) 313–316
61. P.W. Bates, P.C. Fife: *The dynamics of nucleation for the Cahn-Hilliard equation.* S.I.A.M. J. Appl. Math. **53** (1993) 990–108
62. B. Beauzamy: *Introduction to Banach Spaces and their Geometry.* North-Holland, Amsterdam 1985
63. Ph. Bénilan: *Solutions intégrales d'équations d'évolution dans un espace de Banach.* C.R. Acad. Sci. Paris, Série A **274** (1972) 47–50
64. Ph. Bénilan: *Equations d'Évolution dans un Espace de Banach Quelconque et Applications.* Thèse, Orsay 1972
65. Ph. Bénilan: *Opérateurs accrétifs et semi-groupes dans les espaces L^p.* In: Functional Analysis and Numerical Analysis (H. Fujita, ed.). Japan Society for Promotion of Science, 1978, pp. 15–53
66. Ph. Bénilan, M.G. Crandall: *Regularizing effects of homogeneous evolution equations.* In: Contributions to Analysis and Geometry (a supplement to Amer. J. Math.) (D.N. Clark, G. Pecelli, R. Sacksteder, eds.). John Hopkins Univ. Press, Baltimore 1981, pp. 23–30
67. Ph. Bénilan, M.G. Crandall, P. Sacks: *Some L^1 existence and dependence results for semilinear elliptic equations under nonlinear boundary conditions.* Appl. Math. Optim. **17** (1988) 203–224
68. Ph. Bénilan, K.S. Ha: *Equations d'évolution du type $du/dt + \beta\partial\varphi(u) \ni 0$ dans $L^\infty(\Omega)$.* C.R. Acad. Sci. Paris, Série A **281** (1975) 947–950
69. A. Bensoussan, J.L. Lions, G. Papanicolau: *Asymptotic Analysis for Periodic Structures.* North-Holland, Amsterdam 1978
70. A.E. Berger, H. Brézis, J.W. Rogers: *A numerical method for solving the problem $u_t - \Delta f(u) = 0$.* R.A.I.R.O., Analyse Numérique **13** (1979) 297–312
71. A.E. Berger, J.W. Rogers: *Some properties of the nonlinear semigroup for the problem $u_t - \Delta f(u) = 0$.* Nonlinear Analysis, T.M.A. **8** (1984) 909–939
72. J. Bergh, J. Löfström: *Interpolation Spaces. An Introduction.* Springer, Berlin 1976
73. A. Bermudez, C. Saguez: *Mathematical formulation and numerical solution of an alloy solidification problem.* In: Free Boundary Problems: Theory and Applications (A. Fasano, M. Primicerio, eds.). Pitman, Boston 1983, pp. 237–247
74. F. Bernis: *Existence results for doubly nonlinear higher order parabolic equations on unbounded domains.* Math. Ann. **279** (1988) 373–394
75. J.F. Blowey, C.M. Elliott: *The Cahn-Hilliard gradient theory for phase separation with non-smooth free energy.* Euro J. Applied Math. **2** (1991) 233–280
76. Bor-Luh Lin, Pei-Kee Lin, S.L. Troyanski: *Characterizations of denting points.* Proceedings of A.M.S. **102** (1988) 526–528
77. A. Bossavit: *Stefan models for eddy currents in steel.* In: Free Boundary Problems: Theory and Applications (A. Fasano, M. Primicerio, eds.). Pitman, Boston 1983, pp. 349–364
78. A. Bossavit: *Free boundaries in induction heating.* Control and Cybernetics **14** (1985), 69–96
79. A. Bossavit: *Mixed methods for a vectorial Stefan problem.* In: Free Boundary Problems: Theory and Applications (K.-H. Hoffmann, J. Sprekels, eds.). Longman, Harlow 1990, pp. 25–37
80. A. Bossavit: *Électromagnétisme, en vue de la modélisation.* Springer, Paris 1993

81. A. Bossavit, A. Damlamian: *Homogenization of the Stefan problem and application to composite magnetic media.* I.M.A. J. Appl. Math. **27** (1981) 319–334
82. A. Bossavit, A. Damlamian, M. Frémond (eds.): *Free Boundary Problems: Theory and Applications.* Pitman, Boston 1985
83. A. Bossavit, J.C. Vérité: *A mixed FEM-BIEM method to solve eddy currents problems.* I.E.E.E. Trans. on Magnetism **MAG-18** (1982) 431–435
84. K.A. Brakke: *The Motion of a Surface by its Mean Curvature.* Princeton University Press, Princeton 1978
85. H. Brézis: *Equations et inéquations non linéaires dans les espaces vectoriels en dualité.* Ann. Inst. Fourier **18** (1968) 115–175
86. H. Brézis: *On some degenerate nonlinear parabolic equations.* In: Nonlinear Functional Analysis (F.E. Browder, ed.). Proc. Symp. Pure Math. **XVIII**. A.M.S., Providence 1970, pp. 28–38
87. H. Brézis: *Monotonicity methods in Hilbert spaces and some applications to nonlinear partial differential equations.* In: Contributions to Nonlinear Functional Analysis (E.H. Zarantonello, ed.). Academic Press, New York 1971, pp. 101–156
88. H. Brézis: *Intégrales convexes dans les espaces de Sobolev.* Israel J. Math. **13** (1972) 9–23
89. H. Brézis: *Problemes unilateraux.* J. Math. Pures et Appl. **51** (1972) 1–168
90. H. Brézis: *Opérateurs Maximaux Monotones et Semi-Groupes de Contractions dans les Espaces de Hilbert.* North-Holland, Amsterdam 1973
91. H. Brézis: *Asymptotic behavior of some evolution systems.* In: Nonlinear Evolution Equations (M.C. Crandall, ed.). Academic Press, New York 1978, pp. 141–154
92. H. Brézis: *Analyse Fonctionelle. Théorie et Applications.* Masson, Paris 1983
93. H. Brézis: *Convergence in $\mathcal{D}'$ and in L^1 under strict convexity.* In: Boundary Value Problems for Partial Differential Equations and Applications (J.L. Lions, C. Baiocchi, eds.). Dunod, Paris 1993, pp. 43–52
94. H. Brézis, A. Pazy: *Convergence and approximation of semigroups of nonlinear operators in Banach spaces.* J. Funct. Anal. **9** (1972) 63–74
95. H. Brézis, W.A. Strauss: *Semi-linear second-order elliptic equations in L^1.* J. Math. Soc. Japan **25** (1973) 565–590
96. F. Brezzi, G. Gilardi: *Functional Analysis, Functional Spaces, Boundary Value Problems.* In: Finite Element Handbook, Chaps. 1, 2, 3 (D.A. Kardestuncer, ed.). McGraw-Hill, New York 1987, pp. I.3–I.121
97. J.C. Brice: *The Growth of Crystals from Liquids.* North-Holland, Amsterdam 1973
98. M. Brokate, J. Sprekels: *Hysteresis and Phase Transitions.* Springer, Heidelberg 1996
99. F. Browder: *Nonlinear operators and nonlinear equations of evolution in Banach spaces.* In: Proc. Sympos. Pure Math., **XVIII** Part II. A.M.S., Providence 1976
100. W.F. Brown, jr.: *Magnetostatics Principles in Ferromagnetism.* North-Holland, Amsterdam 1962
101. W.F. Brown, jr.: *Micromagnetics.* Krieger, Huntington 1978
102. R.A. Brown, S.H. Davis (eds.): *Free Boundaries in Viscous Flows.* Springer, New York 1993
103. R.E. Bruck: *Asymptotic convergence of nonlinear contraction semigroups in Hilbert spaces.* J. Funct. Anal. **18** (1975) 15–26
104. G. Buttazzo: *Semicontinuity, Relaxation and Integral Representation in the Calculus of Variations.* Longman, Harlow 1989
105. G. Buttazzo, A. Visintin: *Motion by Mean Curvature and Related Topics.* De Gruyter, Berlin 1994
106. L.A. Caffarelli: *The regularity of free boundaries in higher dimensions.* Acta Math. **139** (1977) 155–184
107. L.A. Caffarelli: *Some aspects of the one-phase Stefan problem.* Indiana Univ. Math. J. **27** (1978) 73–77
108. L.A. Caffarelli, L.C. Evans: *Continuity of the temperature in the two-phase Stefan problem.* Arch. Rational Mech. Anal. **81** (1983) 199–220

109. L.A. Caffarelli, A. Friedman: *Continuity of the temperature in the Stefan problem.* Indiana Univ. Math. J. **28** (1979) 53–70
110. G. Caginalp: *Surface tension and supercooling in solidification theory.* In: Applications of Field Theory to Statistical Mechanics (L. Garrido, ed.). Springer, Berlin 1985, pp. 216–226
111. G. Caginalp: *An analysis of a phase field model of a free boundary.* Arch. Rational Mech. Anal. **92** (1986) 205–245
112. G. Caginalp: *Stefan and Hele-Shaw type models as asymptotic limits of phase field equations.* Pys. Rev. **A 39** (1989) 5887–5896
113. G. Caginalp: *The dynamics of a conserved phase field system: Stefan-like, Hele-Shaw, and Cahn-Hilliard models as asymptotic limits.* I.M.A. J. Appl. Math. **44** (1990) 77–94
114. G. Caginalp, W. Xie: *Phase field and sharp-interface alloy models.* Physical Review E **48** (1993) 1897–1909
115. J.W. Cahn. *Theory of crystal growth and interface motion in crystalline materials.* Acta Met. **8** (1960) 554–562
116. J.W. Cahn: *On spinodal decomposition.* Acta Metall. **9** (1961) 795–801
117. J.W. Cahn, J.E. Hilliard: *Free energy of a nonuniform system. I. Interfacial free energy. III. Nucleation in a two-component incompressible fluid.* J. Chem. Phys. **28** (1957) 258–267; **31** (1959) 688–699
118. J.W. Cahn, J.E. Taylor: *Surface motion by surface diffusion.* Acta Metall. Mater. **42** (1994) 1045–1063
119. H.B. Callen: *Thermodynamics and an Introduction to Thermostatistics.* Wiley, New York 1985
120. J.R. Cannon: *The One-Dimensional Heat Equation.* Encyclopedia of Mathematics and its Applications. Vol. 23. Addison Wesley, Menlo Park 1984
121. J.R. Cannon, E. DiBenedetto: *The steady state Stefan problem with convection, with mixed temperature and non-linear heat flux boundary conditions.* In: Free Boundary Problems (E. Magenes, ed.). Istituto di Alta Matematica, Roma 1980, pp. 231–266
122. J.R. Cannon, D. Henry, D. Kotlow: *Continuous differentiability of the free boundary for weak solutions of the Stefan problem.* Bull. Amer. Math. Soc. **80** (1974) 45–48
123. J.R. Cannon, C.D. Hill: *On the infinite differentiability of the free boundary in a Stefan problem.* J. Math. Anal. Appl. **22** (1968) 385–387
124. R.W. Carroll, R.E. Showalter: *Singular and Degenerate Cauchy Problems.* Academic Press, New York 1976
125. C. Castaing, M. Valadier: *Convex Analysis and Measurable Multifunctions.* Springer, Berlin 1977
126. J.M. Chadam, H. Rasmussen (eds.): *Emerging Applications in Free Boundary Problems.* Longman, Harlow 1993
127. J.M. Chadam, H. Rasmussen (eds.): *Free Boundary Problems Involving Solids.* Longman, Harlow 1993
128. J.M. Chadam, H. Rasmussen (eds.): *Free Boundary Problems in Fluid Flow with Applications.* Longman, Harlow 1993
129. B. Chalmers: *Principles of Solidification.* Wiley, New York 1964
130. X. Chen, F. Reitich: *Local existence and uniqueness of solutions of the Stefan problem with surface tension and kinetic undercooling.* J. Math. Anal. Appl. **154** (1992) 350–362
131. Y.G. Chen, Y. Giga, S. Goto: *Uniqueness and existence of viscosity solutions of generalized solutions of mean curvature flow equation.* J. Differential Geom. **33** (1991) 749–786
132. M. Chipot: *Variational Inequalities and Flow in Porous Media.* Springer, New York 1984
133. M. Chipot, G. Michaille: *Uniqueness results and monotonicity properties for strongly non-linear elliptic variational inequalities.* Ann. Scuola Norm. Sup. Pisa **16** (1989) 137–166
134. Y. Choquet-Bruhat: *Distributions. Théorie et Problèmes.* Masson, Paris 1973
135. J.W. Christian: *The Theory of Transformations in Metals and Alloys. Part 1: Equilibrium and General Kinetic Theory.* Pergamon Press, London 1975

136. I. Cioranescu: *Geometry of Banach Spaces, Duality Mappings and Nonlinear Problems.* Kluwer, Dordrecht 1990
137. F.H. Clarke: *Optimization and Nonsmooth Analysis.* Wiley, New York 1983
138. F.H. Clarke: *Methods of Dynamics and Nonsmooth Optimization.* S.I.A.M. Publications, Philadelphia 1989
139. P. Colli: *On some doubly nonlinear evolution equations in Banach spaces.* Japan J. Indust. Appl. Math. **9** (1992) 181–203
140. P. Colli, G. Savaré: *Time discretization of Stefan problems with singular heat flux.* Preprint 1996
141. P. Colli, J. Sprekels: *On a Penrose-Fife model with zero interfacial energy leading to a phase-field system of relaxed Stefan type.* Ann. Mat. Pura Appl. **169** (1995) 269–289
142. P. Colli, J. Sprekels: *Stefan problems and the Penrose-Fife phase field model.* Adv. Math. Sci. Appl. (to appear)
143. P. Colli, A. Visintin: *On a class of doubly nonlinear evolution problems.* Comm. in P.D.E.s **15** (1990) 737–756
144. F. Colombini, E. De Giorgi, L.C. Piccinini: *Frontiere Orientate di Misura Minima e Questioni Collegate.* Editrice Tecnico Scientifica, Pisa 1972
145. M.G. Crandall: *Semigroups of nonlinear tranformations in Banach spaces.* In: Contributions to Nonlinear Functional Analysis (E.H. Zarantonello, ed.). Academic Press, New York 1971, pp. 157–179
146. M.G. Crandall: *An introduction to evolution governed by accretive operators.* In: Dynamical Systems, Vol. I (L. Cesari, J. Hale, J.P. La Salle, eds.). Academic Press, New York 1976, pp. 131–165
147. M.G. Crandall: *Nonlinear semigroups and evolution governed by accretive operators.* In: Proceedings of Symposium in Pure Math., Part I (F. Browder, ed.). A.M.S., Providence 1986, pp. 305–338
148. M.G. Crandall, H. Ishii, P.L. Lions: *User's guide to viscosity solutions of second order partial differential equations.* Bull. Amer. Math. Soc. **27** (1992) 1–67
149. M.G. Crandall, T. Liggett: *Generation of semi-groups of nonlinear transformations on general Banach spaces.* Amer. J. Math. **93** (1971) 265–298
150. M.G. Crandall, P.L. Lions: *Viscosity solutions of Hamilton-Jacobi equations.* Trans. Amer. Math. Soc. **277** (1983) 1–42
151. M.G. Crandall, A. Pazy: *Semigroups of nonlinear contractions and dissipative sets.* J. Funct. Anal. **3** (1969) 376–418
152. M.G. Crandall, A. Pazy: *Nonlinear evolution equations in Banach spaces.* Israel J. Math. **11** (1972) 57–94
153. M.G. Crandall, M. Pierre: *Regularizing effect for $u_t = \Delta\varphi(u)$.* Trans. A.M.S. **274** (1982) 159–168
154. J. Crank: *Free and Moving Boundary Problems.* Clarendon Press, Oxford 1984
155. J. Crank, J.R. Ockendon: *Proceedings of an I.M.A. conference on crystal growth.* I.M.A. J. of Appl. Math. **35** (1985) 115–264
156. A.B. Crowley: *Numerical solution of alloy solidification problem revisited.* In: Free Boundary Problems: Theory and Applications (A. Bossavit, A. Damlamian, M. Frémond, eds.). Pitman, Boston 1985, pp. 122–131
157. A.B. Crowley, J.R. Ockendon: *On the numerical solution of an alloy solidification problem.* Int. J. Heat Mass Transfer **22** (1979) 941–947
158. C.W. Cryer: *A Bibliography of Free Boundary Problems.* M.R.C. Rep. No. 1793, Math. Res. Cent., Univ. of Wisconsin 1977
159. P. Čižek, V. Janovsky: *Hele-Shaw flow model of the injection by a point source.* Proc. Royal Soc. Edinburgh. **A91** (1981) 147–159
160. B. Dacorogna: *Weak Continuity and Weak Lower Semi-Continuity of Nonlinear Functionals.* Springer, Berlin 1982
161. B. Dacorogna: *Direct Methods in the Calculus of Variations.* Springer, Berlin 1989

162. C.M. Dafermos, M. Slemrod: *Asymptotic behaviour of nonlinear contraction semigroups.* J. Funct. Anal. **13** (1973) 97–106
163. G. Dal Maso: *An Introduction to Γ-Convergence.* Birkäuser, Boston 1993
164. A. Damlamian: *Some results on the multi-phase Stefan problem.* Comm. in P.D.E.s **2** (1977) 1017–1044
165. A. Damlamian: *Homogenization for eddy currents.* Delft Progress Report **6** (1981) 268–275
166. A. Damlamian: *How to homogenize a nonlinear diffusion equation: Stefan's problem.* S.I.A.M. J. Math. Anal. **12** (1981) 306–313
167. A. Damlamian: *On the Stefan problem: the variational approach and some applications.* Banach Center Publ. **15** (1985) 253–275
168. A. Damlamian: *Asymptotic behavior of solutions to a multi-phase Stefan problem.* In: Free Boundary Problems: Theory and Applications (K.-H. Hoffmann and J. Sprekels, eds.). Longman, Harlow 1990, pp. 811–817
169. A. Damlamian, N. Kenmochi: *Asymptotic behavior of solutions to a multi-phase Stefan problem.* Japan J. Appl. Math. **3** (1986) 15–36
170. A. Damlamian, N. Kenmochi: *Periodicity and almost periodicity of the solutions to a multi-phase Stefan problem.* Nonlinear Analysis, T.M.A. **12** (1988) 921–934
171. A. Damlamian, N. Kenmochi: *Nonlinear Evolution Problems in Hilbert Spaces.* Gakkotosho, Tokyo (Monograph in preparation)
172. A. Damlamian, J. Spruck, A. Visintin: *Curvature Flows and Related Topics.* Gakkotosho, Tokyo 1995
173. I.I. Danilyuk: *On the Stefan problem.* Russian Math. Surveys **40** (1985) 157–223
174. G. Da Prato: *Applications Croissantes et Équations d'Évolution dans les Espaces de Banach.* Academic Press, London 1976
175. A. Datzeff: *Sur le Problème Linéaire de Stefan.* Gauthier-Villars, Paris 1970
176. R. Dautray, J.L. Lions: *Analyse Mathématique et Calcul Numérique pour les Sciences et les Techniques.* (9 volumes.) Masson, Paris 1987-88
177. M.M. Day: Normed linear spaces. Springer, Berlin 1973
178. R. Defay, I. Prigogine, A. Bellemans, D.H. Everett: *Surface Tension and Adsorption.* Longmans, London 1966
179. E. De Giorgi: *Γ-convergenza e G-convergenza.* Boll. Un. Mat. Ital. **5-B** (1977) 213–220
180. E. De Giorgi, T. Franzoni: *Su un tipo di convergenza variazionale.* Atti Accad. Naz. Lincei Cl. Sci. Mat. Fis. Natur. **58** (1975) 842–850
181. E. De Giorgi, A. Marino, M. Tosques: *Problemi di evoluzione in spazi metrici e curve di massima pendenza.* Atti Accad. Naz. Lincei Cl. Sci. Mat. Fis. Natur. **68** (1980) 180–187
182. S.R. De Groot: *Thermodynamics of Irreversible Processes.* Amsterdam, North-Holland 1961
183. S.R. De Groot, P. Mazur: *Nonequilibrium Thermodynamics.* Amsterdam, North-Holland 1962
184. M.C. Delfour and G. Sabidussi (eds.): *Shape Optimization and Free Boundaries.* Kluwer, Dordrecht 1992
185. J.I. Diaz: *Nonlinear Partial Differential Equations and Free Boundary Problems. Vol. I, Elliptic Equations.* Pitman, Boston 1985
186. J.I. Diaz, M.A. Herrero, A. Liñán, J.L. Vázquez (eds.): *Free Boundary Problems: Theory and Applications.* Longman, Harlow 1995
187. E. DiBenedetto: *Regularity results for the n-dimensional two-phase Stefan problem.* Boll. Un. Mat. Ital. Suppl. (1980) 129–152
188. E. DiBenedetto: *Continuity of weak solutions to certain singular parabolic equations.* Ann. Mat. Pura Appl. **121** (1982) 131–176
189. E. DiBenedetto, A. Friedman: *The ill-posed Hele-Shaw model and the Stefan problem for supercooled water.* Trans. A.M.S. **282** (1984) 183–204
190. E. DiBenedetto, A. Friedman: *Conduction-convection problems with change of phase.* J. Differential Equations **62** (1986) 129–185

191. E. DiBenedetto, R.E. Showalter: *Implicit degenerate evolution equations and applications.* S.I.A.M. J. Math. Anal. **12** (1981) 731–751
192. E. DiBenedetto, V. Vespri: *On the singular equation $\beta(u)_t = \Delta u$.* Arch. Rational Mech. Anal. **132** (1995) 247–309
193. J. Diestel: *Geometry of Banach Spaces – Selected Topics.* Springer, Berlin 1975
194. J. Diestel: *Sequences and Series in Banach Spaces.* Springer, New York 1984
195. J. Diestel, J.J. Uhl, jr.: *Vector Measures.* A.M.S., Providence 1977
196. R.J. DiPerna: *Measure-valued solutions to conservation laws.* Arch. Rational Mech. Anal. **88** (1985) 223–270
197. R. Dobrushin, R. Kotecký, S. Shlosman: *Wulff Construction. A Global Shape from Local Interaction.* A.M.S., Providence 1992
198. J.D.P. Donnelly: *A model for non-equilibrium thermodynamic processes involving phase changes.* J. Inst. Math. Appl. **24** (1979) 425–438
199. A.L. Dontchev, T. Zolezzi: *Well-Posed Optimization Problems.* Springer, Berlin 1993
200. R.H. Doremus: *Rates of Phase Transformations.* Academic Press, Orlando 1985
201. J. Douglas, J.R. Cannon, C.D. Hill: *A multi-boundary Stefan problem and the disappearence of phases.* J. Math. Mech. **17** (1967) 21–35
202. N. Dunford, J. Schwartz: *Linear Operators.* Vol. I. Interscience, New York 1958
203. G. Duvaut: *Résolution d'un problème de Stefan (fusion d'un bloc de glace à zéro degrés).* C.R. Acad. Sci. Paris, Série I **276-A** (1973) 1461–1463
204. G. Duvaut: *The solution of a two-phase Stefan by a variational inequality.* In: Moving Boundary Problems in Heat Flow and Diffusion (J.R. Ockendon, W.R. Hodgkins, eds.). Clarendon Press, Oxford 1975, pp. 173–181
205. G. Duvaut, J.L. Lions: *Les Inéquations en Mécanique et en Physique.* Dunod, Paris 1972
206. R.E. Edwards: *Functional Analysis. Theory and Applications.* Holt, Rinehart and Winston, New York 1965
207. I. Ekeland, R. Temam: *Analyse Convexe et Problèmes Variationnelles.* Dunod Gauthier-Villars, Paris 1974
208. C.M. Elliott: *On a variational inequality formulation of an electrochemical machining moving boundary problem and its approximation by the finite element method.* J. Inst. Maths. Applics. **25** (1980) 121–131
209. C.M. Elliott: *Error analysis of the enthalpy method for the Stefan problem.* I.M.A. J. Numer. Anal.. **7** (1987) 61–71
210. C.M. Elliott: *The Cahn-Hilliard model for the kinetics of phase separation.* In: Mathematical Models for Phase Change Problems (J.-F. Rodrigues, ed.). Birkhäuser, Basel 1989, pp. 35–73
211. C.M. Elliott, V. Janovsky: *A variational inequality approach to Hele-Shaw flow with a moving boundary.* Proc. Roy. Soc. Edin. **88A** (1981) 93–107
212. C.M. Elliott, J.R. Ockendon: *Weak and Variational Methods for Moving Boundary Problems.* Pitman, Boston, 1982
213. C.M. Elliott, S. Zheng: *On the Cahn-Hilliard equation.* Arch. Rational Mech. Anal. **96** (1986) 339–357
214. G.W. Evans: *A note on the existence of a solution to a Stefan problem.* Quart. Appl. Math. **IX** (1951) 185–193
215. L.C. Evans: *Application of nonlinear semigroups to certain partial differential equations.* In: Nonlinear Evolution Equations (M.C. Crandall, ed.). Academic Press, New York 1978, pp. 163–188
216. L.C. Evans: *Weak Convergence Methods for Nonlinear Partial Differential Equations.* A.M.S., Providence 1990
217. L.C. Evans, R.F. Gariepy: *Some remarks on convexity and strong convergence.* Proc. Royal Soc. Edinburgh. **A106** (1987) 53–61
218. L.C. Evans, R.F. Gariepy: *Measure Theory and Fine Properties of Functions.* CRC Press, Boca Raton (1992)

219. L.C. Evans, M. Soner, P.E. Souganidis: *Phase transitions and generalized motion by mean curvature.* Comm. Pure Appl. Math. **45** (1992) 1097–1123
220. L.C. Evans, J. Spruck: *Motion of level sets by mean curvature I.* J. Diff. Geom. **33** (1991) 635–681
221. A. Fasano: *Las Zonas Pastosas en el Problema de Stefan.* Cuad. Inst. Mat. Beppo Levi, No. 13, Rosario 1987
222. A. Fasano: *Modelling the solidification of polymers: an example of an E.C.M.I. cooperation.* In: Proc. I.C.I.A.M. 91 (R.E. O'Malley, ed.). S.I.A.M., Philadelphia 1991, pp. 99–118
223. A. Fasano, M. Primicerio: *Convergence of Huber's method for heat conduction problems with change of phase.* Z. Angew. Math. Mech.. **53** (1973) 341–348
224. A. Fasano, M. Primicerio: *Free boundary problems for nonlinear parabolic equations with nonlinear free boundary conditions.* J. Math. Anal. Appl. **72** (1979) 247–273
225. A. Fasano, M. Primicerio (eds.): *Free Boundary Problems: Theory and Applications.* Pitman, Boston 1983
226. A. Fasano, M. Primicerio: *Phase-change with volumetric heat sources: Stefan's scheme vs. enthalpy formulation.* Boll. Un. Mat. Ital. Suppl. **4** (1985) 131–149
227. A. Fasano, M. Primicerio: *Mushy regions with variable temperature in melting processes.* Boll. Un. Mat. Ital. **4-B** (1985) 601–626
228. A. Fasano, M. Primicerio: *A parabolic-hyperbolic free boundary problem.* S.I.A.M. J. Math. Anal. **17** (1986) 67–73
229. A. Fasano, M. Primicerio, S. Kamin: *Regularity of weak solutions of one-dimensional two-phase Stefan problems.* Ann. Mat. Pura Appl. **115** (1977) 341–348
230. A. Fasano, M. Primicerio, L. Rubinstein: *A model problem for heat conduction with a free boundary in a concentrated capacity.* J. Inst. Maths. Applics. **26** (1980) 327–347
231. H. Federer: *Geometric Measure Theory.* Springer, Berlin 1969
232. G. Fichera: *Sul problema elastostatico di Signorini con ambigue condizioni al contorno.* Rendic. Accad. Naz. Lincei **34** (1963)
233. G. Fichera: *Boundary value problems of elasticity with unilateral constraints.* In: Mechanics of Solids (C. Truesdell, ed.). Handbuch der Physik, vol. VIa/2. Springer, Berlin 1972, pp. 391–424
234. P.C. Fife, G.S. Gill: *A phase-field description of mushy zones.* Physica D **35** (1989) 267–275
235. R. Finn: *Equilibrium Capillary Surfaces.* Springer, New York 1986
236. G. Fix: *Numerical methods for alloy solidification.* In: Moving Boundary Problems (D.G. Wilson, A.D. Solomon, P.T. Boggs, eds.). Academic Press, New York 1978, pp. 109–128
237. G. Fix: *Numerical simulation of free boundary problems using phase field methods.* In: The Mathematics of Finite Element and Applications (J.R. Whiteman, ed.). Academic Press, London 1982
238. G. Fix: *Phase field methods for free boundary problems.* In: Free Boundary Problems: Theory and Applications (A. Fasano, M. Primicerio, eds.). Pitman, Boston 1983, pp. 580–589
239. W.H. Fleming, R. Rishel: *An integral formula for total gradient variation.* Arch. Math. **11** (1960) 218–222
240. M.C. Flemings: *Solidification Processing.* McGraw-Hill, New York 1973
241. I. Fonseca, L. Tartar: *The gradient theory of phase transitions for systems with two potential wells.* Proc. R. Soc. Edinburgh **111** (1989) 89–102
242. R. Fosdick, E. Dunn, M. Slemrod (eds.): *Shock Induced Transitions and Phase Structures in General Media.* Springer, New York 1993
243. M.I. Freidlin, A.D. Wentzell: *Random Perturbations of Dynamical Systems.* Springer, New York 1984
244. M. Frémond: *Variational formulation of the Stefan problem, coupled Stefan problem, frost propagation in porous media.* In: Proc. Conf. Computational Methods in Nonlinear Mechanics (J.T. Oden, ed.). University of Texas, Austin 1974, pp. 341–349
245. M. Frémond, A. Visintin: *Dissipation dans le changement de phase. Surfusion. Changement de phase irréversible.* C.R. Acad. Sci. Paris, Série II **301** (1985) 1265–1268

246. E. Fried, M.E. Gurtin: *Continuum theory of thermally induced phase transitions based on an order parameter.* Physica D **68** (1993) 326–343
247. A. Friedman: *Free boundary problems for parabolic equations. I, II, III.* J. Math. Mech. **8** (1959) 499–517; **9** (1960) 19–66; **9** (1960) 327–345;
248. A. Friedman: *Partial Differential Equations of Parabolic Type.* Prentice-Hall, Englewood Cliffs 1964
249. A. Friedman: *The Stefan problem in several space variables.* Trans. Amer. Math. Soc. **133** (1968) 51–87
250. A. Friedman: *One dimensional Stefan problems with non-monotone free boundary.* Trans. Amer. Math. Soc. **133** (1968) 89–114
251. A. Friedman: *Foundations of Modern Analysis.* Holt, Rinehart and Winston, New York 1970
252. A. Friedman: *Analyticity of the free boundary for the Stefan problem.* Arch. Rational Mech. Anal. **61** (1976) 97–125
253. A. Friedman: *Variational Principles and Free Boundary Problems.* Wiley, New York 1982
254. A. Friedman (ed.): *Mathematics in Industrial Problems.* Parts 1 – 6. Springer, New York 1988 – 1993
255. A. Friedman, D. Kinderlehrer: *A one phase Stefan problem.* Indiana Univ. Math. J. **25** (1975) 1005–1035
256. A. Friedman, J. Spruck (eds.): *Variational and Free Boundary Problems.* Springer, New York 1993
257. S. Fučík: *Solvability of Nonlinear Equations and Boundary Value Problems.* Reidel, Dordrecht 1980
258. R.M. Furzeland: *A comparative study of numerical methods for moving boundary problems.* J. Inst. Math. Appl. **26** (1980) 411–429
259. M. Gage: *An isoperimetric inequality with applications to curve shortening.* Duke Math. J. **50** (1983) 1225–1229
260. H. Gajewski, K. Gröger, K. Zacharias: *Nichtlineare Operatorgleichungen und Operatordifferentialgleichungen.* Akademie-Verlag, Berlin 1974
261. I.M. Gel'fand et al.: *Generalized Functions.* Vols. I–V. Academic Press, New York 1964
262. P. Germain: *Cours de Mécanique des Milieux Continus.* Masson, Paris 1973
263. A. Ghizzetti (ed.): *Theory and Applications of Monotone Operators.* Oderisi, Gubbio 1969
264. M. Giaquinta, S. Hildebrandt: *Calculus of Variations.* Vols. I, II. Springer, Berlin 1996
265. I.I. Gihman, A.V. Skorohod: *Stochastic Differential Equations.* Springer, Berlin 1972
266. D. Gilbarg, N.S. Trudinger: *Elliptic Partial Differential Equations of Second Order.* (2nd ed.) Springer, Berlin 1983
267. J.R. Giles: *Convex Analysis with Application in Differentiation of Convex Functions.* Pitman, Boston 1982
268. E. Giusti: *Minimal Surfaces and Functions of Bounded Variation.* Birkhäuser, Boston 1984
269. P. Glansdorff, I. Prigogine: *Thermodynamic Theory of Structure, Stability and Fluctuations.* Wiley-Interscience, London 1971
270. J. Goldstein: *Semigroups of Operators and Applications.* Oxford University Press, Oxford 1985
271. I.G. Götz, B.B. Zaltzman: *Nonincrease of mushy region in a nonhomogeneous Stefan problem.* Quart. Appl. Math. **XLIX** (1991) 741–746
272. O. Grange, F. Mignot: *Sur la résolution d'une équation et une inéquation paraboliques non linéaires.* J. Funct. Anal. **11** (1972) 77–92
273. M. Grayson: *The heat equation shrinks embedded plane curves to points.* J. Diff. Geometry **26** (1987) 285–314
274. P. Grisvard: *Elliptic Problems in Nonsmooth Domains.* Pitman, Boston 1985
275. M.E. Gurtin: *On a theory of phase transitions with interfacial energy.* Arch. Rational Mech. Anal. **87** (1985) 187–212
276. M.E. Gurtin: *On the two-phase Stefan problem with interfacial energy and entropy.* Arch. Rational Mech. Anal. **96** (1986) 199–241

277. M.E. Gurtin: *Toward a nonequilibrium thermodynamics of two phase materials.* Arch. Rational Mech. Anal. **100** (1988) 275–312
278. M.E. Gurtin: *Multiphase thermomechanics with interfacial structure. 1. Heat conduction and the capillary balance law.* Arch. Rational Mech. Anal. **104** (1988) 195–221
279. M.E. Gurtin: *On diffusion in two-phase systems: the sharp interface versus the transition layer.* In: P.D.E.s and Continuum Models of Phase Transitions (M. Rascle, D. Serre, M. Slemrod, eds.). Springer, Heidelberg 1989, pp. 99–112
280. M.E. Gurtin: *On thermomechanical laws for the motion of a phase interface.* Zeit. Angew. Math. Phys. **42** (1991) 370–388
281. M.E. Gurtin: *Thermomechanics of Evolving Phase Boundaries in the Plane.* Clarendon Press, Oxford 1993
282. M.E. Gurtin: *The dynamics of solid-solid phase transitions. 1. Coherent interfaces.* Arch. Rational Mech. Anal. **123** (1993) 305–335
283. M.E. Gurtin: *Thermodynamics and supercritical Stefan equations with nucleations.* Quart. Appl. Math. **LII** (1994) 133–155
284. M.E. Gurtin, G. McFadden (eds.): *On the Evolution of Phase Boundaries.* Springer, New York 1991
285. M.E. Gurtin, H.M. Soner: *Some remarks on the Stefan problem with surface structure.* Quart. Appl. Math. **L** (1992) 291–303
286. K.S. Ha: *Sur les semi-groupes non linéaires dans les espaces $L^\infty(\Omega)$.* J. Math. Soc. Japan **31** (1979) 593–622
287. J. Hadamard: *Lectures on Cauchy's Problem in Linear Partial Differential Equations.* Yale University Press, New Haven 1923
288. J.K. Hale: *Asymptotic Behavior of Dissipative Systems.* A.M.S., Providence 1988
289. E.-I. Hanzawa: *Classical solution of the Stefan problem.* Tohoku Math. **33** (1981) 297–335
290. E. Hewitt, K. Stromberg: *Real and Abstract Analysis.* Springer, Berlin 1965
291. J.M. Hill: *One-Dimensional Stefan Problem: An Introduction.* Longman, Harlow 1987
292. E. Hille, R.S. Phillips: *Functional Analysis and Semi-Groups.* A.M.S., Providence 1957
293. J.-B. Hiriart-Urruty, C. Lemarechal: *Convex Analysis and Minimization Algorithms.* Springer, Berlin 1993
294. P.C. Hohenberg, B.I. Halperin: *Theory of dynamic critical phenomena.* Rev. Mod. Phys. **49** (1977) 435–479
295. K.-H. Hoffmann, J. Sprekels (eds.): *Free Boundary Problems: Theory and Applications.* Longman, Harlow 1990
296. K.-H. Hoffmann, J. Sprekels (eds.): *Free Boundary Value Problems.* Birkäuser, Boston 1990
297. R.B. Holmes: *A Course on Optimization and Best Approximation.* Springer, Berlin 1972
298. R.B. Holmes: *Geometric Functional Analysis and its Applications.* Springer, New York 1975
299. D. Hömberg: *A mathematical model for the phase transitions in eutectoid carbon steel.* I.M.A. J. Appl. Math. **54** (1995) 31-57
300. L. Hörmander: *The Analysis of Linear Partial Differential Operators I.* Springer, Berlin 1983
301. J. Horváth: *Topological Vector Spaces and Distributions.* Addison-Wesley, Reading 1966
302. K. Huang: *Statistical Mechanics.* Wiley, New York 1987
303. S. Huang: *Regularity of the enthalpy for two-phase Stefan problem in several space variables.* Preprint I.M.A. of Minneapolis
304. A. Huber: *Hauptaufsätze über das Fortschreiten der Schmelzgrenze in einem linearen Leiter.* Z. Angew. Math. Mech. **19** (1939) 1–21
305. G. Huisken: *Flow by mean curvature of convex surfaces into spheres.* J. Differ. Geom. **20** (1984) 237–266
306. J. Hulshof, J.L. Vázquez: *Self-similar solutions of the second kind for the modified porous medium equation.* Euro J. Applied Math. **5** (1994) 391–403
307. T. Ilmanen: *Elliptic regularization and partial regularity for motion by mean curvature.* Memoirs A.M.S. **520** (1994)
308. A.D. Ioffe, V.M. Tikhomirov: *Theory of Extremal Problems.* Noth-Holland, Amsterdam 1979

309. J.W. Jerome: *Nonlinear equations of evolution and a generalized Stefan problem.* J. Differential Equations **26** (1977) 240–261
310. L.S. Jiang: *The two-phase Stefan problem. I, II.* Chinese Math. **4** (1963) 686–702; **5** (1964) 36–53
311. S. Kamenomostskaya: *On the Stefan problem.* Math. Sbornik **53** (1961) 489–514 (Russian)
312. S. Kaminin: *On the existence of solutions of the Verigin problem.* J. Vichis. Mat. Mat. Phiz. **2** (1962) 833–858 (Russian)
313. T. Kato: *Accretive operators and nonlinear evolution equations in Banach spaces.* In: Proc. Sympos. Pure Math., **XVIII** Part II (F. Browder, ed.). A.M.S., Providence 1976, pp. 138–161
314. H. Kawarada, N. Kenmochi, N. Yanagihara (eds.): *Nonlinear Mathematical Problems in Industry.* Gakkotosho, Tokyo 1993
315. J.L. Kelley, I. Namioka et al.: *Linear Topological Spaces.* Van Nostrand, Princeton 1963
316. N. Kenmochi: *Systems of nonlinear P.D.E.s arising from dynamical phase transitions.* In: Modelling and Analysis of Phase Transition and Hysteresis Phenomena (A. Visintin, ed.). Springer, Heidelberg 1994, pp. 39–86
317. N. Kenmochi: *Uniqueness of the solution to a nonlinear system arising in phase transition.* In: Nonlinear Analysis and Applications (N. Kenmochi, M. Niezgódka, P. Strzelecki, eds.). Gakkotosho, Tokyo 1996, pp. 261–271
318. N. Kenmochi, M. Niezgódka: *Systems of nonlinear parabolic equations for nonlinear phase change problems.* Adv. Math. Sci. Appl. **3** (1993/94) 89–117
319. N. Kenmochi, M. Niezgódka: *Viscosity approach to modelling non-isothermal diffusive phase separation.* Japan J. Indust. Appl. Math. **13** (1996) 135–169
320. N. Kenmochi, M. Niezgódka, P. Strzelecki (eds.): *Nonlinear Analysis and Applications.* Gakkotosho, Tokyo 1996
321. D. Kinderlehrer, L. Nirenberg: *Regularity in free boundary value problems.* Ann. Scuola Norm. Sup. Pisa **4** (1977) 373–391
322. D. Kinderlehrer, L. Nirenberg: *The smoothness of the free boundary in the one-phase Stefan problem.* Comm. Pure Appl. Math. **31** (1978) 257–282
323. D. Kinderlehrer, P. Pedregal: *Gradient Young measures generated by sequences in Sobolev spaces.* J. Geometric Analysis **4** (1994) 59–90
324. D. Kinderlehrer, G. Stampacchia: *An Introduction to Variational Inequalities and their Applications.* Academic Press, New York 1980
325. C. Kittel, H. Krömer: *Thermal Physics.* Freeman, San Francisco 1980
326. R. Kluge: *Nichtlineare Variationsungleichungen und Extremalaufgaben.* VEB Deutscher Verlag der Wissenschaften, Berlin 1979
327. A.N. Kolmogorov, S.V. Fomin: *Elements of the Theory of Functions and Functional Analysis.* Vols. I, II. Graglock Press, Rochester 1957
328. I.I. Kolodner: *Free boundary problem for the heat equation with applications to problems with change of phase.* Comm. Pure Appl. Math., **10** (1957) 220–231
329. G. Köthe: *Topological Vector Spaces.* Vols. I, II. Springer, Berlin 1969
330. A. Kufner, O. John, S. Fučik: *Function Spaces.* Academia, Prague and Noordhoff, Leyden 1977
331. W. Kurz, D.J. Fisher: *Fundamentals of Solidification.* Trans Tech, Aedermannsdorf 1989
332. A.G. Kusraev, S.S. Kutateladze: *Subdifferentials: Theory and Applications.* Kluwer, Dordrecht 1995
333. O.A. Ladyženskaja: *The Boundary Value Problems of Mathematical Physics.* Springer, New York 1985 (Russian edition: Nauka, Moscow 1973)
334. O.A. Ladyženskaja, V.A. Solonnikov, N.N. Ural'ceva: *Linear and Quasilinear Equations of Parabolic Type.* Trans. Math. Monographs **23**. A.M.S., Providence 1968 (Russian edition: Nauka, Moscow 1967)
335. O.A. Ladyženskaja, N.N. Ural'ceva: *Equations aux Dérivées Partielles de Type Elliptique.* Academic Press, New York 1968. (Russian edition: Nauka, Moscow 1964)

336. V. Lakshmikantham, S. Leela: *Nonlinear Differential Equations in Abstract Spaces.* Pergamon Press, Oxford 1981
337. G. Lamé, B.P. Clayperon: *Mémoire sur la solidification par refroidissement d'un globe solide.* Ann. Chem. Phys. **47** (1831) 250–256
338. L. Landau, E. Lifshitz: *On the theory of dispersion of magnetic permeability in ferromagnetic bodies.* Physik. Z. Sowietunion **8** (1935) 153–169
339. L. Landau, E. Lifshitz: *Electrodynamique des Milieux Continus.* MIR, Moscow 1969
340. L. Landau, E. Lifshitz: *Statistical Physics.* Pergamon Press, Oxford 1969
341. J.S. Langer: *Instabilities and pattern formation in crystal growth.* Rev. Mod. Phys. **52** (1980) 1–28
342. Ph. Laurençot: *Solutions to a Penrose-Fife model of phase-field type.* J. Math. Anal. Appl. **185** (1994) 262–274
343. B.H. Lavenda: *Thermodynamics of Irreversible Processes.* Macmillan, London 1978
344. M.M. Lavret'ev, V.G. Romanov, S.P. Shishatskiĭ: *Ill-Posed Problems of Mathematical Physics and Analysis.* A.M.S., Providence 1986
345. M.M. Lavret'ev, V.G. Romanov, V.G. Vasiliev: *Multi-Dimensional Inverse Problems for Differential Equations.* Springer, Berlin 1970
346. L.S. Leĭbenzon: *Handbook on Petroleum Mechanics.* G.N.T.I., Moscow 1931 (Russian)
347. C.C. Lin, L.A. Segel: *Mathematics Applied to Deterministic Problems in the Natural Sciences.* S.I.A.M., Philadelphia 1988
348. J.L. Lions: *Problèmes aux Limites dans les Équations aux Dérivées Partielles.* Presses de l'Université de Montréal, Montréal 1965
349. J.L. Lions: *Sur Quelques Problèmes de Calcul des Variations.* Istituto Naz. di Alta Mat., Symposia Mathematica **22** (1968) 125–144
350. J.L. Lions: *Sur le Contrôle Optimal des Systèmes Gouvernés par des Équations aux Dérivées Partielles.* Dunod, Gauthier-Villars, Paris 1968
351. J.L. Lions: *Quelques Méthodes de Résolution des Problèmes aux Limites non Linéaires.* Dunod, Paris 1969
352. J.L. Lions: *Perturbations Singulières dans les Problèmes aux Limites et en Contrôle Optimal.* Springer, Berlin 1973
353. P.L. Lions: *Generalized Solutions of Hamilton-Jacobi Equations.* Pitman, New York 1982
354. P.L. Lions: *The concentration-compactness principle in the calculus of variations: The locally compact case.* Parts I, II Ann. Inst. H. Poincaré **A1** (1984) 109–145; **A1** (1984) 223–283
355. P.L. Lions: *The concentration-compactness principle in the calculus of variations: The limit case.* Parts I, II Rev. Mat. Ibero. **1** (1) (1985) 145–201; **1** (2) (1985) 45–121
356. J.L. Lions, E. Magenes: *Non-Homogeneous Boundary Value Problems and Applications.* Vols. I, II. Springer, Berlin 1972 (French edition: Dunod, Paris 1968)
357. J.L. Lions, G. Stampacchia: *Inéquations variationnelles non coercives.* C.R. Acad. Sci. Paris, Série A **261** (1965) 25–27
358. J.L. Lions, G. Stampacchia: *Variational inequalities.* Comm. Pure Appl. Math. **20** (1967) 493–519
359. J. Li-Shang: *Existence and differentiability of the solution of a two-phase Stefan problem for quasilinear parabolic equations.* Chinese Math. Acta **7** (1965) 481–496
360. F. London: *Superfluids, vol. 1. Macroscopic Theory of Superconductivity.* Wiley, New York 1950
361. S. Luckhaus: *Solutions of the two-phase Stefan problem with the Gibbs-Thomson law for the melting temperature.* Euro. J. Appl. Math. **1** (1990) 101–111
362. S. Luckhaus: *The Stefan problem with the Gibbs-Thomson law.* Preprint 1991
363. S. Luckhaus: *The Stefan problem with surface tension.* In: Variational and Free Boundary Problems (A. Friedman, J. Spruck, eds.). Springer, New York 1993, pp. 153–157
364. S. Luckhaus: *Solidification of alloys and the Gibbs-Thomson law.* Preprint 1994
365. S. Luckhaus, L. Modica: *The Gibbs-Thompson relation within the gradient theory of phase transitions.* Arch. Rational Mech. Anal. **107** (1989) 71–83

366. S. Luckhaus, T. Sturzenhecker: *Implicit time discretization for the mean curvature flow equation.* Calc. Var. **3** (1995) 253–271
367. S. Luckhaus, A. Visintin: *Phase transition in a multicomponent system.* Manuscripta Math. **43** (1983) 261–288
368. A. Lunardi: *Analytic Semigroups and Optimal Regularity in Parabolic Problems.* Birkhäuser, Basel 1995
369. C.H.P. Lupis: *Chemical Thermodynamics of Materials.* PTR Prentice-Hall, Englewood Cliffs, 1983
370. H. Lüth: *Surfaces and Interfaces of Solid Materials.* Springer, Berlin 1995
371. E. Magenes: *Topics in parabolic equations: Some typical free boundary problems.* In: Boundary Value Problems for Nonlinear Partial Differential Equations (H.G. Garnier, ed.). Reidel, Dordrecht 1976, pp. 239–312
372. E. Magenes (ed.): *Free Boundary Problems.* Istituto di Alta Matematica, Roma 1980
373. E. Magenes: *Problemi di Stefan bifase in più variabili spaziali.* Le Matematiche **36** (1981) 65–108
374. E. Magenes: *On a Stefan problem on a boundary of a domain.* In: Partial Differential Equations and Related Subjects (M. Miranda, ed.). Longman, Harlow 1992, 209–226
375. E. Magenes: *Some new results on a Stefan problem in a concentrated capacity.* Rend. Acc. Naz. Lincei, Matem. Applic. **3** (1992) 23–34
376. E. Magenes: *On a Stefan problem in a concentrated capacity.* In: Topics in P.D.E.s and Applications (P. Marcellini, G. Talenti, E. Vesentini, eds.). Marcel Dekker, New York 1996
377. E. Magenes, C. Verdi, A. Visintin: *Semigroup approach to the Stefan problem with non-linear flux.* Rend. Acc. Naz. Lincei **75** (1983) 24–33
378. U. Massari: *Esistenza e regolarità delle ipersuperfici di curvatura media assegnata in R^N.* Arch. Rational Mech. Anal. **55** (1974) 357–382
379. U. Massari, L. Pepe: *Su di una impostazione parametrica del problema dei capillari.* Ann. Univ. Ferrara **20** (1974) 21–31
380. U. Massari, L. Pepe: *Successioni convergenti di ipersuperfici di curvatura media assegnata.* Rend. Sem. Mat. Univ. Padova **53** (1975) 53–68
381. V.G. Maz'ja: *Sobolev Spaces.* Springer, Berlin 1985
382. J.A. McGeough: *Principles of Electrochemical Machining.* Chapman and Hall, London 1974
383. J.A. McGeough, H. Rasmussen: *On the derivation of the quasi-steady model in electrochemical machining.* J. Inst. Maths. Applics. **13** (1974) 13–21
384. J.A. McGeough, P.G. Saffman: *The effect of surface tension on the shape of fingers in Hele-Shaw cell.* J. Fluid Mech. **102** (1981) 455–469
385. A.M. Meirmanov: *On the classical solvability of the Stefan problem.* Soviet Math. Dokl. **20** (1979) 1426–1429
386. A.M. Meirmanov: *On the classical solution of the multidimensional Stefan problem for quasilinear parabolic equations.* Math. U.S.S.R. Sbornik **40** (1981) 157–178
387. A.M. Meirmanov: *An example of nonexistence of a classical solution of the Stefan problem.* Soviet Math. Dokl. **23** (1981) 564–566
388. A.M. Meirmanov: *The Stefan Problem.* De Gruyter, Berlin 1992 (Russian edition: Nauka, Novosibirsk 1986)
389. A.M. Meirmanov: *The Stefan problem with surface tension in the three dimensional case with spherical symmetry: non-existence of the classical solution.* Euro J. Applied Math. **5** (1994) 1–20
390. G.H. Meyer: *The numerical solution of multidimensional Stefan problems; a survey.* In: Moving Boundary Problems. Proc. Sympos. and Workshop, Gatlinburg, Tenn. 1977. Academic Press, New York 1978, pp. 73–89
391. G.H. Meyer: *Numerical methods for free boundary problems: 1981 survey.* In: Free Boundary Problems: Theory and Applications (A. Fasano, M. Primicerio, eds.). Pitman, Boston 1983, pp. 590–600

392. G. Minty: *Monotone (nonlinear) operators in Hilbert space.* Duke Math. J. **29** (1962) 341–346
393. I. Miyadera: *Nonlinear Semigroups.* A.M.S., Providence 1992 (Japanese edition: Kinokuniya, Tokyo 1977)
394. S. Mizohata: *The Theory of Partial Differential Equations.* Cambridge University Press, Cambridge 1973
395. L. Modica: *Gradient theory of phase transitions with boundary contact energy.* Ann. Inst. H. Poincaré. Analyse non linéaire **4** (1987) 487–512
396. L. Modica: *Gradient theory of phase transitions and the minimal interface criterion.* Arch. Rational Mech. Anal. **98** (1987) 123–142
397. L. Modica, S. Mortola: *Un esempio di Γ-convergenza.* Boll. Un. Mat. Ital. **5** (1977) 285–299
398. V.N. Monachov: *Boundary Problems with Free Boundaries for Systems of Elliptic Equations.* Nauka, Novosibirsk 1977 (Russian)
399. J.J. Moreau: *Fonctionnelles sous-différentiables.* C.R. Acad. Sci. Paris, Série A **257** (1963) 4117–4119
400. J.J. Moreau: *Proximité et dualité dans un espace hilbertien.* Bull. Soc. Math. Fr. **93** (1965) 273–299
401. J.J. Moreau: *Fonctionnelles Convexes.* Séminaires sur les équations aux derivées partielles. Collège de France, Paris 1967
402. J.J. Moreau: *La notion de sur-potentiel et les liaisons unilatérales en élastotatique.* C.R. Acad. Sci. Paris, Série A **267** (1968) 954–957
403. C. Morrey: *Functions of several variables and absolute continuity.* Duke Math. J. **6** (1940) 187–215
404. C. Morrey: *Multiple Integrals in the Calculus of Variations.* Springer, Berlin 1966
405. A. Mukherjea, K. Pothoven: *Real and Functional Analysis.* Vols. I, II. Plenum Press, New York 1986
406. I. Müller: *Thermodynamics.* Pitman, Boston 1985
407. W.W. Mullins, R.F. Sekerka: *Morphological stability of a particle growing by diffusion and heat flow.* J. Appl. Phys. **34** (1963) 323–329
408. W.W. Mullins, R.F. Sekerka: *Stability of a planar interface during solidification of a dilute alloy.* J. Appl. Phys. **35** (1964) 441–451
409. F. Murat: *Compacité par compensation.* Ann. Scuola Norm. Sup. Pisa **5** (1978) 489–507
410. L.E. Murr: *Interfacial Phenomena in Metals and Alloys.* Addison-Wesley, Reading 1975
411. M. Muskat: *Two fluid systems in porous media. The encroachment of water into an oil sand.* Physics **5** (1934) 250–264
412. M. Muskat: *The Flow of Homogeneous Fluids Through Porous Media.* McGraw-Hill, New York 1937
413. J. Naumann: *Einfürung in die Theorie Parabolischer Variationsungleichungen.* Teubner, Leipzig 1984
414. J. Nečas: *Les Méthodes Directes en Théorie des Équations Elliptiques.* Masson, Paris, and Academia, Prague 1967
415. J. Nečas: *Introduction to the Theory of Nonlinear Elliptic Equations.* Wiley Interscience, New York 1986
416. J. Nečas, I. Hlaváček: *Mathematical Theory of Elastic and Elastico-Plastic Bodies: An Introduction.* Elsevier, Amsterdam 1982
417. P. Neittanmäki (ed.): *Numerical Methods for Free Boundary Problems.* Birkhäuser, Basel 1991
418. W.-M. Ni, L.A. Peletier, J.L. Vázquez (eds.): *Degenerate Diffusion.* Springer, New York 1992
419. G. Nicolis, I. Prigogine: *Self-Organization in Nonequilibrium Systems.* Wiley, New York 1977
420. M. Niezgódka: *Stefan-like problems.* In: Free Boundary Problems: Theory and Applications (A. Fasano, M. Primicerio, eds.). Pitman, Boston 1983, pp. 321–347

421. M. Niezgódka (ed.): *Free Boundary Problems: Theory and Applications*. Gakkotosho, Tokyo (to appear)
422. M. Niezgódka, I. Pawlow (eds.): *Recent Advances in Free Boundary Problems*. Control and Cybernetics **14** (1985), 1–307
423. R.H. Nochetto: *Numerical methods for free boundary problems*. In: Free Boundary Problems: Theory and Applications (K.-H. Hoffmann and J. Sprekels, eds.). Longman, Harlow 1990, pp. 555–566
424. R.H. Nochetto: *Finite element methods for parabolic free boundary problems*. In: Advances in Numerical Analysis. Vol. I (W. Light, ed.). Oxford Academic Press, Oxford 1991, pp. 34–95
425. R.H. Nochetto, M. Paolini, C. Verdi: *An adaptive finite element method for two-phase Stefan problems in two space dimensions. Part I: Stability and error estimates*. Math. Comp., Supplement **57** (1991) 73–108, S1–11
426. R.H. Nochetto, M. Paolini, C. Verdi: *An adaptive finite element method for two-phase Stefan problems in two space dimensions. Part II: Implementation and numerical experiments*. S.I.A.M. J. Sci. Statist. Comput. **12** (1991) 1207–1244
427. R.H. Nochetto, M. Paolini, C. Verdi: *Towards a unified approach for the adaptive solution of evolution phase changes*. In: Variational and Free Boundary Problems (A. Friedman, J. Spruck, eds.). Springer, New York 1993, pp. 171–193
428. R.H. Nochetto, M. Paolini, C. Verdi: *A fully discrete adaptive nonlinear Chernoff formula*. S.I.A.M. J. Numer. Anal. **30** (1993) 991–1014
429. R.H. Nochetto, C. Verdi: *Approximation of degenerate parabolic problems using numerical integration*. S.I.A.M. J. Numer. Anal. **25** (1988) 784–814
430. R.H. Nochetto, C. Verdi: *An efficent linear scheme to approximate parabolic free boundary problems: Error estimates and implementation*. Math. Comp. **51** (1988) 27–53
431. A. Novick-Cohen, L.A. Segel: *Nonlinear aspects of the Cahn-Hilliard equation*. Physica D **10** (1984) 278–298
432. J.R. Ockendon, W.R. Hodgkins (eds.): *Moving Boundary Problems in Heat Flow and Diffusion*. Clarendon Press, Oxford 1975
433. C. Olech: *Integrals of set-valued functions and linear optimal control problems*. In: Colloque sur la Théorie Mathématique du Contrôle Optimal. C.B.R.M., Vander Louvain 1970, pp. 109–125
434. C. Olech: *Existence theory in optimal control*. In: Control Theory and Topics in Functional Analysis. Vol. I. International Atomic Energy Agency, Vienna 1976, pp. 291–328
435. O.A. Oleĭnik: *A method of solution of the general Stefan problem*. Soviet Math. Dokl. **1** (1960) 1350–1353
436. O.A. Oleĭnik, M. Primicerio, E.V. Radkevich: *Stefan-like problems*. Meccanica **28** (1993) 129–143
437. S. Osher, J.A. Sethian: *Fronts propagating with curvature dependent speed: Algorithms based on Hamilton-Jacobi formulations*. J. Comput. Physics **79** (1988) 12–49
438. N. Owen: *Nonconvex variational problems with singular general perturbations*. Trans. A.M.S. **310** (1989) 393–404
439. R. Pamplin (ed.): *Crystal Growth*. Pergamon Press, Oxford 1975
440. P.D. Panagiotopoulos: *Inequality Problems in Mechanics and Applications*. Birkhäuser, Basel 1985
441. P.D. Panagiotopoulos: *Hemivariational Inequalities. Applications in Mechanics and Engineering*. Springer, Heidelberg 1993
442. M. Paolini, G. Sacchi, C. Verdi: *Finite element approximation of singular parabolic problems*. Internat. J. Numer. Methods Engrg. **26** (1988) 1989–2007
443. D. Pascali, S. Sburlan: *Nonlinear Mappings of Monotone Type*. Ed. Academiei, Bucharest 1978
444. N.H. Pavel: *Nonlinear Evolution Operators and Semigroups*. Springer, Berlin 1987

445. L.E. Payne: *Improperly Posed Problems in Partial Differential Equations.* S.I.A.M. Publications, Philadelphia 1975
446. A. Pazy: *On the asymptotic behaviour of semigroups of nonlinear contractions in Hilbert space.* J. Funct. Anal. **27** (1978) 202–307
447. A. Pazy: *Semigroups of nonlinear contractions and their asymptotic behaviour.* In: Nonlinear Analysis and Mechanics: Heriot-Watt Symposium. Vol. III (R.J. Knops, ed.). Pitman, London 1979, pp. 36–134
448. A. Pazy: *The Lyapunov method for semigroups of nonlinear contractions in Banach spaces.* J. d'Analyse Mathém. **40** (1981) 239–262
449. A. Pazy: *Initial value problems for nonlinear differential equations in Banach spaces.* In: Nonlinear Partial Differential Equations and their Applications. Collège de France Seminar, Vol. IV (H. Brézis, J.L. Lions, D. Cioranescu, eds.). Pitman, Boston 1983, pp. 154–172
450. A. Pazy: *Semigroups of Linear Operators and Applications to Partial Differential Equations.* Springer, New York 1983
451. O. Penrose, P.C. Fife: *Thermodynamically consistent models of phase-field type for the kinetics of phase transitions.* Physica D **43** (1990) 44–62
452. O. Penrose, P.C. Fife: *On the relation between the standard phase-field model and a "thermodynamically consistent" phase-field model.* Physica D **69** (1993) 107–113
453. R.R. Phelps: *Convex Functions, Monotone Operators and Differentiability.* Springer, Heidelberg 1989
454. P.I. Plotnikov, V.N. Starovoitov: *The Stefan problem with surface tension as a limit of phase field model.* Differential Equations **29** (1993) 395–404
455. I. Prigogine: *Thermodynamics of Irreversible Processes.* Wiley-Interscience, New York 1967
456. M. Primicerio: *Problemi a contorno libero per l'equazione della diffusione.* Rend. Sem. Mat. Univers. Politecn. Torino **32** (1973-74) 183–206
457. M. Primicerio: *Problemi di diffusione a frontiera libera.* Boll. Un. Matem. Ital. **18-A** (1981) 11–68
458. M. Primicerio: *Mushy regions in phase-change problems.* In: Applied Functional Analysis (R. Gorenflo, K.-H. Hoffmann, eds.). Lang, Frankfurt 1983, pp. 251–269
459. M. Primicerio, M. Ughi: *Phase-change problems with mushy regions.* In: Free Boundary Problems: Theory and Applications (A. Bossavit, A. Damlamian, M. Frémond, eds.). Pitman, Boston 1985, pp. 61–71
460. E. Radkevitch: *Gibbs-Thomson law and existence of the classical solution of the modified Stefan problem.* Soviet Dokl. Acad. Sci. **43** (1991) 274–278
461. E. Radkevich, A.C. Melikulov: *Free Boundary Problems.* F.A.N., Tashkent 1992 (Russian)
462. P.A. Raviart: *Sur la résolution des certaines équations paraboliques non linéaires.* J. Funct. Anal. **5** (1970) 299–328
463. M. Reed, J. Simon: *Functional Analysis.* Vol. I. Academic Press, New York 1980
464. M. Renardy, R. Rogers: *An Introduction to Partial Differential Equations.* Springer, New York 1993
465. S. Richardson: *Hele-Shaw flow with a free boundary produced by the injections of a fluid into a narrow channel.* J. Fluid. Mech. **56** (1972) 609–618
466. S. Richardson: *Some Hele-Shaw flow with time-dependent free boundaries.* J. Fluid. Mech. **102** (1981) 263–278
467. A.W. Roberts, D.E. Varberg: *Convex Functions.* Academic Press, New York 1973
468. R.T. Rockafellar: *Convex Analysis.* Princeton University Press, Princeton 1969
469. R.T. Rockafellar: *Conjugate Duality and Optimization.* S.I.A.M., Philadelphia 1974
470. J.-F. Rodrigues: *An evolutionary continuous casting problem of Stefan type.* Quart. Appl. Math. **XLIV** (1986) 109–131
471. J.-F. Rodrigues: *Obstacle Problems in Mathematical Physics.* North-Holland, Amsterdam 1987
472. J.-F. Rodrigues: *The variational inequality approach to the one-phase Stefan problem.* Acta Applicandae Mathematicae **8** (1987) 1–35

473. J.-F. Rodrigues (ed.): *Mathematical Models for Phase Change Problems.* Birkhäuser, Basel 1989
474. J.-F. Rodrigues: *Variational methods in the Stefan problem.* In: Modelling and Analysis of Phase Transition and Hysteresis Phenomena (A. Visintin, ed.). Springer, Heidelberg 1994, pp. 147–212
475. J.-F. Rodrigues (ed.): *Free boundary problems news.* Newsletter of the E.S.F. Program "Mathematical treatment of free boundary problems", since 1993
476. J.C.W. Rogers: *The Stefan problem with surface tension.* In: Free Boundary Problems: Theory and Applications. Vol. I (A. Fasano, M. Primicerio, eds.). Pitman, Boston 1983, pp. 263–274
477. A. Romano: *Thermomechanics of Phase Transitions in Classical Field Theory.* World Scientific, Singapore 1993
478. V.G. Romanov: *Inverse Problems of Mathematical Physics.* V.N.U. Science Press, Utrecht 1987
479. L. Rubinstein: *On the determination of the position of the boundary which separates two phases in the one-dimensional problem of Stefan.* Dokl. Acad. Nauk USSR **58** (1947) 217–220
480. L. Rubinstein: *The Stefan Problem.* A.M.S., Providence 1971 (Russian edition: Zvaigzne, Riga 1967)
481. L. Rubinstein: *The Stefan problem: Comments on its present state.* J. Inst. Maths. Applics. **24** (1979) 259–277
482. L. Rubinstein: *On mathematical modelling of growth of an individual monocomponent crystal from melt in a non-uniform temperature field.* Control and Cybernetics **12** (1983) 5–18
483. L. Rubinstein: *On mathematical models of solid-liquid zones in two phase monocomponent system and in binary alloys.* In: Free Boundary Problems: Theory and Applications (A. Fasano, M. Primicerio, eds.). Pitman, Boston 1983, pp. 275–282
484. L. Rubinstein: *Mathematical modelling of growth of an individual monocomponent crystal from its melt in a non-homogeneous temperature field.* In: Free Boundary Problems: Theory and Applications (A. Bossavit, A. Damlamian, M. Frémond, eds.). Pitman, Boston 1985, pp. 166–178
485. L. Rubinstein, A. Fasano, M. Primicerio: *Remarks on the analyticity of the free boundary for the one-dimensional Stefan problem.* Ann. Mat. Pura Appl. **125** (1980) 295–311
486. L. Rubinstein, H. Geiman, M. Shachaf: *Heat transfer with a free boundary moving within a concentrated capacity.* I.M.A. J. Appl. Math. **28** (1982) 131–147
487. I. Rubinstein, B. Martuzans: *Free Boundary Problems Related to Osmotic Mass Transfer through Semipermeable Membranes.* Gakkotosho, Tokyo 1995
488. I. Rubinstein, L. Rubinstein: *Partial Differential Equations in Classical Mathematical Physics.* Cambridge University Press, Cambridge 1993
489. W. Rudin: *Real and Complex Analysis.* McGraw-Hill, New York 1970
490. W. Rudin: *Functional Analysis.* McGraw-Hill, New York 1973
491. T. Rzezuchowski: *Strong convergence of selections implied by weak.* Bull. Austral. Math. Soc. **39** (1989) 201–214
492. T. Rzezuchowski: *Impact of dentability on weak convergence in L^1.* Boll. Un. Mat. Ital. **6-A** (1992) 71–80
493. P.G. Saffman, G.I. Taylor: *The penetration of fluid into a porous medium or Hele-Shaw cell.* Proc. Roy. Soc. **A 245** (1958) 312–329
494. J. Sanchez-Hubert, E. Sanchez-Palencia: *Introduction aux Méthodes Asymptotiques et à l'Homogénéisation des Milieux Continus.* Masson, Paris 1992
495. E. Sanchez-Palencia: *Non Homogeneous Media and Vibration Theory.* Springer, Berlin 1980
496. D. Schaeffer: *A new proof of infinite differentiability of the free boundary in the Stefan problem.* J. Differential Equations **20** (1976) 266–269
497. L. Schwartz: *Théorie des Distributions.* Vols. I, II. (2nd ed.) Hermann, Paris 1957, 1959

498. L. Schwartz: *Distributions à valeurs vectorielles.* Parts I, II. Ann. Inst. Fourier **7** (1957) 1–141; **8** (1958) 1–209
499. G. Sestini: *Esistenza di una soluzione in problemi analoghi a quello di Stefan.* Rivista Mat. Univ. Parma **3** (1952) 3–23; **8** (1958) 1–209
500. J.A. Sethian: *Curvature and evolution of fronts.* Comm. Math. Phys. **101** (1985) 487–499
501. M.J. Sewell: *Maximum and Minimum Principles.* Cambridge Univ. Press, Cambridge 1987
502. M. Shillor: *Existence and continuity of a weak solution to the problem of a free boundary in a concentrated capacity.* Proc. Roy. Soc. Edinburgh **A100** (1985) 271–280
503. R.E. Showalter: *Mathematical formulation of the Stefan problem.* Int. J. Eng. Sc. **20** (1982) 909–912
504. R.E. Showalter: *Monotone Operators and Nonlinear P.D.E.s.* (Monograph in preparation)
505. R.E. Showalter, N.J. Walkington: *A hyperbolic Stefan problem.* Quart. Appl. Math. **XLV** (1987) 769–781
506. J. Simon: *Compact sets in the space $L^p(0,T;B)$.* Ann. Mat. Pura Appl. **146** (1987) 65–96
507. L. Simon: *Lectures on Geometric Measure Theory.* Australian National University, Canberra **3** 1983
508. V.P. Skripov: *Metastable Liquids.* Wiley, Chichester 1974
509. D.R. Smart: *Fixed Point Theorems.* Cambridge University Press, Cambridge 1974
510. P. Smith: *Convexity Methods in Variational Calculus.* Wiley, New York 1985
511. S.L. Sobolev: *Some Applications of Functional Analysis in Mathematical Physics.* A.M.S., Providence 1991 (First Russian edition: Leningrad Univ., Leningrad 1950)
512. M. Soner: *Convergence of the phase-field equation to the Mullins-Sekerka problem with kinetic undercooling.* Arch. Rational Mech. Anal. **131** (1995) 139–197
513. J. Sprekels, S. Zheng: *Global smooth solutions to a thermodynamically consistent model of phase-field type in higher space dimensions.* J. Math. Anal. Appl. **176** (1993) 200–223
514. G. Stampacchia: *Formes bilinéaires coercives sur les ensembles convexes.* C.R. Acad. Sci. Paris, Série A **258** (1964) 4413–4416
515. G. Stampacchia: *Variational inequalities.* In: Theory and Applications of Monotone Operators (A. Ghizzetti, ed.). Oderisi, Gubbio 1969, pp. 101–192
516. J. Stefan: *Über einige Probleme der Theorie der Wärmeleitung.* Sitzungber., Wien, Akad. Mat. Natur. **98** (1889) 473–484. Also ibid. pp. 614–634, 965–983, 1418–1442
517. M. Struwe: *Variational Methods.* Springer, Berlin 1990
518. J. Szekely: *Some mathematical physical and engineering aspects of melting and solidification problems.* In: Free Boundary Problems: Theory and Applications. Vol. II (A. Fasano, M. Primicerio, eds.). Pitman, Boston 1983, pp. 283–292
519. I. Tamanini: *Boundary of Caccioppoli sets with Hölder continuous normal vectors.* J. Reine Angew. Math. **334** (1982) 27–39
520. I. Tamanini: *Regularity results for almost minimal oriented hypersurfaces in R^n.* Quaderni Università di Lecce **1** (1984)
521. L. Tartar: *Interpolation non linéaire et régularité.* J. Funct. Anal. **9** (1972) 469–489
522. L. Tartar: *Compensated compactness and applications to partial differential equations.* In: Nonlinear Analysis and Mechanics: Heriott-Watt Symposium, Vol. IV (R.J. Knops, ed.). Pitman, London 1979, pp. 136–212
523. L. Tartar: *The compensated compactness method applied to systems of conservation laws.* In: Systems of Nonlinear Partial Differential Equations (J. Ball, ed.). Reidel, Dordrecht 1983, pp. 263–285
524. L. Tartar: *H-measures, a new approach for studying homogenisation, oscillation and concentration effects in partial differential equations.* Proc. Roy. Soc. Edinburgh **A 115** (1990) 193–230
525. D.A. Tarzia: *Una revisión sobre problemas de frontera móvil y libre para la ecuación del calor. El problema de Stefan.* Math. Notae **29** (1981/82) 147–241
526. D.A. Tarzia: *The Two-Phase Stefan Problem and Some Related Conduction Problems.* S.B.M.A.C., Gramado 1987

527. D.A. Tarzia: *A Bibliography on Moving-Free Boundary Problems for the Heat Diffusion Equation. The Stefan Problem.* Prog. Naz. M.P.I. Ital., Firenze 1988
528. A.B. Tayler: *A mathematical formulation of Stefan problems.* In: Moving Boundary Problems in Heat Flow and Diffusion (J.R. Ockendon, W.R. Hodgkins, eds.). Clarendon Press, Oxford 1975, pp. 120–137
529. J.E. Taylor: *Mean curvature and weighted mean curvature.* Acta Metall. Mater. **40** (1992) 1475–1485
530. J.E. Taylor, J.W. Cahn: *Linking anisotropic sharp and diffuse surface motion laws via gradient flows.* J. Stat. Phys. **77** (1994) 183–197
531. R. Temam: *Infinite Dimensional Mechanical Systems in Mechanics and Physics.* Springer, Berlin 1988
532. R. Thom: *Stabilité Structurelle et Morphogénèse. Essai d'une Théorie Générale des Modèles.* Benjamin, Reading 1972
533. V.M. Tikhomirov: *Fundamental Principles of the Theory of Extremal Problems.* Wiley, Chichester 1982
534. A.N. Tikhonov, A.A. Samarskiĭ: *Equations of Mathematical Physics.* Pergamon Press, Oxford 1963 (2nd Russian edition: G.I.I.T.L., Moscow 1953)
535. F. Treves: *Topological Vector Spaces, Distributions and Kernels.* Academic Press, New York 1967
536. H. Triebel: *Interpolation Theory, Function Spaces, Differential Operators.* Verlag der Wissenschaften, Berlin 1978
537. D. Turnbull: *Phase Changes.* Solid State Physics **3** (1956) 225–306
538. D. Turnbull, J.C. Fisher: *Rate of nucleation in condensed systems.* J. Chem. Phys. **17** (1949) 71–73
539. A.R. Ubbelohde: *The Molten State of Matter.* Wiley, Chichester 1978
540. M. Valadier: *Différents cas où, grâce a une propriété d'extrémalité, une suite de fonctions intégrables faiblement convergente converge fortement.* Séminaire d'Analyse Convexe, Montpellier 1989, exposé n. 5
541. M. Valadier: *Young measures.* In: Methods of Nonconvex Analysis (A. Cellina, ed.). Springer, Heidelberg 1990, pp. 152–188
542. M. Valadier: *Oscillations et compacité forte dans L^1.* Séminaire d'Analyse Convexe, Montpellier 1991, exposé n. 7
543. M. Valadier: *Young measures, weak and strong convergence and the Visintin-Balder theorem.* Set-Valued Anal. **2** (1994) 357–367
544. C. Verdi: *Optimal error estimates for an approximation of degenerate parabolic problems.* Numer. Funct. Anal. Optim. **9** (1987) 657–670
545. C. Verdi: *BV regularity of the enthalpy for semidiscrete two-phase Stefan problem.* Rend. Istit. Lomb. **126** (1992) 29–42
546. C. Verdi: *Numerical aspects of free boundary and hysteresis problems.* In: Modelling and Analysis of Phase Transition and Hysteresis Phenomena (A. Visintin, ed.). Springer, Heidelberg 1994, pp. 213–284
547. C. Verdi, A. Visintin: *A mathematical model of the austenite-pearlite transformation in plain steel based on the Scheil additivity rule.* Acta Metall. **35** (1987) 2711–2717
548. N.N. Verigin: *On the pressured forcing of binder solutions into rocks in order to increase the solidity and imperviousness to water of the foundations of hydrotechnical constructions.* Izv. Acad. Nauk U.S.S.R. (Ser. Tech.) **5** (1954) 674–687 (Russian)
549. A. Visintin: *Strong convergence results related to strict convexity.* Comm. in P.D.E.s **9** (1984) 439–466
550. A. Visintin: *Stefan problem with phase relaxation.* I.M.A. J. Appl. Math. **34** (1985) 225–245
551. A. Visintin: *Study of the eddy-current problem taking account of Hall's effect.* Applicable Analysis, **19** (1985), 217–226
552. A. Visintin: *On Landau-Lifshitz equations in ferromagnetism.* Japan J. Appl. Math. **2** (1985) 69–84

Bibliography 317

553. A. Visintin: *A new model for supercooling and superheating effects.* I.M.A. J. Appl. Math. **36** (1986) 141-157
554. A. Visintin: *Stefan problem with a kinetic condition at the free boundary.* Ann. Mat. Pura Appl. **146** (1987), 97-122
555. A. Visintin: *Coupled thermal and electromagnetic evolution in a ferromagnetic body.* Z. Angew. Math. Mech. **67** (1987) 409–417
556. A. Visintin: *Supercooling and superheating effects in heterogeneous systems.* Quart. Appl. Math. **XLV** (1987), 239–263
557. A. Visintin: *Mathematical models of solid-solid phase transformations in steel.* I.M.A. J. Appl. Math. **39** (1987) 143-157
558. A. Visintin: *Surface tension effects in phase transitions.* In: Material Instabilities in Continuum Mechanics and Related Mathematical Problems (J. Ball, ed.). Clarendon Press, Oxford 1988, pp. 505–537
559. A. Visintin: *The Stefan problem with surface tension.* In: Mathematical Models of Phase Change Problems (J.F. Rodrigues, ed.). Birkhäuser, Basel 1989, pp. 191–213
560. A. Visintin: *Non-convex functionals related to multi-phase systems.* S.I.A.M. J. Math. Anal. **21** (1990) 1281–1304
561. A. Visintin: *Generalized coarea formula and fractal sets.* Japan J. Appl. Math. **8** (1991) 175–201
562. A. Visintin: *Pattern evolution.* Ann. Scuola Norm. Sup. Pisa Cl. Sci. **XVII** (1990) 195–225 (Errata corrige: Ibid. **XVIII** (1991) 319–320)
563. A. Visintin: *Remarks on the Stefan problem with surface tension.* In: Boundary Value Problems for Partial Differential Equations and Applications (C. Baiocchi, J.L. Lions, eds.). Dunod, Paris 1993.
564. A. Visintin: *Differential Models of Hysteresis.* Springer, Heidelberg 1994
565. A. Visintin (ed.): *Modelling and Analysis of Phase Transition and Hysteresis Phenomena.* Springer, Heidelberg 1994
566. A. Visintin: *Two-scale Stefan problem with surface tension.* In: Nonlinear Analysis and Applications (N. Kenmochi, M. Niezgódka, eds.). Gakkotosho, Tokyo (to appear)
567. A. Visintin: *Two-scale model of phase transitions.* Preprint, 1996
568. A. Visintin: *Nucleation and mean curvature flow.* Preprint, 1996
569. A. Visintin: *Stefan Problem with nucleation and mean curvature flow.* Preprint, 1996
570. A.I. Vol'pert, S.I. Hudjaev: *Analysis in Classes of Discontinuous Functions and Equations of Mathematical Physics.* Nijhoff, Dordrecht 1985
571. I.I. Vrabie: *Compactness Methods for Nonlinear Evolutions.*
572. W. von Wahl: *On the Cahn-Hilliard equation $u' + \Delta^2 u - \Delta f(u) = 0$.* Delft Progress Report **10** (1985) 391–310
573. W. Walter: *Differential and Integral Inequalities.* Springer, Berlin 1970
574. S.-L. Wang, R.F. Sekerka, A.A. Wheeler, B.T. Murray, S.R. Coriell, R.J. Braun, G.B. McFadden: *Thermodynamically-consistent phase-field models for solidification.* Physica D **69** (1993) 189–200
575. M.C. Weinberg (ed.): *Nucleation and Crystallization In Liquid and Glasses* American Ceramic Society, Westerville 1993
576. J.A. Wheeler: *Permafrost thermal design for the Trans-Alaska pipeline.* In: Moving Boundary Problems (D.G. Wilson, A.D. Solomon, P.T. Boggs, eds.). Academic Press, New York 1978, pp. 267–284
577. D.G. Wilson, A.D. Solomon, V. Alexiades: *Progress with simple binary alloy solidification problems.* In: Free Boundary Problems: Theory and Applications (A. Fasano, M. Primicerio, eds.). Pitman, Boston 1983, pp. 306–320
578. D.G. Wilson, A.D. Solomon, P.T. Boggs (eds.): *Moving Boundary Problems.* Academic Press, New York 1978
579. D.G. Wilson, A.D. Solomon, J.S. Trent: *A Bibliography on Moving Boundary Problems with Key Word Index.* Oak Ridge National Laboratory 1979

580. J. Wloka: *Partial Differential Equations.* Cambridge University Press, Cambridge 1987
581. D.J. Wollkind, R.D. Notestine: *A nonlinear stability analysis of the solidification of a pure substance.* I.M.A. J. Appl. Math. **27** (1981) 85–104
582. D.J. Wollkind, R.D. Notestine, R.N. Maurer: *A continuum model appropriate for nonlinear analysis of the solidification of a pure metal.* In: Applied Nonlinear Analysis (V. Lakshmikantham, ed.). Academic Press, New York 1979, pp. 657–668
583. P.D. Woodruff: *The Solid-Liquid Interface.* Cambridge Univ. Press, Cambridge 1973
584. L.C. Woods: *The Thermodynamics of Fluid Systems.* Clarendon Press, Oxford 1975
585. L.C. Wrobel, C.A. Brebbia (eds.): *Computational Modelling of Free and Moving Boundary Problems.* Vols. I, II. Computational Mechanics Publ., Southampton 1991, 1993
586. L.C. Wrobel, C.A. Brebbia (eds.): *Computational Modelling of Free and Moving Boundary Problems in Heat and Fluid Flow.* Computational Mechanics Publ., Southampton 1993
587. L.C. Wrobel, C.A. Brebbia, B. Sarler (eds.): *Computational Modelling of Free and Moving Boundary Problems.* Vol. III. Computational Mechanics Publ., Southampton 1995
588. M. Yamaguchi, T. Nogi: *The Stefan Problem.* Sangyo-Tosho, Tokyo 1977 (Japanese)
589. K. Yosida: *Functional Analysis.* Springer, Berlin 1971
590. L. Young: *Lectures on the Calculus of Variations and Optimal Control Theory.* Saunders, Philadelphia 1968
591. E. Zeidler: *Nonlinear Functional Analysis and its Applications.* Vols. I – IV. Springer, New York 1985
592. E. Zeidler: *Applied Functional Analysis: Applications to Mathematical Physics.* Springer, New York 1995
593. E. Zeidler: *Applied Functional Analysis: Main Principles and their Applications.* Springer, New York 1995
594. S. Zheng: *Global existence for a thermodynamically consistent model of phase field type.* Differential Integral Eq. **5** (1992) 241–253
595. A. Ziabicki: *Generalized theory of nucleation kinetics. I. General formulations; II. Athermal nucleation involving spherical clusters.* J. Chem. Phys. **48** (1968) 4368–4374; **48** (1968) 4374–4380
596. W.P. Ziemer: *Interior and boundary continuity of weak solutions of degenerate parabolic equations.* Trans. A.M.S. **271** (1982) 733–748
597. W.P. Ziemer: *Weakly Differentiable Functions.* Springer, New York 1989
598. T. Zolezzi: *On weak convergence in L^∞.* Indiana Univ. Math. J. **23** (1973-74) 765–766

Index

$BV(\Omega)$, 263
$C^k(\bar{\Omega};B)$, 261
$C^{k,\nu}(\bar{\Omega};B)$, 261
$\Gamma(B)$, 275
$\Gamma_0(B)$, 275
σ-finite measure, 248

a priori estimates, 19
absolute minima, 168
absolute minimizer, 186, 188
absolute temperature, 147
accretiveness, 285
activation energy, 192
adiabatic nucleation, 204, 229
almost minimal boundary, 291
almost-minimal boundary, 222
analytical formulation, 7
anisotropy energy, 242
approximate limit, 172
approximation methods, 18
Ascoli-Arzelà theorem, 21
Ascoli-Arzelà theorem, 268
asymptotic behaviour, 14, 283

Banach lattice, 287
binary alloy problem, 152
Bloch and Néel walls, 243
boundary and initial conditions, 10

Cahn model, 198
Cahn-Hilliard equation, 120, 194
Cattaneo law, 152
Cauchy problem, 17, 281, 286
classification of P.D.E.s, 10, 11
Clausius-Duhem inequality, 148
coarea formula, 188
coerciveness, 19
compactness, 20, 21, 201, 268
compactness and closure, 13
compactness by strict convexity, 84
compactness by strict monotonicity, 257
compensated compactness, 273
competing terms, 232
concentrated capacities, 121

conjugate function, 275
constitutive law, 25
contact angle condition, 168, 172, 210
continuity, 13
continuous casting, 121
continuous phase transition, 193, 204
convex conjugate, 255
convex function, 274
convex functional, 252
convex set, 274
convexity, 22, 245
convolution, 230, 239
coupled systems, 121
Crank-Nicolson scheme, 218
critical cap, 180
crystal growth, 121
cyclical monotonicity, 280

De la Vallée Poussin theorem, 270
dentability, 259
dependence on the data, 166
Dirac measure, 232, 262
direct method, 15, 21, 167, 289
discontinuous coalescence, 210
discontinuous phase transition, 192, 204
displacement current, 27
distributions, 262
dry friction, 200
Dunford-Pettis theorem, 250, 270

Eberlein-Šmulian theorem, 270
effective domain, 274
effective magnetic field, 244
Egoroff theorem, 271
Ehrling theorem, 269
electro-machining, 122
energy balance, 203
energy conservation, 148
energy estimate, 19
entropy, 160, 173
entropy balance, 148
epigraph, 274
equation + lower semicontinuity technique, 39, 72, 76, 80, 81, 84, 85

equiintegrability, 270
essential boundary, 291
Euler-Lagrange method, 21
exchange energy, 242
existence, 12, 17, 34, 59, 60, 70, 74, 77, 159, 166–168, 186, 216, 234, 238, 240, 244, 245
explicit solution, 12
exposed point, 248
extremal point, 248

fast heat diffusion, 229
Fatou lemma, 252, 254
ferromagnetism, 122, 242
fine structure, 245
fluctuations, 190
formal procedure, 20
Fourier law, 26, 152, 203
Fréchet differential, 277
Fréchet space, 262, 263
Fréchet-Riesz-Kolmogorov theorem, 215
free boundary problem, 12, 26, 245
free energy, 24, 147, 155, 156, 181, 185, 190, 204, 233
free surfaces, 120
front motion, 204
Fréchet-Riesz-Kolmogorov theorem, 268
fully nonlinear P.D.E.s, 11

Gâteaux differential, 193, 277
Gaussian kernel, 230
generalized Poincaré inequality, 293
Gibbs-Thomson law, 168, 170, 172, 179, 200, 204, 212, 222, 234
glass, 184, 192
global minimizer, 178, 180
gradient flow, 24, 149, 150, 193
grains, 184
graph norm, 261
Gronwall lemma, 293

Hahn-Banach theorem, 275
heat equation, 156
heat operator, 10
Hele-Shaw problem, 122
heterogeneous nucleation, 181, 182
Hilbert triplet, 294
homogeneous nucleation, 179, 181
hyperbolic equation, 152
hypercooling, 213
hysteresis, 197, 199, 207, 236, 243, 245

impurity, 181

indicator function, 274
infinite dimension, 8, 251
integral solution, 281, 286
integral transformation, 58
interaction distance, 230, 233
isothermal nucleation, 181
isothermal phase transition, 236
iterative methods, 17

Kirchhoff transformation, 156

Landau-Lifshitz equation, 243, 247
latent heat, 230
lateral growth, 198
Lax-Milgram theorem, 293
Lebesgue spaces, 261
Legendre transformation, 147, 273
Legendre-Fenchel transformation, 274, 275
length scale, 182
level method, 289
lift operator, 266
limit procedure, 20
Lions-Aubin theorem, 21, 269
Lipschitz domain, 260
local equilibrium, 155, 204, 212
local minimizer, 178
locally convex topological space, 262
locally uniformly convex Banach space, 253
lower semicontinuity, 158
lower semicontinuous functional, 252
lower semicontinuous functions, 274
Lyapunov function, 195

m-accretiveness, 22, 63, 285
macroscopic length scale, 235
magnetic energy, 242
magnetic field energy, 242
magnetostatic equations, 242
maximal monotone graph, 22, 31, 196, 197, 256, 279
Maxwell equations, 27, 244
mean curvature, 180
mean curvature flow, 120, 191, 196, 197, 200, 201, 236, 247
mean field, 230, 239
mesoscopic scale, 181, 230, 235
metastability, 155, 245
metastable states, 164, 185, 190
micromagnetics, 242
microstructure, 9, 120
minimization, 159, 288
model, 7
model problem, 9

modified free energy, 186
modified Gibbs-Thomson law, 231
monotonicity, 14, 20, 22, 279
Mullins-Sekerka problem, 225
multi-index, 260
multi-stability, 23, 243
multi-valued mapping, 248
mushy region, 234

nonconvex constraint, 243
nonconvex functional, 155, 160
nonconvexity, 21, 216
nonequilibrium thermodynamics, 153
nonlinear mean curvature flow, 196
nonlinear P.D.E.s, 10
nucleant, 182
nucleation, 185, 202, 204, 208, 212
nucleation temperature, 183, 226, 234
nucleus, 230
numerical approximation, 15

Onsager relations, 146

penalization, 19
perimeter, 157, 233, 243, 290
periodic solution, 284
phase, 29, 181
phase nucleation, 205
phase-field model, 120, 237
Poisson process, 192
polymers, 121, 202
principal part, 10
pseudo-parabolic equation, 73, 77
pure phase, 188

qualitative properties, 13
quasilinear P.D.E.s, 11

radial symmetry, 179, 191
Radon-Nikodým property, 287
range, 279
rate independence, 200
recalescence, 192, 230, 234
reduced boundary, 291
regularity, 14, 77, 282, 292
regularization, 18
relative minima, 168
relative minimizer, 186, 243
Rellich-Kondrachov theorem, 21, 220, 269
reversibility, 205
Riesz operator, 294

scale transformation, 235

scaling, 235
second principle, 148
semi-inner product, 285
semicontinuity, 20
semigroup of ω-contractions, 287
semigroups of nonlinear contractions, 22, 62, 282, 286
semilinear P.D.E.s, 11
singular evolution, 197
Sobolev inclusion, 166
Sobolev spaces, 28, 264
soil freezing, 121
solid-solid transitions, 121
space interaction term, 247
space interpolation, 220, 267, 269
space of measures, 263
space-distributed system, 9, 25
stable states, 164, 185, 190
Stefan condition, 204
Stefan-Gibbs-Thomson problem, 208, 212
Stefan-type model, 230
strict convexity, 22
strictly convex function, 274
strong formulation, 212, 245
strong monotonicity, 22
strong solution, 16, 281, 286
strongly measurable functions, 261
structural stability, 14
subdifferential, 25, 147, 193, 255, 276, 278, 279
superheating, 207, 212, 229
surface tension, 30, 204
symmetry breaking, 23, 164

T-accretiveness, 63, 288
test functions, 262
thermodynamical equilibrium, 207
time discretization, 8, 18
time scale, 192
total variation, 263, 290
trace, 266, 267
transposition, 23, 294
triply nonlinear equation, 80
two-dimensional nucleation, 199, 236
two-scale model, 29, 235

undercooling, 120, 207, 212, 229, 235
uniform integrability, 270
uniqueness, 13, 16, 22, 59, 60, 238, 240

variational equation, 22
variational inequality, 22, 28, 33, 58, 66
variational problem, 28

vectorial Stefan-type problem, 242
Vitali theorem, 272
Volmer-Weber-Becker-Döring theory, 183
volume discontinuity, 183

Weak compactness in $L^1(\Omega)$, 270
weak compactness theorem, 268
weak formulation, 33, 208, 225
weak solution, 16, 281, 286
Weiss domains, 243
well-posedness, 15
Wiener process, 191

Young measures, 259

Progress in Nonlinear Differential Equations and Their Applications

Editor
Haim Brezis
Département de Mathématiques
Université P. et M. Curie
4, Place Jussieu
75252 Paris Cedex 05
France
and
Department of Mathematics
Rutgers University
New Brunswick, NJ 08903
U.S.A.

Progress in Nonlinear Differential Equations and Their Applications is a book series that lies at the interface of pure and applied mathematics. Many differential equations are motivated by problems arising in such diversified fields as Mechanics, Physics, Differential Geometry, Engineering, Control Theory, Biology, and Economics. This series is open to both the theoretical and applied aspects, hopefully stimulating a fruitful interaction between the two sides. It will publish monographs, polished notes arising from lectures and seminars, graduate level texts, and proceedings of focused and refereed conferences.

We encourage preparation of manuscripts in some form of TeX for delivery in camera-ready copy, which leads to rapid publication, or in electronic form for interfacing with laser printers or typesetters.

Proposals should be sent directly to the editor or to: Birkhäuser Boston, 675 Massachusetts Avenue, Cambridge, MA 02139

PNLDE 1 Partial Differential Equations and the Calculus of Variations, Volume I
Essays in Honor of Ennio De Giorgi
F. Colombini, A. Marino, L. Modica, and S. Spagnolo, editors

PNLDE 2 Partial Differential Equations and the Calculus of Variations, Volume II
Essays in Honor of Ennio De Giorgi
F. Colombini, A. Marino, L. Modica, and S. Spagnolo, editors

PNLDE 3 Propagation and Interaction of Singularities in Nonlinear Hyperbolic Problems
Michael Beals

PNLDE 4 Variational Methods
Henri Berestycki, Jean-Michel Coron, and Ivar Ekeland, editors

PNLDE 5 Composite Media and Homogenization Theory
Gianni Dal Maso and Gian Fausto Dell'Antonio, editors

PNLDE 6   Infinite Dimensional Morse Theory and Multiple Solution Problems
          Kung-ching Chang

PNLDE 7   Nonlinear Differential Equations and their Equilibrium States, 3
          N.G. Lloyd, W.M. Ni, L.A. Peletier, J. Serrin, editors

PNLDE 8   Introduction to Γ–Convergence
          Gianni Dal Maso

PNLDE 9   Differential Inclusions in Nonsmooth
          Mechanical Problems: Shocks and Dry Frictions
          Manuel D. P. Monteiro Marques

PNLDE 10 Periodic Solutions of Singular Lagrangian Systems
          Antonio Ambrosetti and Vittorio Coti Zelati

PNLDE 11 Nonlinear Waves and Weak Turbulence
          with Applications in Oceanography and
          Condensed Matter Physics
          *N. Fitzmaurice, D. Gurarie, F. McCaugham,
          and W. A. Woyczyński, editors*

PNLDE 12 Semester on Dynamical Systems: Euler
          International Math Institute, St. Petersburg, 1991
          Kuksin, Lazutkin, and Poeschel, editors

PNLDE 13 Ginzburg-Landau Vortices
          F. Bethuel, H. Brezis, and F. Hélein

PNLDE 14 Variational Methods for Image Segmentation
          Jean-Michel Morel and Sergio Solomini

PNLDE 15 Topological Nonlinear Analysis: Degree,
          Singularity, and Variations
          Michele Matzeu and Alfonso Vignoli, editors

PNLDE 16 Analytic Semigroups and Optimal Regularity
          in Parabolic Problems
          A. Lunardi

PNLDE 17 Blowup for Nonlinear Hyperbolic Equations
          Serge Alinhac

PNLDE 18 The Heat Kernel Lefschetz Fixed Point Formula
          for the Spin-c Dirac Operator
          J. Duistermaat

PNLDE 19   Chaos and Nonlinear Dynamical Systems
           I. Hoveijn

PNLDE 20   Topics in Geometry: Honoring the Memory
           of Joseph D'Atri
           Simon Gindikin, editor

PNLDE 21   Partial Differential Equations and Mathematical
           Physics: The Danish-Swedish Analysis Seminar,
           1995
           Lars Hörmander and Anders Melin, editors

PNLDE 22   Partial Differential Equations and Functional
           Analysis: In Memory of Pierre Grisvard
           *J. Cea, D. Chenais, G. Geymonat, and
           J.-L. Lions, editors*

PNLDE 23   Geometry of Harmonic Maps
           Yuanlong Xin

PNLDE 24   Minimax Theorems
           Michel Willem

PNLDE 25   Variational Methods for Discontinuous Structures
           Raul Serapioni and Franco Tomarelli

PNLDE 26   Algebraic Aspects of Integrable Systems: In Memory
           of Irene Dorfman
           A. S. Fokas and I. M. Gelfand

PNLDE 27   Topological Nonlinear Analysis II
           Degree, Singularity, and Variations
           Michele Matzeu and Alfonso Vignoli

PNLDE 28   Models of Phase Transitions
           Augusto Visintin